CW00503042

Disaster Risk, Resilient Agriculture and Livelihood

This volume discusses important issues associated with agricultural disaster risk, resilient agriculture, and livelihood. It highlights the role of sustainable development goals in reducing the impact of climate change on agriculture. The contributions found in this volume discuss methodological and innovative resilience approaches to various natural hazards including flood, landslide, environmental challenges, strategies of disaster risk management, livelihood, ecosystem services, and agricultural sustainability. It explores the relationship between climatic change and agricultural transformation. While throwing light on the role of ecosystem services in disaster risk reduction, the book explores the impact of land degradation and change on growth of agricultural production and food production. The book will be useful for students and researchers of geography, environmental sciences, disaster management, and environmental geology. It will also be useful for geographers, environmentalists, hydrologists, geomorphologists, planners, and professionals working on related ideas.

Asraful Alam is an Assistant Professor and Head of the Department of Geography, Serampore Girls' College, University of Calcutta, West Bengal, India. He received his MA and PhD degrees in Geography from Aligarh Muslim University, Aligarh and Aliah University, Kolkata, India, respectively, and also completed a PG diploma in remote sensing and geographic information systems (GIS). Alam completed his post doctorate (PDF) from the Department of Geography, University of Calcutta, Kolkata, India. Previously, he was an Assistant Coordinator in the PG Department of Geography, Calcutta Women's College, University of Calcutta, Kolkata, India. Dr. Alam is one of the Project Directors of a collaborative major project on "Pradhan Mantri Ujjwala Yojana: An Impact Assessment in Relation to the Life of Women in Assam and West Bengal" sponsored by Indian Council of Social Science Research, Ministry of Education, New Delhi. His research interests include population geography, agricultural geography, climatology, health geography, remote sensing and GIS, and developmental studies. He has contributed various research papers published in various reputed national and international journals and edited book volumes. He has authored jointly edited books entitled *Habitat, Ecology and*

Ekistics: Case Studies of Human-Environment Interactions in India, Agriculture, Food and Nutrition Security: Case Study of Availability and Sustainability in India, Agriculture, Environment and Sustainable Development: Experiences and Case Studies, Life and Living through Newer Spectrum of Geography, Self-Reliance (Atmanirbhar) and Sustainable Resource Management in India, and *Climate Change, Agriculture and Society—Approaches toward Sustainability, Population, Sanitation and Health: A Geographical Study Towards Sustainability and Agriculture and Climatic Issues in South Asia Geospatial Applications*. He was a convener in the National Seminar on Self-reliance (Atmanirbhar), Sustainable Development and Environment 25–26 March 2022 sponsored by the Indian Council of Social Science Research (ICSSR) organized by the Department of Geography, Serampore Girls' College. Alam has served as an editorial board member in peer-reviewed international journals *PLOS ONE, Earth Science, Scientific Journal of Health Science Research,* and *Frontiers in Geochemistry*.

Rukhsana is an Assistant Professor and former Head of the Department of Geography at Aliah University, Kolkata. She has 15 years of teaching as well as research experience. She obtained her MA and PhD degrees in Geography from Aligarh Muslims University. She was awarded nine academic awards and fellowships like International Young Geographer Award 2009, the AMU-JRF award 2006–07, and UGC-JRF 2007–2009 award, etc. She has published more than 40 papers at the national as well as international levels in reputed journals and 23 chapters in edited books. Dr Rukhsana has published eleven books and presented a number of research papers at national and international levels. She has attended XXV FIG International Congress 2014, Malaysia and ICGGS-2018, Thailand Bangkok. Dr Rukhsana has successfully completed three major research project sponsored by ICSSR, Ministry of Education, New Delhi, in the agriculture field. Four PhDs have been successfully awarded under her supervision. She has served as a reviewer for many reputed international journals. Her specialization in research is agriculture development and planning, urban expansion and planning, environment, rural development, RS, and GIS. She has completed various training programs, workshops from different organizations, and a number of training courses run by IIRS Indian Institute of Remote Sensing, ISRO, Department of Space, Govt of India. She is engaged in various professional activities and served the university in various posts. Presently she is working as project director on a major research project and project director of a collaborative major project also sponsored by Indian Council of Social Science Research, Ministry of Education, New Delhi.

Disaster Risk, Resilient Agriculture and Livelihood

Methods and Applications

Edited by Asraful Alam and Rukhsana

LONDON AND NEW YORK

First published 2025
by Routledge
4 Park Square, Milton Park, Abingdon, Oxon OX14 4RN

and by Routledge
605 Third Avenue, New York, NY 10158

Routledge is an imprint of the Taylor & Francis Group, an informa business

British Library Cataloguing-in-Publication Data
A *catalogue* record for this book is available from the British Library

ISBN: 978-1-032-16236-2 (hbk)
ISBN: 978-1-032-23132-7 (pbk)
ISBN: 978-1-003-27591-6 (ebk)

DOI: 10.4324/9781003275916

Typeset in Times New Roman
by SPi Technologies India Pvt Ltd (Straive)

Contents

Figures

Tables

Contributors

Mbanga Lawrence Akei is part of the Department of Geography and Planning, Faculty of Arts, University of Bamenda, Cameroon.

Ershad Ali is part of the Department of Geography & Applied Geography, University of North Bengal, West Bengal, India.

Md. Julfikar Ali is part of the Department of Geography, Aliah University, 17 Gorachand Road, Park Circus Campus, Kolkata 700014, West Bengal, India.

Nso Ngang Andre works with the Institute of Agricultural Research for Development (IRAD), Yaounde Cameroon.

Najib Ansari is Research Schoar, Department of Geography, Aliah University, 17 Gorachand Road, Park Circus Campus.

A. K. M. Anwaruzzaman is part of the Department of Geography, Aliah University, Kolkata.

Kaldjob Mbeh Christian Bernard works with the Institute of Agricultural Research for Development (IRAD), Yaounde Cameroon & Department of public economy, Faculty of Economic Sciences and Management, University of Yaounde II SOA (UYII).

Pradip Chouhan is part of Department of Geography, University of Gour Banga, Malda, West Bengal.

Kailash Chandra Das is Professor, Department of Migration & Urban Studies, International Institute for Population Sciences (IIPS), Mumbai.

Douya Emmanuel is part of the Department of public economy, faculty of Economic Sciences and Management, University of Yaounde II SOA (UYII).

Augustine Toh Gam is part of Department of Geography and Planning, Faculty of Arts, University of Bamenda, Cameroon.

Arijit Ghosh is a PhD Research Scholar, Department of Geography, Sidho-Kanho-Birsha University, Purulia, West Bengal, India.

Woheeul Islam is part of the Department of Geography, Aliah University, 17 Gorachand Road, Park Circus Campus, Kolkata, West Bengal, India.

Gurucharan Karmakar works with SALAM, International Institute for Population Sciences (IIPS), Mumbai.

Soleman Khan is part of the Department of Geography, Aliah University, Kolkata.

Sumaiya Khatun is part of the Department of Geography, University of Calcutta, Kolkata, West Bengal, India

Bamou Tankoua Lydie is part of the Department of Public Economy, Faculty of Economic Sciences and Management, University of Yaounde II SOA (UYII).

Pravakar Mishra works with the National Center for Coastal Research, NIOT Campus, Pallikaranai, Chennai, India.

Biraj Kanti Mondal is Assistant Professor, Department of Geography, School of Sciences Netaji Subhas Open University, Kolkata, West Bengal, India.

Pratap Kumar Mohanty is part of the Department of Marine Sciences, Berhampur University, Berhampur, India.

Md. Mustaquim is part of the Department of Geography, Aliah University, 17 Gorachand Road, Park Circus Campus, Kolkata, West Bengal, India.

Suika Rita Nyuyfoni is part of the Department of Geography and Planning, The University of Bamenda, P.O. Box 39, Bambili, Northwest Region, Cameroon.

Sivaraj Paramasivam is part of Amrita School of Agricultural Sciences, Amrita Vishwa Vidyapeetham, Coimbatore, Tamil Nadu, India.

Umakanta Pradhan works with the National Center for Coastal Research, NIOT Campus, Pallikaranai, Chennai, India.

Tata Ngome Precillia works with the Institute of Agricultural Research for Development (IRAD).

Margubur Rahaman is Senior Research Fellow, Department of Migration & Urban Studies, International Institute for Population Sciences (IIPS), Mumbai.

Md. Juel Rana is Assistant Professor of G.B. Pant Social Science Institute (a constituent institute of the University of Allahabad), Jhusi, Prayagraj.

Alisha Safder is part of the Department of Geography, Aliah University, 17 Gorachand Road, Park Circus Campus.

Amiya Saha is Senior Research Fellow, Family & Generations, International Institute for Population Sciences (IIPS), Mumbai.

Apurba Sarkar is part of Department of Geography, University of Gour Banga, Malda, West Bengal.

Bipul Chandra Sarkar is part of the Department of Geography, Ananda Chandra College, Jalpaiguri, West Bengal, India.

Lakshminarayan Satpati is Professor of Geography & Director, UGC-HRDC, University of Calcutta, Kolkata, West Bengal, India.

Parthasarathy Seethapathy is part of Amrita School of Agricultural Sciences, Amrita Vishwa Vidyapeetham, Coimbatore, Tamil Nadu, India.

Sabirul Sk is part of an Interdisciplinary Program in Climate Studies, Indian Institute of Technology Bombay, Mumbai, India.

S. Bala Subramaniyam is an Independent (GIS + XR) Consultant and Researcher, India.

Suiven John Paul Tume is part of the Department of Geography and Planning, The University of Bamenda, P.O. Box 39, Bambili, Northwest Region, Cameroon.

G. Vivekanathapatmanaban is part of the Department of Agricultural Extension and Rural Sociology, Tamil Nadu Agricultural University, Coimbatore, Tamil Nadu, India.

Mofor Gilbert Zechia is part of the Department of Geography, Higher Teachers Training College (HTTC), University of Bamenda, Cameroon.

Acknowledgements

We would like to express our gratitude to the teachers and colleagues who have helped us in a variety of career paths, as we did in our prefaces to the earlier edited book. We would like to express our gratitude to the many colleagues at Aliah University, Serampore Girls' College, and University of Calcutta who have collaborated with us, given their knowledge, and taught us everything about the research, writing, and teaching. We must also express our gratitude to each and every one of our esteemed authors who contributed heir valuable papers to make a great success of this project. We really appreciate the hard work put in by the entire Taylor & Francis team in getting this volume ready.

Dr. Asraful Alam
Dr. Rukhsana

Part I

Climate Change, Vulnerability and Agricultural Transformation

1 Disaster Risk, Resilient Agriculture, and Livelihood

An Overview

Rukhsana, Asraful Alam, and Sumaiya Khatun

1.1 Introduction

Many rural areas are at risk for disasters caused by climate change as a result of the rising number of extreme weather occurrences against the backdrop of climate change and widespread poverty. Poor households may experience long-lasting consequences from atmospheric shocks or stressors, which can contribute to the development or reinforcement of poverty traps, which are self-reinforcing mechanisms that provide considerable obstacles to exiting poverty (Leichenko & Silva, 2014).

The agriculture sector is complex and inconsistent. Despite the fact that agriculture (which also includes the relatively small hunting, fishing, and forestry industries) only accounted for 2.8% of global income in 2012, estimates indicate that 1.3 billion (or 19%) of the 7.1 billion people on the Earth were directly worked in farming. Today's majority of farmers may be found in middle- and low-income countries, where agriculture contributes significantly to their economy and provides employment for a considerable portion of their populations. The significant increase in global food production over the past 40 years has been a tremendous achievement, but it has also brought about serious environmental problems. The hazards associated with pesticides and chemical fertilizers, desertification, and the cumulative effects of salinization on land productivity and soil erosion are a few of these (FAO, 2010).

According to several studies, great strides have been made in lowering poverty and hunger as well as improving the security of food and nutrition. Technology development and increasing productivity have improved resource management and improved food security. But serious problems still persist. Approximately 795 million people still go hungry, and over 2 billion people have vitamin deficiencies. Additionally, increasing pressures on natural resources and climate change, both of which endanger global food security, may compromise the sustainability of large-scale agricultural systems (FAO, 2010). More than 820 million people in the world endure daily hunger, and this number has been steadily rising over the past three years, according to the most recent State of Food Safety and Nutrition in the World Report. It is surprising how persistent and widespread hunger remains on the planet despite all of

DOI: 10.4324/9781003275916-2

development's great achievements, including significantly improved food production. Additionally, about 2 billion people suffer from some form of food insecurity, such as a lack of access to enough nutritious food. Women, children, and native communities are particularly vulnerable to starvation. Undernutrition also increases the risk of overweight and obesity, which are on the rise globally and are approaching epidemic proportions (FAO, 2019; Rukhsana & Alam, 2021).

The resilience concept has advanced significantly in the disaster sector in recent years. Most academics and professionals define disaster resilience as a social system's ability to achieve ecological, social, and economic objectives while managing environmental dangers throughout time in a mutually reinforcing manner (Béné et al., 2015). This concept places a strong emphasis on the livelihood capitals (natural, social, economic, human, and physical) of the sustainable livelihood framework that enable community welfare, which continues to be the overarching purpose of disaster resilience (Barua et al., 2014). The three dimensions that make up disaster resilience are, "a) the capacity to absorb shocks and stressors, b) the capacity to adapt to shocks and stressors, and c) the capacity to transform in the face of shocks and stressors" (Constas et al., 2014). Risk management techniques that help people and/or families deal with environmental dangers without incurring long-term, detrimental consequences on their quality of life are known as absorptive capacity (Béné et al., 2015; Smith et al., 2015).

There is a growing understanding that farmers' and local knowledge is a valuable resource that can redirect modern agriculture towards more sustainable and resilient paths of development in the face of the numerous contemporary challenges facing agriculture, such as climate change, food security, and resource depletion (Šūmane et al., 2018). Resilience is the ability of an agricultural system to change and adapt such that it remains viable over the long term, "in recent years agricultural sustainability has been linked with the concept of resilience, which emphasises dynamics, disequilibrium and unpredictability in agricultural development" (Darnhofer, 2014; Walker et al., 2004). Agricultural systems need to be adaptable and robust in order to deal with climate change. In the face of extreme external shifts like climate change and fluctuating prices, resilient agriculture systems are more likely to preserve economic, ecological, and social benefits (Reddy, 2015).

Disaster resilience, on the contrary, is the capacity of people, communities, organizations, and countries to deal with and recover from risks, shocks, or pressures without jeopardizing long-term development prospects. According to the Hyogo Framework for Action (2005), the ability of individuals to organize themselves in order to learn from previous disasters and lower their risks from upcoming ones determines disaster resilience at the international, regional, national, and local levels. A smaller subset of the greater idea of resilience is disaster resilience, which is described as "the capacity of individuals, communities, states, and their institutions to absorb and recover from shocks while positively adapting and transforming their structures and means of living in the face of long-term changes and uncertainty" (OECD, 2013).

Knowing the sustainable development objective of meeting the demands of disaster resilience is necessary to better comprehend the idea of disaster resilience. The idea that the economy, society, and environment are the three pillars that best support sustainability is the basis of sustainable development. According to the Brundtland Report, sustainable development "meets the needs of the present without compromising the ability of future generations to meet their own needs." In developing nations, immediate concerns for the poorest residents' survival may lead emphasis to be directed toward fulfilling their basic local needs (Stefen et al., 2015). As a result, the idea of sustainable development includes both time and geographical variables. Therefore, unlike the Millennium Development Goals, which focused on action in developing countries, the Sustainable Development Goals (SDGs) address the universal need for development that meets everyone's needs.

To create more livable, resilient, and sustainable housing cities, we have finally included the 2030 Agenda for Sustainable Development. By implementing numerous city planning initiatives like the Swachh Bharat Mission (SBM), Smart Cities Mission (SCM), Pradhan Mantri Awas Yojana-Housing for All for the Urban Poor Population (PMAY-U), Atal Mission for Rejuvenation and Urban Transformation (AMRUT), and many others, the government of contemporary developing countries like India has modernized their cities. Different planning and policy efforts must be evaluated, put into action, and then modified using certain frameworks in order to achieve urban resilience and sustainability. Without solid strategic development planning, resilience and sustainability in landscape and urban planning cannot be achieved in the future (Zanotti et al., 2020; Ribeiro & Goncalves, 2019). In both developing and developed nations, many planning and policy initiatives for urban land use have failed in recent decades to fully realize urban resilience and sustainable development (Keith et al., 2020; Zanotti et al., 2020). The dynamics, tactics, and post-disaster restoration processes of livelihoods may be explored through the perspective of "resilience thinking," which also helps to enhance the adaptive capacity1 to deal with future changes (Folke et al., 2003).

1.2 Part I: Climate Change, Disaster Risk and Agricultural Transformation

Climate change and agriculture are interrelated. Climate change has an influence on agriculture in many different ways, including changes in average temperatures, rainfall, and other aspects. Pests and illnesses on agricultural land are exacerbated by a lack of rain and changes in atmospheric constituents like carbon dioxide and ground-level ozone. Globally, a more in-depth geographic examination of climate change and its repercussions is becoming more and more necessary. The modern world has to pay attention to and find novel answers for two critical concerns: agriculture transformation and climate issues. In order to solve these issues and promote sustainable agricultural practices, geospatial technology is crucial. Geospatial technology also makes it easier to establish policies and monitor and assess sustainable agriculture practices. It is feasible to evaluate the efficacy of conservation strategies, land-use regulations, and the effects of agricultural operations on ecosystems by using satellite images and other geospatial data. This information is crucial for developing policies that are supported by facts and for advancing sustainable agriculture practices at the municipal, regional, and federal levels (Rukhsana et al., 2021; Molla and Rukhsana, 2023; Rukhsana and Mandal 2021).

Food security and rural livelihoods are suffering greatly as a result of the frequent disasters caused by climate change and climatic variability. The region's food security and assets used for generating a living are under greater danger due to the expected increases in temperature and rainfall. Globally, more people are being impacted by natural disasters, with floods being the most severe, pervasive, and common ones that are caused by climate change (Guha-Sapir et al., 2016). According to scientists, climatic variability is brought on by human climate change (CC). According to climate forecasts, for instance, the frequency of days with exceptionally heavy rainfall has increased by 1 to 2% per decade in the driest parts of the world. This means that more floods will occur during the ensuing decades due to rising rainfall. Human security is affected by rising temperatures and fluctuating precipitation levels, and current research on how climate change affects agriculture in developing nations reveals that many African agricultural economies face a serious danger to their food security (IPCC, 2014).

Throughout the Earth's origin, climate change has happened naturally, but throughout the 20th and 21st centuries, its frequency and intensity have increased mostly owing to human activity. Numerous aspects of human life are being impacted by climate change, including socioeconomics, severe occurrences, decreasing food supply, illnesses, etc., in addition to effects on natural systems. The mountain ecosystems are among the most vulnerable ecosystems in the world, and the implications of climate change there are severe (Bhatta et al., 2015).

Changes in land use and cover, which are related to human-nature interactions, have a considerable impact on the effects of human activities and global environmental variance (Liu et al., 2014; Rukhsana & Molla, 2023). Activities involving land usage and land cover demonstrate how humans have altered the Earth's surface. As a result of changes in land use/land cover, the planet's energy balance and biogeochemical cycles have been significantly altered (Foley et al., 2005; Alkama & Cescatti, 2016; Turner et al., 2007). This has an impact on the properties of the extent of the land and the availability of ecosystem services. According to a recent study (Song et al., 2018), 60% of worldwide land changes are directly related to human activity. Kennedy et al.'s (2019) most recent estimate states that only 5% (6.96 million km^2) of the planet's terrestrial area is still undisturbed by human activity. Urbanization is characterized by migrated population to the urban extent from the rural world, and the expansion of urban land use (Wu et al., 2015) is one common manifestation of all these human-related modifications on global lands (Theobald et al., 2020).

1.2.1 Disaster Risk

Agriculture is more susceptible to natural disasters than other sectors since it depends on the environment for the growth of animals and plants (Zhang et al., 2022). Disasters can have a significant impact on people's ability to sustain their livelihoods and to face additional burdens. Poor people may become more vulnerable as a result of impacts like asset loss, which can create a "downward spiral of deepening poverty and increasing risk" (Davies et al., 2009).

Agricultural production is plagued with risks that may cause a negative impact on output levels and result in substantial losses. There are a variety of risks and uncertainties that affect agricultural activities due to fluctuating economic and biophysical conditions. Risk occurs from a variety of sources in the agricultural sector. Weather poses the greatest threat to productivity risk. Extreme weather events, including hailstorms, cyclones, droughts, floods, storm surges, etc. are mostly uncertain and farmers are generally unprepared for such calamities (Ullah et al., 2015). The effects of global climate change are progressively becoming more severe, and the recurring appearance of extreme weather events severely impacts agricultural production. Urbanization and global warming have altered the frequency and severity of weather-causing elements as well as the exposure of crop-bearing bodies, which has significant effects on how resilient agricultural output is to natural disasters (Zhang et al., 2022). Agriculture is significantly impacted by severe floods and droughts. Fast drought-flood cycles can possibly be impacted by climate change effects such as increased precipitation uncertainty, changes in snow water equivalent, and fast snowmelt, particularly in an area where there is an abundance of snow. For instance, Afghanistan experienced a multi-year snow drought during the winters of 2017–2018. According to estimates, 9.8 million people experienced food insecurity by September 2018 due to the drought (Ward et al., 2020).

In order to effectively reduce the risks of natural disasters and regulate agricultural production, it is important to analyze and evaluate agricultural multi-hazard meteorological risks of disasters based on historical disaster data and an overview of disaster events (Zhang et al., 2022). The insurance is able to compensate for a variety of risks that might arise from both climatic and non-climatic factors. Insurance also offers opportunities to create public–private collaborations and minimizes the reliance on public resources during the post-disaster reconstruction and relief stages (Alam et al., 2020). According to research on agricultural insurance, farmers who improve their agricultural operations and seek to maximize their earnings have a higher rate of formal credit adoption (Budhathoki et al., 2019). The efficiency of the current insurance products in terms of disaster risk reduction (DRR) and climate change adaptation (CCA) seems to be limited, even in places where insurance is available. DRR is "the process of reducing exposure, lessening underlying vulnerabilities, better management of resources and improved preparedness towards future hazards" (Setiadi et al., 2010). These descriptions demonstrate that CCA and DRR both address the root reasons of vulnerability to a disaster risk (Alam et al., 2020). Disaster risk reduction (DRR) operations in agriculture have been utilized to alleviate the consequences of persistent shortages of food and avert widespread famines. Programs include risk awareness and assessment, social protection measures, infrastructure investment, early warning systems, education and training, and environmental management. DRR strives to increase a community's ability to cope with the effects of hazards shocks, and disasters before they occur (Davies et al., 2009). A sustainable adaptation strategy to inevitable disasters or changes should not only aim to lessen a social-ecological system's vulnerability, but also to increase its resilience and capacity for adaptation to potential threats and hazards in the future (Lei et al., 2014).

1.3 Part II: Disaster, Population, and Livelihoods

Population health concerns are already being impacted by climate change, including the impacts of high temperatures on people's capacity to labor in the field, with significant consequences for non-mechanized farming-based livelihoods. Between 2000 and 2016, the worldwide capacity of rural labor decreased by more than 5%. The areas of the globe most at risk from malnutrition are also those where food yields have decreased the most as a result of climate change (Watts et al., 2017). In addition to being one of the main contributors to climate change, agriculture is also very vulnerable to it. The obstacles facing the expansion of the agricultural industry as a whole include comprehending how the weather varies over time and changing management strategies to produce better harvests. Agriculture's climate sensitivity is unknown since rainfall, temperature, crops and cropping methods, soils, and management techniques vary according to area. Compared to the expected changes in temperature and precipitation, the inter-annual variability in temperature and precipitation were significantly larger. Crop losses might rise if the expected climate change results in a rise in climatic variability. Because the effects of global warming will be varied, different crops will react differently. Some 75% of the world's population lives in the tropics and two thirds of them make their living from agriculture, the tropics are more dependent on it. Given the low levels of technology, the broad variety of pests, illnesses, and weeds, the deterioration of the soil, the unequal distribution of the land, and the rapid population expansion, any influence on tropical agriculture will have an impact on their way of life. Six major crops—rice, wheat, maize, sorghum, soybeans, and barley—are cultivated on 40% of the world's arable land and account for nearly 70% of animal feed and 55% of non-meat calories (FAO, 2006).

Extreme weather occurrences are frequently thought to disrupt people's daily lives by destroying their homes, belongings, and other capital as well as their income and stockpiles, leaving them in permanent poverty and subject to ongoing shocks and pressures (Daly et al., 2020). Extreme weather events pose a severe danger to the local economy and asset bases in coastal areas in developing countries (Abdullah et al., 2016). When individuals are able to respond swiftly to stressors and shocks, or when they are robust to such shocks, their livelihoods are sustainable (Chamber & Conway, 1992). This prevents the depletion of the natural resource base on which they rely. Research on post-disaster recovery places an emphasis on the processes of strengthening victim and community resilience, as well as of their adaptation and livelihood reconstruction techniques, in light of the rising frequency and intensity of nature-triggered disasters (such as cyclones and floods) globally (Adger et al., 2011).

The largest issue confronting the Earth in a period of frequent climate change is how to produce enough food to sustain a growing population. High-yield cultivars, technological advancements, irrigation water, and fertilizers have all contributed to the current global food surplus. The achievement of food and nutrition security must be given priority in developing countries in order to have a strong global food system, achieve sustainable agriculture, and control and conserve the environment. One of the main ways to do this is to restructure relevant policies (Rukhsana & Alam, 2022; Mollah & Rukhsana, 2022).

1.3.1 Livelihoods

Livelihoods are a collection of resources that comprise abilities, assets, and activities that are essential to sustain daily living (Zhao et al., 2019). People's livelihoods have been widely seen to be disrupted by extreme weather events that destroy their homes, properties, and other capital as well as their income and stocks, leaving them in persistent poverty and subject to ongoing shocks and stresses (Uddin et al., 2021). Livelihoods are regarded as sustainable if they can withstand stress and shocks, recover from them, and preserve and strengthen their assets, activities, and capacities without depleting natural resource (Serrat, 2010).

One of the most common occurrences that destroy populations, and economies, and hinder progress in developed as well as developing countries are natural disasters. In 2015, natural disasters affected over 98 million people globally and cost the economy US$66.5 billion in damages (Patel et al., 2020). Recent global meteorological occurrences have brought attention to the possible negative effects of disasters on livelihoods and populations. A growing body of research indicates that storms, droughts, and floods negatively impact livelihoods by destroying agricultural land or significantly damaging essential infrastructure, which can ultimately cause displacement and migration in numerous parts of the world. Extreme climate-related occurrences may be so detrimental that maintaining a stable way of life may become impossible, forcing the communities to look for safety elsewhere (Krishnamurthy, 2012).

In global climate change research studies interest in improving livelihoods has grown over recent years (Smit & Wandel, 2006). It is necessary to strengthen risk mitigation and adaptation policies in order to be economically viable because extreme weather events are predicted to become more frequent and intense. To stop the widespread eviction of vulnerable populations, such initiatives can be encouraged (Krishnamurthy, 2012). Studies on sustainable livelihoods (SL) have a direct impact on accomplishing the SDGs, a global development agenda proposed by the United Nations (UN) in 2015 ("Tracking Progress on the SDGs," 2018). There is an urgent need to concentrate on lowering people's underlying vulnerabilities by taking proactive steps, involving the community in decision-making, and developing alternative and sustainable livelihoods (Patel et al., 2020).

1.4 Part III: Agriculture, Environment and Livelihoods towards Sustainability

Global attention has been drawn more and more to agricultural sustainability and farmers' sustainable livelihoods, particularly in rural areas, especially in relation to efforts for managing food and water resources (Chapman & Darby, 2016; Makate et al., 2017; Tran et al., 2018; Tran & Tuan, 2020). A system of increasing agricultural output without negatively affecting the environment is known as sustainable agricultural intensification (Struik & Kuyper, 2017). Weekley et al. discussed how to design a sustainable agriculture by

> combining environmental sensors (partially mounted to drones or through satellite vision), digital imaging, and data analysis, allowing subsurface precision soil management (including the soil meso- and microbiota), and plant-input delivery, based on real-time monitoring of the status of the soil, crop, and environment.
>
> (Struik & Kuyper, 2017)

In spite of the significant increases in agricultural output brought about by technological advancement, the poor in agriculturally marginal areas are ignored and left behind. Food insecurity is an undesirable reality in the daily lives of these marginalized poor people, and with the rising trends in population growth, the problem of hunger and food insecurity is increasingly becoming serious and urgent mainly in developing countries. Climate change is likely to further complicate this equation and increase risks to these communities' way of life (Ahmadzai et al., 2021). Through the perception of economic and ecological values, climate change cannot only have a direct impact on farmers' livelihoods but also have an indirect effect on their ability to maintain their livelihoods (Guo et al., 2022). Climate change possesses an influence on people's lives and livelihoods, especially in mountainous areas where a large portion of the population may be ethnic minorities. People are prone to vulnerability as a big portion of the population cannot understand and interpret climate data (Sohail et al., 2022). In recent years, digital agriculture, which uses a variety of tools and management techniques, has advanced significantly with the goal of reducing food insecurity and addressing climate risk. With the advancement of unmanned aerial vehicles (UAV), low-powered long-range wireless sensors, Internet of Things (IoT) gadgets, and robotics, digital agriculture (DA) and smart farming methods shifted toward digitization. A concerted effort from policymakers, researchers, and farmers is necessary to ensure that the benefits of digitalization are realized in a sustainable equitable manner which contributes to the economic development and sustainability of agricultural production (Balasundram et al., 2023).

The Brundtland Commission on Environment and Development pioneered the concept of sustainable livelihoods as a means of integrating socioeconomic and ecological concerns into a coherent, policy-relevant framework. The United Nations Conference on Environment and Development (UNCED) expanded the idea in 1992, particularly in the context of Agenda 21, It stated that sustainable livelihoods could serve as "an integrating factor that allows policies to address development, sustainable resource management, and poverty eradication simultaneously" (Karki, 2021). The term "livelihood security" refers to the protection of ownership of, or access to, resources and income-generating activities, as well as reserves and assets to manage risks, lessen shocks, and meet contingencies (Acharya, 2006). "A livelihood is sustainable when it can cope with and recover from stresses and shocks and maintain or enhance its capabilities and assets both now and in the future, while not undermining the natural resource base" (Karki, 2021). Various capital sources enable people to engage in livelihood activities, such as agricultural production, fishing, aquaculture, tourism, and migration, to accomplish certain livelihood goals. To carry out livelihood activities and achieve targeted livelihood outcomes, institutions and policies are crucial in this analytical framework (D. D. Tran et al., 2021).

1.5 Conclusion

This chapter offers a comprehensive analysis of the interconnections between these key elements of disaster risk, resilient agriculture, and livelihood and their effects on communities and agricultural systems. The chapter demonstrates the challenges posed by climate change-induced disasters, the requirement for resilient agricultural practices, and their direct impact on livelihoods. Catastrophic events including floods, droughts, and storms are much more frequent and intense as a result of climate change. These natural calamities put agricultural systems at serious risk, increasing crop failure, livestock losses, and decreased output. As a consequence, farmers and rural communities are particularly more vulnerable as their main source of income and food security are at risk. To address these challenges, it is important to build resilience in agricultural practices. Adopting techniques and strategies that help farmers more effectively respond to and rebuild from disasters is an essential part of resilient agriculture.

Combining traditional knowledge, local practices, and awareness with scientific advancements is crucial for the creation of context-specific resilience solutions. To create frameworks that promote sustainable agriculture, disaster risk reduction, and livelihood prosperity, governments, international organizations, and local communities need to collaborate together. Adequate investments in research, infrastructure, and capacity building are crucial in order to foster resilience at different scales. By comprehending the risks, implementing resilient farming practices, and embracing holistic methods, Communities may better endure the effects of climate change, safeguard their livelihoods, and create sustainable futures.

Bibliography

Abdullah, ANM, Zander, KK, Myers, B, Stacey N, Garnett ST (2016). A short-term decrease in household income inequality in the Sundarbans, Bangladesh, following Cyclone Aila. *Natural Hazards*, 83(2), 1103–1123. https://doi.org/10.1007/s11069-016-2358-1

Acharya, SS (2006). Sustainable agriculture and rural livelihoods. *Agricultural Economics Research Review*, 19, 205–217.

Adger, WN, Brown, K, Nelson, DR, Berkes, F, Eakin, H, Folke, C, Galvin, K, Gunderson, L, Goulden, M, O'Brien, K, Ruitenbeek, J, Tompkins, EL, (2011). Resilience implications of policy responses to climate change. *Wiley Interdisciplinary Reviews: Climate Change*, 2(2011), 757–766. https://doi.org/10.1002/wcc.133

Ahmadzai, H, Tutundjian, S, Elouafi, I (2021). Policies for sustainable agriculture and livelihood in marginal lands: A review. *Sustainability*, 13(16), Article 16. https://doi.org/10.3390/su13168692

Alam, ASAF, Begum, H, Masud, MM, Al-Amin, AQ, Filho, WL (2020). Agriculture insurance for disaster risk reduction: A case study of Malaysia. *International Journal of Disaster Risk Reduction*, 47, 101626. https://doi.org/10.1016/j.ijdrr.2020.101626

Alkama, R, Cescatti, A (2016) Biophysical climate impacts of recent changes in global forest cover. *Science*, 351(6273), 600–604.

Balasundram, SK, Shamshiri, RR, Sridhara, S, Rizan, N (2023). The role of digital agriculture in mitigating climate change and ensuring food security: An overview. *Sustainability*, 15(6), Article 6. https://doi.org/10.3390/su15065325

Barua, A, Katyaini, S, Mili, B, Gooch, P (2014) Climate change and poverty: Building resilience of rural mountain communities in South Sikkim, Eastern Himalaya, India. *Regional Environmental Change*, 14(1), 267–280.

Béné, C, Frankenberger, T, Nelson, S (2015) Design, monitoring and evaluation of resilience interventions: Conceptual and empirical considerations (No. 459) (Vol. 2015) London.

Bhatta, LD, van Oort, BEH, Stork, NE, Baral, H (2015). Ecosystem services and livelihoods in a changing climate: Understanding local adaptations in the Upper Koshi, Nepal. *International Journal of Biodiversity Science, Ecosystem Services & Management*, 11(2), 145–155

Budhathoki, NK, Lassa, JA, Pun, S, Zander, KK (2019). Farmers' interest and willingness-to-pay for index-based crop insurance in the lowlands of Nepal. *Land Use Policy*, 85, 1–10.

Chamber, R, Conway, G (1992), Sustainable rural livelihoods: Practical concepts for the 21st century. IDS Discussion Paper 296, Institute of Development Studies, Brighton, Sussex, UK.

Chapman, A, Darby, S (2016). Evaluating sustainable adaptation strategies for vulnerable mega-deltas using system dynamics modelling: Rice agriculture in the Mekong Delta's An Giang Province, Vietnam. *Science of the Total Environment*, 559, 326–338. https://doi.org/10.1016/j.scitotenv.2016.02.162

Constas, MA, Frankenberger, TR, Hoddinott, J, Mock, N, Romano, D, Béném, C, Maxwell, D (2014) A common analytical model of resilience measurement for development. Causal framework and methodological options. Resilience Measurement Technical Working Group Technical Series No. 2. Rome.

Daly, P, Mahdi, S, McCaughey, J, Mundzir, I, Halim, Ardiansyah,Nizamuddin, A, Srimulyani, E, (2020) Rethinking relief, reconstruction and development: Evaluating the effectiveness and sustainability of post-disaster livelihood aid. *International Journal of Disaster Risk Reduction*, 49(2020) 101650. https://doi.org/10.1016/j.ijdrr.2020.101650

Darnhofer, I (2014). Resilience and why it matters for farm management. *European Review of Agricultural Economics*, 41(3), 461–484. https://doi.org/10.1093/erae/jbu012

Davies, M, Guenther, B, Leavy, J, Mitchell, T, Tanner, T (2009). Climate change adaptation, disaster risk reduction and social protection: Complementary roles in agriculture and rural growth? *IDS Working Papers*, 2009(320), 01–37. https://doi.org/10.1111/j.2040-0209.2009.00320_2.x

FAO (2010) Characterisation of small farmers in Asia and the Pacific. Asia and Pacific Commission on Agricultural Statistics, Twenty-Third Session, Siem Reap, 26–30 Apr 2010.

FAO (2019) State of food security and nutrition in the World. www.fao.org/3/ca5162en/ca5162en.pdf. Accessed 20/11/2019

Foley, JA, DeFries, R, Asner, GP, Barford, C, Bonan, G, Carpenter, SR, Chapin, FS, Coe, MT, Daily, GC, Gibbs, HK, Helkowski, JH, Holloway, T, Howard, EA, Kucharik, CJ, Monfreda, C, Patz, JA, Prentice, IC, Ramankutty, N, Snyder, PK (2005) Global consequences of land use. *Science*, 309(5734), 570–574.

Folke, C, Colding, J, Berkes, F (2003) Synthesis: Building resilience and adaptive capacity in social–ecological systems, in: F Berkes, J Colding, C Folke (Eds.), *Navigating social–ecological systems: Building resilience for complexity and change*, Cambridge, UK: Cambridge University, 2003, pp. 352–387.

Food and Agriculture Organization of the United Nations (2006) *The state of food insecurity in the world*. Rome: FAO.

Guha-Sapir, D, Below, R, Hoyois, P (2016). EM-DAT: The CRED/OFDA international disaster database, Université Catholique de Louvain, Belgium. www.emdat.be. Accessed 25/06/2018.

Guo, A, Wei, Y, Zhong, F, Wang, P (2022). How do climate change perception and value cognition affect farmers' sustainable livelihood capacity? An analysis based on an improved DFID sustainable livelihood framework. *Sustainable Production and Consumption*, 33, 636–650. https://doi.org/10.1016/j.spc.2022.08.002

IPCC (2014). Impacts, adaptations and vulnerability, fifth assessment report. www.ipcc.ch/report/ar5/wg2. Accessed 13/06/2018.

ISDR (2005). *Hyogo, framework for action 2005–2015: Building the resilience of nations and communities to disasters*, Kobe, Hyogo.

Karki, S (2021). Sustainable livelihood framework: Monitoring and evaluation. *International Journal of Social Sciences and Management*, 8(1), Article 1. https://doi.org/10.3126/ijssm.v8i1.34399

Keith, M, O'Clery, N, Parnell, S, Revi, A (2020). The future of the future city? The new urban sciences and a PEAK Urban interdisciplinary disposition. *Cities*, 105, Article 102820. https://doi.org/10.1016/j.cities.2020.102820

Kennedy, CM, Oakleaf, JR, Theobald, DM, Baruch-Mordo, S, Kiesecker, J (2019) Managing the middle: A shift in conservation priorities based on the global human modification gradient. *Global Change Biology*, 25(3), 811–826.

Krishnamurthy, PK (2012). Disaster-induced migration: Assessing the impact of extreme weather events on livelihoods. *Environmental Hazards*, 11(2), 96–111. https://doi.org/10.1080/17477891.2011.609879

Lei, Y, Wang, J, Yue, Y, Zhou, H, Yin, W (2014). Rethinking the relationships of vulnerability, resilience, and adaptation from a disaster risk perspective. *Natural Hazards*, 70(1), 609–627. https://doi.org/10.1007/s11069-013-0831-7

Leichenko, R, Silva, JA (2014) Climate change and poverty: Vulnerability, impacts, and alleviation strategies. *Wiley Interdisciplinary Reviews: Climate Change*, 5(4), 539–556.

Liu, J, Kuang, W, Zhang, Z et al. (2014) Spatiotemporal characteristics, patterns, and causes of land-use changes in China since the late 1980s. *Journal of Geographical Sciences*, 24, 195–210. https://doi.org/10.1007/s11442-014-1082-6

Makate, C, Makate, M, Mango, N (2017). Sustainable agriculture practices and livelihoods in pro-poor smallholder farming systems in southern Africa. *African Journal of Science, Technology, Innovation and Development*, 9(3), 269–279. https://doi.org/10.1080/20421338.2017.1322350

Molla, SH, Rukhsana (2022). Spatio-temporal analysis of built-up area expansion on agricultural land in Mousuni Island of Indian sundarban region. In: Rukhsana, Alam, A. (eds) *Agriculture, environment and sustainable development*. Cham: Springer. https://doi.org/10.1007/978-3-031-10406-0_6

Molla, SH, Rukhsana (2023). Mapping spatial dynamicity of cropping pattern and long-term surveillance of land-use/land-cover alterations in the Indian Sundarban region. *Arabian Journal of Geosciences*, 16, 379. https://doi.org/10.1007/s12517-023-11444-8

OECD (2013). What does 'Resilience' mean for donors? An OECD factsheet. OECD. www.oecd.org/dac/governancedevelopment/May%2010%202013%20FINAL%20resilience%20PDF.pdf

Patel, SK, Mathew, B, Nanda, A, Mohanty, B, Saggurti, N (2020). Voices of rural people: Community-level assessment of effects and resilience to natural disasters in Odisha, India. *International Journal of Population Studies*, 6(1). https://doi.org/10.18063/ijps.v6i1.1042

Reddy, PP (2015). Introduction. In PP Reddy (Ed.), *Climate resilient agriculture for ensuring food security* (pp. 1–15). India: Springer https://doi.org/10.1007/978-81-322-2199-9_1

Ribeiro, PJG, Goncalves, LAPJ (2019). Urban resilience: A conceptual framework. *Sustainable Cities and Society*, 50, Article 101625. https://doi.org/ 10.1016/j.scs.2019.101625

Rukhsana, Alam A. (2021). Agriculture, food, and nutritional security: An overview. In: Rukhsana, Alam A. (Eds.), *Agriculture, food and nutrition security*. Cham: Springer. https://doi.org/10.1007/978-3-030-69333-6_1

Rukhsana, Alam A. (2022). Agriculture, environment and sustainable development: An overview. In: Rukhsana, Alam A. (Eds.), *Agriculture, environment and sustainable development*. Cham: Springer. https://doi.org/10.1007/978-3-031-10406-0_1

Rukhsana, Alam A., Mandal, I (2021). Impact of microclimate on agriculture in India: Transformation and adaptation. In: Rukhsana, Alam A. (Eds.), *Agriculture, food and nutrition security*. Cham: Springer. https://doi.org/10.1007/978-3-030-69333-6_3

Rukhsana, Molla S.H. (2023). Soil site suitability for sustainable intensive agriculture in Sagar Island, India: A geospatial approach. *Journal of Coastal Conservation*, 27, 14. https://doi.org/10.1007/s11852-023-00943-1

Serrat, O. (2010). *The sustainable livelihoods approach*. Washington, DC: Asian Development Bank.

Setiadi, N, Birkmann, J, Buckle, P (2010). Disaster risk reduction and climate change adaptation: Case studies from South and Southeast Asia. SOURCE Publication No. 14/2010. Publication Series of UNU-EHS. UNU-EHS, Bonn.

Smit, B, Wandel, J (2006). Adaptation, adaptive capacity and vulnerability. *Global Environmental Change*, 16(3), 282–292. https://doi.org/10.1016/j.gloenvcha.2006.03.008

Smith, L, Frankenberger, T, Langworthy, B, Martin, S, Spangler, T, Nelson, S (2015) Ethiopia pastoralist areas resilience improvement and market expansion (PRIME) project impact evaluation baseline survey report. Feed the Future FEEDBACK Project, Washington DC, USA. 1–155.

Sohail, MT, Mustafa, S, Ali, MM, Riaz, S (2022). Agricultural communities' risk assessment and the effects of climate change: A pathway toward green productivity and sustainable development. *Frontiers in Environmental Science*, 10, 948016. https://doi.org/10.3389/fenvs.2022.948016

Song, X-P, Hansen, MC, Stehman, SV, Potapov, PV, Tyukavina, A, Vermote, EF, Townshend, JR (2018) Global land change from 1982 to 2016. *Nature*, 560(7720), 639.

Stefen, W et al. (2015). Planetary boundaries: Guiding human development on a changing planet. *Science* 347, 1259855.

Struik, PC, Kuyper, TW (2017). Sustainable intensification in agriculture: The richer shade of green. A review. *Agronomy for Sustainable Development*, 37(5), 39. https://doi.org/10.1007/s13593-017-0445-7

Šūmane, S, Kunda, I, Knickel, K, Strauss, A, Tisenkopfs, T, Rios, I, Des, I., Rivera, M, Chebach, T, Ashkenazy, A (2018). Local and farmers' knowledge matters! How integrating informal and formal knowledge enhances sustainable and resilient agriculture. *Journal of Rural Studies*, 59, 232–241. https://doi.org/10.1016/j.jrurstud.2017.01.020

Theobald, DM, Kennedy, C, Chen, B, Oakleaf, J, Baruch-Mordo, S, Kiesecker, J (2020) Earth transformed: Detailed mapping of global human modification from 1990 to 2017. *Earth System Science Data*, 12(3), 1953–1972.

Tracking progress on the SDGs. (2018). *Nature Sustainability*, 1(8), Article 8. https://doi.org/10.1038/s41893-018-0131-z

Tran, DD, Huu, LH, Hoang, LP, Pham, TD, Nguyen, AH (2021). Sustainability of rice-based livelihoods in the upper floodplains of Vietnamese Mekong Delta: Prospects and challenges. *Agricultural Water Management*, 243, 106495. https://doi.org/10.1016/j.agwat.2020.106495

Tran, DD, Van Halsema, G, Hellegers, PJGJ, Ludwig, F, Seijger, C (2018). Stakeholders' assessment of dike-protected and flood-based alternatives from a sustainable livelihood perspective in An Giang Province, Mekong Delta, Vietnam. *Agricultural Water Management*, 206, 187–199. https://doi.org/10.1016/j.agwat.2018.04.039

Tran, TA, Tuan, LA (2020). Policy transfer into flood management in the Vietnamese Mekong Delta: A North Vam Nao study. *International Journal of Water Resources Development*, 36(1), 106–126. https://doi.org/10.1080/07900627.2019.1568862

Turner, BL, Lambin, EF, Reenberg, A (2007) The emergence of land change science for global environmental change and sustainability. *Proceedings of the National Academy of Sciences*, 104(52), 20666–20671.

Uddin, MS, Haque, CE, Khan, MN, Doberstein, B, Cox, RS (2021). "Disasters threaten livelihoods, and people cope, adapt and make transformational changes": Community resilience and livelihoods reconstruction in coastal communities of Bangladesh. *International Journal of Disaster Risk Reduction*, 63, 102444. https://doi.org/10.1016/j.ijdrr.2021.102444

Ullah, R, Jourdain, D, Shivakoti, GP, Dhakal, S (2015). Managing catastrophic risks in agriculture: Simultaneous adoption of diversification and precautionary savings. *International Journal of Disaster Risk Reduction*, 12, 268–277. https://doi.org/10.1016/j.ijdrr.2015.02.001

Walker, B, Holling, CS, Carpenter, S, Kinzig, A (2004). Resilience, adaptability and transformability in social–ecological systems. *Ecology and Society*, 9(2). https://doi.org/10.5751/ES-00650-090205

Ward, PJ, De Ruiter, MC, Mård, J, Schröter, K, Van Loon, A, Veldkamp, T, Von Uexkull, N, Wanders, N, AghaKouchak, A, Arnbjerg-Nielsen, K, Capewell, L, Carmen Llasat, M, Day, R, Dewals, B, Di Baldassarre, G, Huning, LS, Kreibich, H, Mazzoleni, M, Savelli, E, … Wens, M (2020). The need to integrate flood and drought disaster risk reduction strategies. *Water Security*, 11, 100070. https://doi.org/10.1016/j.wasec.2020.100070

Watts, N, Amann, M, Ayeb-Karlsson, S, et al. (2017) The lancet countdown on health and climate change: From 25 years of inaction to a global transformation for public health. *The Lancet*, 391, 581–630.

Wu, W, Zhao, S, Zhu, C, Jiang, J (2015) A comparative study of urban expansion in Beijing, Tianjin and Shijiazhuang over the past three decades. *Landscape and Urban Planning*, 134, 93–106.

Zanotti, L, Ma, Z, Johnson, JL, Johnson, DR, Yu, DJ, Burnham, M, Carothers, C (2020). Sustainability, resilience, adaptation, and transformation: Tensions and plural approaches. *Ecology and Society*, 25(3), 4. https://doi.org/10.5751/ES-11642-250304

Zhang, J, Wang, J, Chen, S, Tang, S, Zhao, W (2022). Multi-hazard meteorological disaster risk assessment for agriculture based on historical disaster data in Jilin Province, China. *Sustainability*, 14(12), 7482. https://doi.org/10.3390/su14127482

Zhao, Y, Fan, J, Liang, B, Zhang, L (2019). Evaluation of sustainable livelihoods in the context of disaster vulnerability: A case study of Shenzha county in Tibet, China. *Sustainability*, 11(10), 2874. https://doi.org/10.3390/su11102874

2 Farmers' Mitigation and Adaptation Strategies against Climate Change Impact on Agriculture

Sivaraj Paramasivam, Parthasarathy Seethapathy and G. Vivekanathapatmanaban

2.1 Impact of Climate Change on Agriculture

Probably the most important environmental challenge to combating hunger, malnutrition, disease, and poverty is global climate change, which has an impact on agricultural output, which is a primary cause of these problems. In addition to being a climate-sensitive sector, agriculture is actually a sector that offers a sustainable livelihood for over 60% of Indians. Global climate change will have major ramifications for food and water security in the next decades, and there are strong signs that developing countries like India will bear the brunt of climate change's negative consequences. Furthermore, the rural population of poor nations, for whom agriculture is the primary source of direct and indirect employment, will be the most affected by agricultural susceptibility to global climate change processes, as their income will be the most affected. The monsoon's arrival and performance are critical for India's agricultural growth because most of the country's states rely on it to meet their agricultural and water needs, as well as to conserve and propagate the country's unique biodiversity. According to the studies, summer rainfall accounts for over 70% of total annual rainfall in India and is critical to Indian agriculture. Small climate changes can have a big impact on water resources, especially in arid and semi-arid areas. Agriculture, drinking water, and hydroelectric power generation will all be affected. The agricultural sector is the largest user of water resources, and water supply unpredictability has a significant impact on rural socioeconomics, food security, employment patterns, and livelihood security. It is critical that the policy addresses the loss of livelihoods in developing countries like India, as climate change would add to the stress on ecological and socio-economic systems that are already under severe strain as a result of human-induced changes in land use patterns, urbanization, the elimination of wetlands, and nutrient overloading in water systems. Climate change may have an impact on the global water cycle's ability to support critical food production. Agriculture is the backbone of the Indian economy, and the majority of Indians rely on it for their living. India is a vast country with 15 distinct agro-climatic zones, as well as a wide range of seasons, crops, and

DOI: 10.4324/9781003275916-3

farming practices. Millions of Indians' lives and livelihoods would undoubtedly be affected. While the amount and impact of climate change differ by place, it is predicted to have a stronger impact on agricultural productivity and crop patterns. According to Sinha and Swaminathan (1991), every 2°C increase in temperature caused by climate change reduces rice output by roughly 0.75 tonnes/ha in high yield areas. For example, every 0.5°C increase in winter temperatures reduces wheat output by 0.45 tonnes per hectare. Rao and Sinha (2008) found that without accounting for CO_2 fertilization effects, wheat yields might drop by 28 to 68%, but after accounting for CO_2 fertilization effects, yields could raise by 4 to 34 %. According to Shindell et al. (1998), doubling CO_2 would rise the average temperature in India from 2.33°C to 4.33°C. Despite the impressive effects of increased CO_2 on crop development, it is predicted that crop production in India will decline by 10–40 % by 2100 AD. Climate change may have an impact on the agriculture sector in terms of crop quantity and quality. Climate change has a direct impact on food production around the world, particularly in industrialized countries like India, which is having difficulty obtaining enough food materials for its citizens. Severe climate change is expected to have a significant influence on food production and may pose a threat to food security, necessitating the use of particular agricultural methods to resist it (IPCC, 2007).

Long-term changes in weather patterns occur in a changing climate, which poses a threat to agricultural output due to extremes in temperature, increasing rainfall unpredictability, and rising sea levels, which threaten coastal freshwater reserves and increase the risk of flooding. Climate change will have a detrimental impact on irrigated crop production across regions due to variations in temperature and water availability. Around 54% of India's groundwater wells are shrinking, with 16% shrinking by more than a meter each year. According to Mall and Agarwal (2003), a 20°C increase in temperature resulted in a 15–17% reduction in rice and wheat yields. Rain-fed agriculture will be harmed mostly as a result of rainfall variability and a decrease in the number of wet days (Venkateswarlu and Rama, 2010). By 2080, global agricultural productivity is expected to fall by 3–16% (Anupama, 2014). Climate change may lead to price increases in agricultural commodities, feed supplies, and, as a result, livestock products such as meat and milk.

Climate change will have a major impact on biodiversity in India, which has been severe in recent years, as have insects and diseases that affect agriculture and food supply. Higher temperatures tend to lower crop production and stimulate the spread of weeds and pests. Pest and disease dynamics will be drastically altered, resulting in novel patterns of pests and illnesses affecting crop productivity. Until now, research on the influence of climate change on crop diseases has been limited, with the majority of work focusing on the effects of a single atmospheric component or meteorological condition on the host, pathogen, or their interaction in structured ecosystems. Indirectly, the effects of rising temperatures and CO_2 levels in the atmosphere on pests and illnesses have an effect on agricultural production. These interactions are

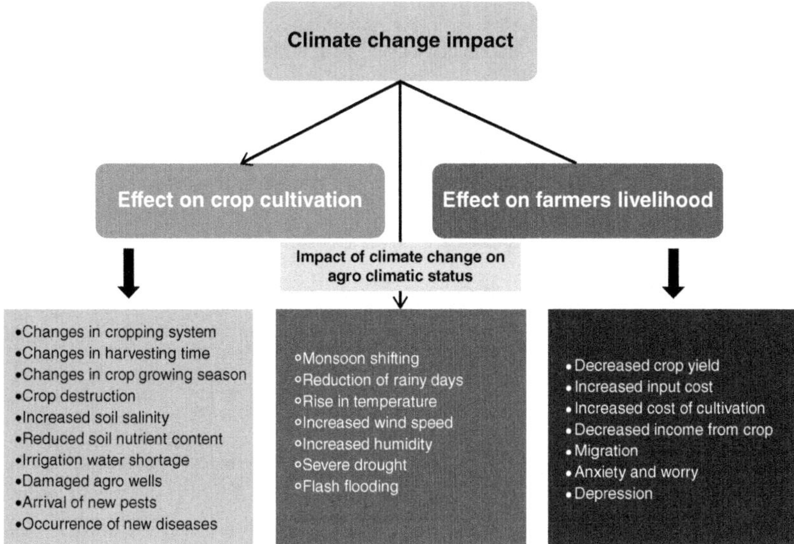

Figure 2.1 Climate change impact on agriculture and farmer's livelihood

complex, and the precise extent to which they affect agricultural productivity is unknown. Increased pathogen virulence or environmental-driven mutations may affect crop resistance to certain diseases as a result of global warming or drought in the country. Numerous studies have been undertaken on the effect of climate change on the plant diseases that follow. Individual factors associated with climate change, such as warming, increased CO_2 levels, decreased rainfall, and irregular rainfall patterns, have been researched for their effects on a variety of infections and illnesses across a wide variety of crops and crop kinds (Figure 2.1).

2.2 Impact of Climate Change on Crops

Agriculture has been mostly unaffected by regional climate change. Changes in agricultural crop life cycle provide critical evidence of crop response to recent regional climate change. Droughts have become more common as a result of climate change, and they are expected to become more frequent and intense throughout Southeast Asia. Growing water usage, population increase, urban expansion, and environmental preservation programs all worsen their consequences in many places. Droughts results in crop failures as well as the loss of cattle pasture grazing land. Some farmers may decide to leave drought-stricken areas and settle elsewhere. Changes in precipitation patterns, including extended periods of excessive rain and drought. Increased average temperatures; hotter summers and warmer winters can cause plant cycles to be disrupted, resulting in premature blossoming, less pollination, and frost damage (Figure 2.2).

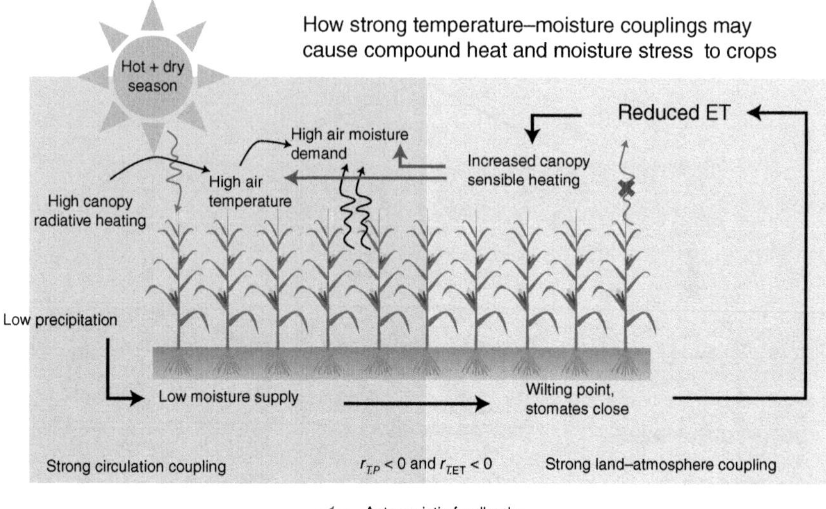

Figure 2.2 Stronger temperature–moisture couplings exacerbate the impact of climate warming on global crop yields (Lesk and Anderson, 2021)

Flooding has increased, causing crop loss, water contamination, and soil erosion. Drought levels are rising, posing a threat to plant life and raising the risk of wildfires. Degraded soils: As a result of monoculture farming, the soil becomes less organically rich and is more prone to erosion and water pollution. Crop factories: Plant viability is decreased in industrial agriculture due to biodiversity loss. Crop output is affected by a wide range of factors, including climate change. The repercussions are induced by a wide range of biotic and abiotic stressors (and their interactions), varied crop responses to stress, and farmers' management adjustments in response to changing socioeconomic and climatic conditions, making their evaluation difficult. Farmers' adaptations have their own climate impacts, such as shifting resource usage or greenhouse gas emissions (Figure 2.3).

2.3 Adaptation and Mitigation Strategies

Climate change adaptation techniques are actions performed to alter farming practices to the current climatic conditions. To handle current risks, increased unpredictability, and emerging trends, manage risk, uncertainty, and build adaptive capacity, it should be a coordinated strategy (Catherine, 2010). By moving planting dates, picking cultivars with varying growth durations, or changing crop rotations, farmers can adapt to rising temperatures, wind speeds, poor rainfall distribution, delayed monsoon, drought, and floods to some extent (Deressa et al., 2008; Saleem and Ifthikar, 2013; Komba and Muchpondwa, 2015).

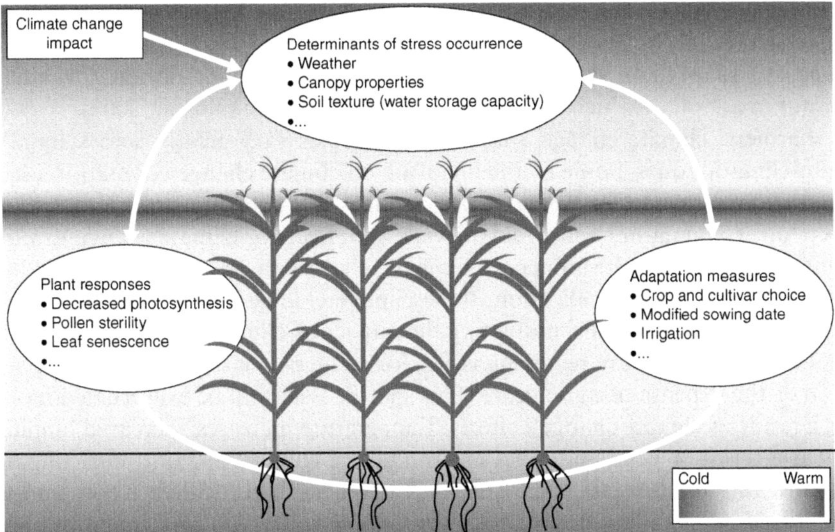

Figure 2.3 Schematic illustration of links between climate change, heat stress occurrence, plant responses to heat stress and farmer's adaptation measures (Siebert and Ewert, 2014)

Early warning systems were also available to help them monitor changes in a climatic phenomenon so that they could alter their farm and other activities accordingly. Weather information disseminated through various media aids farmers in addressing climate change issues. Farmers can use this knowledge to pursue adaptation strategies in a variety of ways, from variety selection to collecting their produce. Conservation of resources, use of organic manures, planting of pest and disease resistant crops, draining of wetland, cover cropping, use of minimum tillage system (zero or minimum), reforestation, early or late sowing varieties, protection of water sheds and mulching, reducing access to eroded and erosion prone areas, mixed farming practices, out migration from climate-risk areas, and use of windbreaks were some of the adoption measures taken by farmers to migrate (Saleem and Ifthikar, 2013).

The "adjustment of natural or human systems to a new or changing environment," according to the IPCC (2001), is the "adjustment of natural or human systems to a new or changing environment." Adaptation to climate change is defined as "modification in natural or human systems in response to present or anticipated climatic stimuli or their effects, with the goal of minimising harm and maximising benefits." According to the IPCC (2001), "a man-made intervention to reduce the sources or boost the sinks of greenhouse gases," In other words, it refers to the actions taken to mitigate the negative effects of natural disasters, as well as linked environmental and technological disasters. In general, the more mitigation there is, the fewer the effects we'll have to adjust to, and the fewer the hazards we'll have to try to prepare for.

Conversely, the greater the degree of preparedness, the fewer the effects of any given degree of climate change are likely to be. At the farm level, there are two stages to adaptation: recognizing the change in climate and deciding whether or not to adapt, or which adaptation method to use (Maddison, 2007).

Farmers' climate change adaptation strategies vary greatly across India's agro-climatic zones. Some of the literature on climate change adaptation used by Indian farmers (Shankara, 2010; Rubina, 2014) demonstrated that they were diverse in nature and could alleviate the effects of climate change to varying degrees. Given the influence of climate change on yield decrease and, ultimately, farm income realization, these climate change adaptation techniques will go a long way toward mitigating the impact of climate change.

Pest and disease are recurring issues for farmers, consuming a sizable portion of the expense of agriculture. This specific issue will be exacerbated more as a result of climate change's effects. Temperature increases, increased humidity as a result of heat, and other environmental changes all contribute to pest growth and disease epidemics. This increased abundance will be accompanied by increased population dynamics, reproduction, transit, and overwintering rates. Crop diversity contributes to the disruption of the pest-pathogen cycle or pest-pathogen crop linkages that become established in mono cropping circumstances. Additionally, because diverse agro ecosystems mirror more natural systems, they can support a higher range of life forms, which are mostly natural antagonists of crop pests and illnesses. Diversifying cereal cropping systems with the addition of oilseed, pulse, and fodder crops is another strategy for controlling pest and disease risk. Crop rotation disrupts pest and disease epidemics by replacing cereal crops with herbaceous crops that are not sensitive to the same causal agents (Nadeem et al., 2019). Minimal tillage may increase soil biodiversity, which may result in increased disease suppression, and stand densities may be changed to allow for more accurate climatological adaptation to disease progression. Climate change, as measured by temperature increases, droughts, and CO_2 levels, also has an effect on the three factors of the plant disease triad (Buttke et al. 2021).Climate change will likely have a more severe impact on a range of apple diseases such as *Alternaria* leaf spot, gummosis, and fruit scab (Shuttleworth, 2021). Another indicator of climate change is the development of the Sanjo Scale as a result of dry spring weather. For every 1°C increase in overall temperature, aphid attacks occur nearly two weeks earlier (Hulle et al., 2010).Numerous climate change-related effects on aphid population demographics will thus have an effect on viral infections. The red mite has established a significant presence in virtually all crops grown in temperate areas, including scale insects, which multiply rapidly when temperatures and host plant receptivity are favorable (Figure 2.4).

Climate change is a phenomenon that scientists have interpreted in a variety of ways. Farmers, too, have an opinion about climate change and how they see the effects of climate change, which varies from farmer to farmer. The literature on farmer perceptions of climate change validated the differences in farmer perceptions of climate change in different parts of the world (Apata et al., 2009;

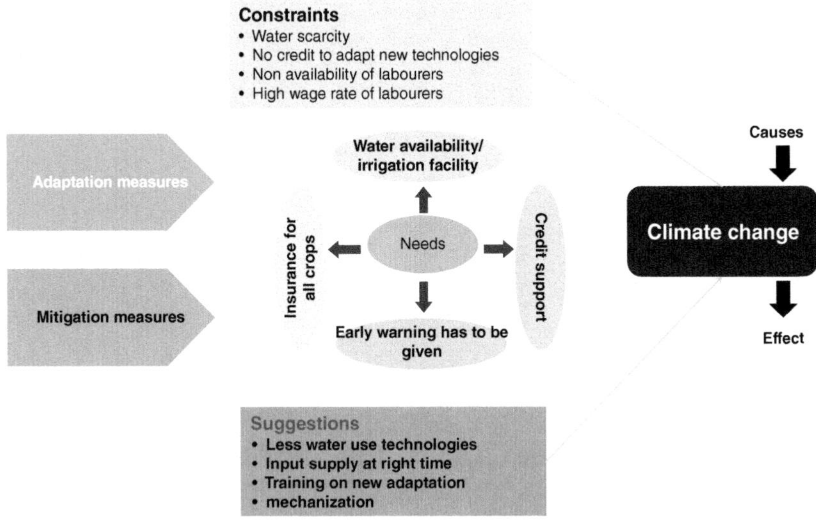

Figure 2.4 Adaptation and mitigation strategies conceptual model on climate change

Gbetibouo and Ringler, 2009; Tiwari et al., 2010; Idrisa et al., 2012; Abid et al., 2015). The way people think about climate change will have a big impact on whether or not they adopt climate change adaptations. Furthermore, several personal and biophysical factors were found to influence these climate change adaptations, as reported in a few studies (Shankara, 2010; Rubina, 2014; Shongwe et al., 2014; Sivaraj et al., 2014). As a result, several factors have an impact on whether or not specific climate change adaptations are adopted. Awareness of climate change issues, exposure to training related to climate change, and interaction with scientists and extension workers might have influenced the adoption of climate change adaptations. Climate change awareness, exposure to climate change training, and interactions with scientists and extension workers may have all encouraged the adoption of climate change adaptations.

2.4 Documentation of Climate Change Adaptations

Through focus group discussions, we documented the location-specific climate change adaptation techniques taken by farmers at the grass-roots level in several farming systems, including wetland, dryland, and garden land eco systems. Tables 2.1–2.3 show the documented adaptation measures in the wet, dry, and garden land farming systems, respectively.

2.5 Coping Strategies against Climate Change

Farmers' ability to adapt to climate change is hindered by their poverty and significant reliance on rainfed agriculture, as well as a lack of economic and technical resources. Given the variety of obstacles, the ability of small and

Table 2.1 Documentation of adaptations followed in the wetland farming system

S. No.	*Adaptation measure*	*Details of the adaptation measures*	*Associated factors*	*Perceived benefits*
1.	**Crop selection and variety selection**	**Crop selection** In wet land conditions, rice, sugarcane and banana were grown as major crops whereas area under sugarcane cultivation was reduced drastically and large number of farmers switched over to rice, bananas, and vegetables in larger area.	Failure of monsoon, water scarcity, poor marketing price for sugarcane, training and awareness about the cultivation of vegetables.	Farmers got good yield in rice with the available water source.
		Changes in rice variety In rice cultivation changing of variety was observed. They moved from the cultivation of long duration varieties (Ponni, Seraga samba, Arachamba) to short duration varieties namely IR-20, NLR-3, ADT- 45.	Awareness about the short duration varieties, water scarcity and rainfall	Short duration varieties were provided the stable income from agriculture
		Changes in banana variety In banana cultivation too farmers had been changing the varieties namely Rasthali, Poovan and local cultivars.	Climatic factors such as temperature, rainfall, wind speed, relative humidity and market demand.	Change of varieties gives good yield and income

2.	**Change in sowing/ planting/transplanting operations**	**Delay in rice transplanting** Under normal conditions transplanting of rice was done in August–September months but the farmers delayed the transplanting of rice to November–December months. It might have been due to the delay in the monsoon during the season.	Awareness and knowledge about weather parameters and decision-making about adaptation measures	Crop failure was avoided during the monsoon and rice crops were protected from extreme climate fluctuations.
		Delaying the planting of banana Banana was normally planted during January–February months. Due to the delay in the monsoon the planting is being taken up in April–May months to save the plants from winds and to grow a healthy crop.	Wind speed, wind velocity, rainfall, suitable harvesting time and market demand	Proper price for banana and protected the plants from the wind problems.
3.	**Diversification in cropping system**	**Cultivation of drought tolerant crops** Sorghum and maize were grown in the fallow lands during drought period (April–June)	Water scarcity and drought during summer,	Farm loss was reduced due to the changing of crop from rice, banana to vegetables and other crops.
		Mixed cropping in vegetables Summer season vegetables namely brinjal, bhendi, tomato were being grown during April–May season.	Market demand and knowledge and experience in vegetable cultivation.	Remunerative income from the vegetable cultivation.
		Fallow crop: sesame In summer season, if farmers receive summer shower then sesame was grown as fallow crop by direct sowing method.	Onset of monsoon, prevalent crop in the field and market demand for the produce	Additional income was obtained from the fallow crops.

(*Continued*)

Table 2.1 (Continued)

S. No.	*Adaptation measure*	*Details of the adaptation measures*	*Associated factors*	*Perceived benefits*
4.	**Changes in the intercultural operations**	**Use of machineries and bio inputs** Harvesting of rice was done with the rice harvester. Usage of bio fertilizers and bio control agents along with the fertilizers and plant protection chemicals for rice and banana was followed by the farmers in wet land.	Trainings and meetings related to application of organic and bio inputs for the crops by the KVK, social participation, extension agency contact and availability of bio inputs in different formulation in the nearby areas.	Plant protection measures were comparatively improved through bio inputs than the usage of chemical fertilizers and plant protection chemical.
5.	**Income diversification**	**Livestock income** Majority of the farmers owned livestock namely sheep, goat, and milch cows.	Bank loans, veterinary hospital and rich experience in animal rearing.	Stable income from the livestock and the farm waste could be used as feed for livestock.
		Farm laborers Some of the farmers were working as farm laborers in farming activities on the nearby by farms.	Low income from agriculture, failure of monsoon, increased cost of cultivation.	Source of revenue for most of the small and marginal farmers.
		Off-farm income Some of the farmers were working as laborers for construction works and in nearby industries.	Lack of credit for agriculture, location of industries, failure of monsoon, low income from farming, crop loss and yield reduction.	Maintenance of their livelihood security.

Table 2.2 Documentation of adaptations followed in the dryland farming system

S. No.	*Adaptation measure*	*Details of the adaptation measures*	*Associated factors*	*Perceived benefits*
1.	**Crop selection**	In dry land farming, the major crops were cotton, millet, and pulses. Crop selection was done based on the onset of monsoon and rain fall. Before the rainfall the millets were sown in the June–July months and cotton was grown during September–November coinciding with the onset of northeast monsoon. Sowing of the crops was entirely based on the seasonal rainfall in the dry land conditions. Pulses namely green gram, black gram and Red gram were grown during November to January months.	Onset of monsoon and rainfall distribution, market demand and farmers' decision.	Normal crop yield was obtained without crop loss. Income from the crops was sustained.
2.	**Diversification in cropping system**	**Inter cropping of red gram with millet and cotton** Red gram was grown as inter crop in cotton cultivation on the border rows. In millet cultivation, red gram was inter cropped with kodo millet and banyard millet.	Educational status and social participation.	Additional income was obtained from the red gram cultivation.
		Mixed cropping of pulses During October–December months the pulses namely black gram and green gram were grown in the mixed cropping system.	Farmers experience and rainfall.	Regular income was obtained with the available land area by the pulses cultivation.

(*Continued*)

Table 2.2 (Continued)

S. No.	*Adaptation measure*	*Details of the adaptation measures*	*Associated factors*	*Perceived benefits*
3.	**Income diversification activities**	**Livestock rearing** More than 90% of the farmers owned sheep, goat, or cows in their homes.	Economic status, awareness and knowledge about livestock rearing.	Stable income for the family.
		Off-farm income Failure of monsoon and drought condition made the farmers move to be laborers in the nearby city and laborers as in MGNREGA Scheme.	Crop failure, monsoon failure and job availability in nearby city.	Maintain their family and sustain their income for their livelihood security.

Table 2.3 Documentation of adaptations followed in the garden land farming system

S. No.	*Adaptation measure*	*Details of the technique*	*Associated factors*	*Perceived benefits*
1.	**Crop selection**	In garden land system of Madurai district the farmers changed to the cultivation of coconut and fodder crops in larger area from the cultivation of vegetables and grapes.	Failure of monsoon, yield reduction, water scarcity and non-availability of farm labor.	Crop loss was reduced and the fodder for livestock could be made available.
		The garden land system was cultivating the vegetables namely onion, chilli, bhendi, bitter gourd, cluster bean etc.,	Climatic variability, educational status, water scarcity and lack of funding	Income from the mango cultivation was comparatively better than vegetable cultivation.
		The farmers changed the larger areas of vegetables into mango groves within 15 years.		
		In drought conditions some of them were grown sorghum and maize during July–September.		
2.	**Moisture and water conservation**	**Micro-irrigation** Drip irrigation was practiced by the garden land farmers through the government subsidy. It was used for onion, bhendi, tomato brinjal to conserve water.	Government support, water scarcity, extension agency contact and social participation.	Water level was maintained in the wells and it was conserved. Comparatively, yield was also improved due to the drip irrigation practices.

(*Continued*)

Table 2.3 (Continued)

S. No.	*Adaptation measure*	*Details of the technique*	*Associated factors*	*Perceived benefits*
3.	**Diversification in cropping system**.	**Inter cropping** Groundnut was grown as intercrop in mango plantations. Fodder crops were grown as intercrop in coconut plantations. Some of the farmers cultivated cucumber in the border of the onion field.	Climatic parameters and water availability	Additional income was obtained from red gram cultivation.
		Mixed cropping of vegetables Onion and chilli, brinjal and bhendi were cultivated under the mixed cropping system.	Crop nature, market demand and farmer experience	Consistent income was obtained with the available land area.
		Introduction of perennial crops Mango and coconut were introduced in the garden land system within 15 years.	Water scarcity, crop nature, drought, and labor scarcity.	Maintainable income was obtained from the perennial crops and for easy maintenance.
4.	**Income diversification activities**	**Livestock rearing** More than 90% of the farmers owned sheep, goat, or cows in their homes.	Economic status, awareness, and knowledge about livestock rearing.	Steady income came from livestock rearing and it helped them to adjust family expenses.
		Off-farm income Farmers moved from agriculture towards industrial works, and as laborers in the city.	Crop failure, monsoon failure and educational status.	Maintain their family and sustained their income for their livelihood security.

marginal farmers to adapt to climate change is now quite limited. Small and marginal farmers' adaptation strategies to climate change are divided into two categories: long-term adjustments based on mitigation technologies that may provide benefits for a long time; and short-term adjustments based on borrowing, reducing total expenditure, selling farm assets, and resorting to crop insurance, which farmers typically use to cushion the effects of weather variability. Farmers may pursue adaptation even if they are unaware of the causal reasons in order to avoid the dangers posed by climate change.

a **Technological mitigation (Long run adjustment)**

1 Change the crop pattern and planting early.
2 Shifting to tree crops.
3 Adopting integrated / mixed farming system.
4 Adopting soil and water conservation measures.

b **Short run adjustment**

1 Reducing their total expenditure
2 Borrowing
3 Crop insurance
4 Selling of land, livestock and nonfarm assets

Suggestions to overcome climate change effects

- Early warning has to be given to the farmers about environmental changes.
- Creating awareness to the farmers about appropriate coping mechanism against climate change.
- Subsidies/compensation has to be given for the crops to make up the cost of cultivation due to weather aberrations.
- Insurance has to be extended to all crops and make individual benefit.
- Providing financial support for soil nutrient enrichment and free soil test.
- Providing incentives/support for increasing the green manuring.
- Support price has to be given to all the crop produce based on cost of cultivation.
- Creating awareness/support for adoption of organic farming technologies.
- Providing short message service (SMS) about new innovations and technologies in regional language.
- Need more information about flower and vegetable cultivation.
- Shifting to organic farming and grow more medicinal plant.
- Insurance has to be extended to all crops for the benefit of farmers.
- Increasing the drip subsidy rate based on crop and area

According to Geethalakshmi and Dheebakaran (2008), agriculture is capable of adapting to limited climatic change through the application of appropriate technology and agronomic modifications. However, this capability varies widely among places. As a result, climate-vulnerable areas must be recognized. It is vital to define the nature of this adaptation and the critical rates of climatic change to which agriculture can adapt under local conditions. To improve our understanding of the significance of climate change and its implications for agriculture and humanity, more research is needed into how agriculture can best adapt to avoid or benefit from annual, seasonal, and intra-seasonal climate variability in different agroclimatic regions of the country. Improved understanding of the effects of climate change on crop yields and physical processes such as soil erosion, salinization, nutrient depletion, insect pests, diseases, and hydrological conditions is required. A range of potentially useful agronomic changes, such as irrigation, crop selection, sowing time fertilization, and so on, is also required. Keeping in mind the rise in population, the need for food grains, and the influence of productivity changes caused by climate change, management methods should be customized to maximize the yield of vital crops even under changing climate scenarios. A new research program should be launched with the goal of identifying or developing cultivars and management approaches that are suitable for changing climates. Despite the fact that many model forecasts on future climate change scenarios are available, more accurate scenarios with finer geographical dimensions are necessary to assess the implications.

The government's responsibility is to improve and promote livestock and poultry among farmers in order to offset the loss of income caused by climate change. Agro-weather alerts are distributed through a variety of media. To promote knowledge and use of meteorological and agricultural advisory services, training and demonstrations should be provided. As a result, policy measures for the promotion of farmer groups at the grass-roots level should be developed in order to enhance the adoption of climate-specific adaptation technology. Farmers in all three farming methods used adaptation measures such as crop or variety selection based on climate variability, planting according to the prevailing monsoon, intercropping, mixed cropping, and cow rearing. To increase the adoption rate, this must be increased. Farmers will be provided with market-driven extension services and market intelligence services to help them effectively market their farm products. Farmers should be offered storage and value-adding possibilities to help them earn a living and reduce the negative effects of climate change on agriculture.

Bibliography

Abid, M., Scheffran, J., Schneider, U.A. and Ashfaq, M. 2015. Farmer's perception and adaptation strategies to climate change and their determinants: The case of Punjab province, Pakistan. *Earth System Dynamics*, 6: 225–243.

Agarwal, P.K. 2003. Impact of climate change on agriculture. *Journal Of Plant Biology*, 30(2), 189–198.

Apata, T.G., Samuel, K.D. and Adeola, A.O. 2009. Analysis of climate change perception and adaptation among Arable food farmers in South West Nigeria. *International Association of Agricultural Economists Conference Proceedings*, Beijing, China.

Anupama, M. 2014. Climate change and its impact on agriculture. *International Journal of Research Publication*, 4(4), 112–115.

Buttke, D., Wild, M., Monello, R., Schuurman, G., Hahn, M. and Jackson, K. 2021. Managing wildlife disease under Climate Change. *EcoHealth*, 18, 1–5.

Catherine, P. 2010. Climate change adaptation enabling people living in poverty to adapt. Oxfam International Research Report, April 2010.

Deressa, T.T., Hassan, R.M., Ringler, C., Alemu, T. and Yesuf, M. 2008. Analysis of the determinants of farmers' choice of adaptation methods and perceptions of climate change in the Nile basin in Ethiopia. IFPRI Discussion Paper No: 798, Washington D.C., 2008.

Geethalakshmi, V. and Dheebakaran, G. 2008. Impact of climate change on agriculture over Tamil Nadu. www.researchgate.net/publication/286657821_Impact_of_climate_change_on_agriculture_over_Tamil_Nadu

Gbetibouo, G.A. and Ringler, C. 2009. Mapping South African farming sector vulnerability to climate change and variability: A subnational assessment. Discussion Paper No. 885, International Food Policy Research Institute, Washington: DC.

Hulle, M., d'Acier, A.C., Bankhead-Dronnet, S. and Harrington, R. 2010. Aphids in the face of global changes. *Comptes Rendus Biologies*, *333*(6–7), 497–503.

Idrisa, Y.L., Ogunbamu, B.O., Ibrahim, A.A. and Bawa, D.B. 2012. Analysis of awareness and adaptation to climate change among farmers in the Savannah agroecological zone of Borno state, Nigeria. *British Journal of Environment and Climate Change*, 2(2), 216–226.

IPCC. 2001. *Climate change: Impacts, adaptation and vulnerability*. Cambridge: Third Assessment Report: Cambridge University Press.

IPCC. 2007. *Climate change 2007: Impacts, Adaptation and Vulnerability*. Cambridge: Cambridge University Press.

Komba, C. and Muchpondwa, E. 2015. Adaptation to climate change by smallholder farmers in Tanzania. Environment for development. *Discussion paper series paper June* 2015, 12–15.

Lesk, C. and Anderson, W. 2021. Variability modulates trends in concurrent heat and drought over global croplands. *Environmental Research Letters* 16 055024.

Maddison, D. 2007. The perception and adaptation to climate change in Africa. CEEPA. Centre for Environmental Economics and Policy in Africa. Discussion Paper No. 10, University of Pretoria, Pretoria, South Africa.

Nadeem, F.; Nawaz, A.; Farooq, M. 2019. Crop rotations, fallowing, and associated environmental benefits. In *Oxford Research Encyclopedia of Environmental Science*. Oxford: Oxford University Press.

Rao, A. and Sinha, A.K. 2008. Climate changes and agriculture. *Nature*, 437, 102–109.

Rubina, S. 2014. Impact of climate change on adaptation and mitigation strategies of Ponnaniyar and Kalingarayan basin farmers: A gender analysis. *Unpub.* M.Sc. (Ag.) Thesis, TNAU, Coimbatore.

Saleem, A. and Ifthikar, M. 2013. Mitigation and adaptation strategies for climate variability: A case of cotton growers in the Punjab, Pakistan. *International Journal of Agricultural Extension*, 1(1), 30–35.

Shankara, M.H. 2010. Farmers perception of climate change and their adaptations. Unpub. M.Sc. (Ag.) Thesis, University of Agricultural Sciences, Bangalore.

Shindell, D.T., Rind, D. and Lonergan, P. 1998. Climate change and the middle atmosphere. Part IV: Ozone response to doubled CO2. *Journal of Climate*, 11, 895–918, doi:10.1175/1520-0442(1998)011<0895:CCATMA>2.0.CO;2

Shongwe, P., Masuku, Micah B. and Manyatsi, Absalom M. 2014. Factors influencing the choice of climate change adaptation strategies by households: A case of Mpolonjeni area development programme (ADP) in Swaziland. *Journal of Agricultural Studies*, 2(1), 86–98.

Shuttleworth, L.A. 2021. Alternative disease management strategies for organic apple production in the United Kingdom. *CABI Agriculture and Bioscience*, 2(1), 1–15.

Sinha, A.K. and Swaminathan, M.S. 1991. Long-term climate variability and changes. *Journal of Indian Geographical Union*, 7(3), 125–134.

Sivaraj, P., Philip, H. and Chinnadurai, M. 2014. Extent of awareness on climate change among small and marginal paddy farmers of Tamil Nadu. *Journal of Extension Education*, 26(2), 5246–5250.

Siebert, S. and Ewert, F.. 2014. Future crop production threatened by extreme heat. *Environmental Research Letters*, 9(4), 1–4. DOI: 10.1088/1748-9326/9/4/041001

Tiwari, K.R., Awasthi, K.D., Bala, M.K. and Situalla, B.K. 2010. Local people perception on climate change, its impact and adaptation practices in Himalaya to Terai region of Nepal. *Himalayan Journal of Development and Democracy*, 01 56–63.

Venkateswarlu, B. and Rama, Rao C.A. 2010. Rainfed agriculture: Challenges of climate change. *Agriculture Today Yearbook*, 43–45.

3 Rainfall Anomaly Index Valuation of Agricultural Production in Jakiri Sub-Division, Northwest Region, Cameroon

Suiven John Paul Tume and Suika Rita Nyuyfoni

3.1 Introduction

Climate variability and change is one of the greatest threats facing humanity, with far-reaching and devastating impacts on people, ecosystem services, natural resource-base, and the natural environment as a whole (Shames *et al.*, 2019). Climate stresses the food system from land preparation, growing crops till the produced food reach the final consumers. Agriculture, on the other hand, contributes to climate variability and change through the emission of Greenhouse Gases from poor farming techniques like slash and burn (Intergovernmental Panel on Climate Change-IPCC, 2019). Climate variability and change have adverse physiological effects on crops through inadequate water when available water cannot meet crop requirements at various critical stages such as germination, flowering and maturity (Feenstra *et al.*, 1998). Inadequate crop water during the growth stage makes crops vulnerable to pests and diseases attacks (Tume *et al.*, 2020).

Most Sub-Saharan African countries are vulnerable to the effect of climate variability and change due to high reliance on rain-fed agriculture and natural resources, which constitute a large part of local livelihoods (Hulme *et al.*, 2005; IPCC, 2007). Agriculture contributes about 70% of the GDP of some African economies (Toulmin and Huq, 2006; Tume *et al.*, 2019). Climate variability is projected to reduce yields from rain-fed agriculture by up to 50% by 2020 for some of the countries in Africa (Campbell *et al.*, 2011). In Cameroon, mean annual rainfall has decreased by about 2.9 mm per month (2.2%) per decade since 1960 (McSweeney *et al.*, 2012). Cameroon experienced particularly low rainfall in 2003 and 2005. Rainfall has continued to dwindle since 2010 (Tume *et al.*, 2019). Human activity has caused a variety of changes in different forcing agents in the Bamenda Highlands, where Jakiri Sub-Division is located.

Previous studies on the impact of climate on agriculture focused on stakeholder signatures to climate change adaptation to the agrarian sector of the Bui Plateau. In the study, Tume and Tani (2018) revealed that despite the glaring consequences of climate and environmental change like overgrazing, bush fires, water scarcity and increasing temperature, the Jakiri Council only sensitizes farmers on the dangers of their actions. These authors also revealed that

DOI: 10.4324/9781003275916-4

the Jakiri Council allocates only 1% of its annual budget for climate action. Again, Tani and Tume (2019) focused on the role of municipal councils in climate change mitigation in the Northwest Region of Cameroon. The authors revealed that municipal authorities are more interested in enriching their pockets and bank accounts rather than taking climate action that affects poor rural farmers. This chapter shows how such administrative lapses indirectly have accelerated the vulnerability of crop production to climate variability and change in Jakiri Sub-Division.

Several characteristics and indices are used in assessing rainfall variability. Some include inter-annual rainfall, rainfall seasonality, rainfall Coefficient of Variation (CV), Standardized Precipitation Index (SPI) and many others (World Meteorological Organization (WMO) and Global Water Partnership (GWP), 2016; Tume and Fogwe, 2018). In this chapter, the Rainfall Anomaly Index (RAI) is used, and it portrays episodes of extreme rainfall (floods) and meteorological droughts which affect crop production in Jakiri Sub-Division. In this light, this chapter falls within the context of the current global development agenda as it addresses some global development goals, viz, SDG 2: end hunger, achieve food security and improved nutrition, and promote sustainable agriculture; SDG 13: take urgent action to combat climate change and its impacts.

3.2 Method of Study

Jakiri Sub-Division is part of the Bui Division of the Northwest Region of Cameroon and has an estimated population of 194,088 (Jakiri Council, 2012), distributed in 58 villages. The main activity of this municipality is agriculture, which takes place in different agro-ecological zones. The lower frontier is an extension of the Ndop plain, while the upland area is part of the Kilum-Ijim mountain forest. The area lies approximately on latitude 6°N of the equator and longitude 10°E of the Greenwich Meridian (Figure 3.1). It has an altitude of about 1,700 m above sea level and a surface area of about 675 km^2 (Jakiri Council, 2012). The area is bounded to the north by Kumbo, west by Elak-Oku, south by Babessi, to the south-east by Bangorain in the west region and the east by Mbiame. The area is made up of Jakiri town at the centre constituting the main centre for commercial and administrative functions, Vekovi, Nkar, Sop, Wainamah, Yer, and Shiy, which act as nodes where growth and development could proceed, and a host of other villages such as Mbokijah, Mbokam, Ber, Mantum, Wahsi, Rontong, Kinsenjam, Limbo, Kwanso, Ndzerem Nyam, Ntohti, Sarkong amongst others. Jakiri Sub-Division is structured into three agro-ecological zones, namely: the low altitude, the transitional, and the high-altitude. This partition is because of the climatic and altitudinal characteristics, which vary over the geographical space, thus, giving a notion that these zones portray different climatic characteristics. Rainfall incidence across these agro-ecological zones is not uniform, which in turn determines agricultural productivity in the area.

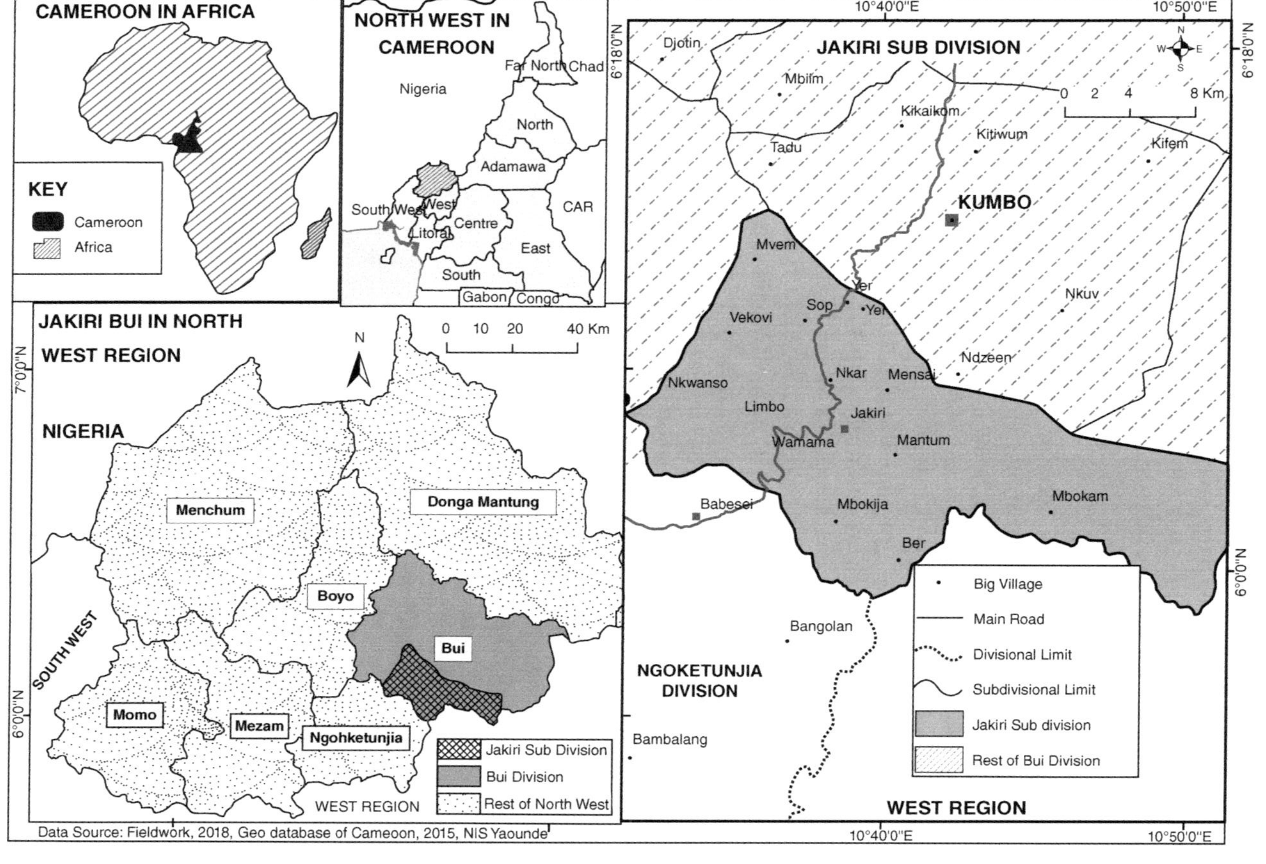

Figure 3.1 Location of Jakiri Sub-Division

Source: Jakiri Council, 2012

Primary sources of data collection used field observations and the administration of questionnaires. Field visits were undertaken at Wahsi-Ber rice fields, Mbokam and Limbo cassava production basins, and Tan farming areas. The questionnaire was designed to capture farmers' perceptions of climate variability and changes in crops. The purposive sampling used was important because it permitted the researcher to reach out to those who have lived in Jakiri Sub-Division for at least 20 years as they were capable to give reliable information on rainfall variability. A total of 211 questionnaires were administered to farmers (Table 3.1).

These questionnaires were age selective, to farmers from all age groups to obtain diverse and concrete information to understand rainfall variability and its implications on crop productivity. Some farmers groups also provided data on approximate cassava production from 1990 to 2018.

The independent variable for this study is climate and the element of climate considered for the study is rainfall. Rainfall data were obtained from the Upper Nun Valley Development Authority (UNVDA) Ndop (1990–2018), which has the same climatic characteristics as the lowland ecological zone. Rainfall data for the mid-altitude ecological zone (Jakiri) was obtained from the National Meteorological Service, Bonanjo-Douala (1961–2018), while data for the highland ecological zone was obtained from the Mount Oku Wildlife Sanctuary (1986–2018), which has the same ecological characteristics like Vekovi and its environs. Other secondary data sources included the UNVDA statistics on rice production for the Wahsi-Ber plain and data on cassava production from farmers groups in Mbokam and Limbo. Variations in rainfall were treated as anomalies to establish trends, illustrated in graphs, fitted with R^2 and linear equations (Coefficient of Determination to show the percentage change). The climatic index used in this chapter is the Rainfall Anomaly Index (RAI), designed by Van Rooy (1965). It considers the rank of the precipitation values to calculate positive and negative rainfall anomalies using the following equations:

$$RAI = +3\frac{\mathrm{RF}-\mathrm{MRF}}{\mathrm{MH10}-\mathrm{MRF}} \text{(Positive anomalies)}$$

$$RAI = -3\frac{\mathrm{RF}-\mathrm{MRF}}{\mathrm{ML10}-\mathrm{MRF}} \text{(Negative anomalies)}$$

Table 3.1 Distribution of questionnaire according to agro-ecological zones

Agro-ecological zones	*Total population*	*Questionnaires administered*
Lowland	49,952	39
Mid-altitude	97,708	77
Highland	46,428	95
Total	194,088	211

Source: Fieldwork, 2019

Where:

RAI = the rainfall anomaly index
RF = the rainfall for the year in question
M_{RF} = the mean actual annual rainfall for the total length of the period
MH_{10} and ML_{10} = the mean of ten highest and lowest values of rainfall (RF) respectively of the period

RAI is dimensionless and it has also been used to determine variation in rainfall for the three agro-ecological zones in the Jakiri Sub-Division. The RAI classification is used to determine extremes conditions of rainfall in a particular area for decadal and annual time frames. The results got from the range and class description determine the variation in rainfall for that region which may be negative or positive. This range is from ≥3.0 (extremely wet) to ≤−3.00 (extremely dry) (Table 3.2).

RAI normalized precipitation values are based on the weather history of a particular location. The only input parameter is precipitation. It addresses agricultural and meteorological droughts that affect agriculture, water resources, and other sectors. RAI is flexible in that it can be analysed at various timescales (World Meteorological Organization-WMO, 2016). It is easy to calculate, with a single input (precipitation) that can be analysed on a monthly, seasonal and annual timescale. For this study, the annual time scale is used. RAI classification ranges from ≥3.0 (extremely wet) to ≤−3.00 (extremely dry). These extreme conditions are not favourable for rain-fed tropical crop production because extreme wetness is associated with flooding that destroys agricultural land, while extreme dry conditions are associated with severe water deficits that cannot support agricultural production. Positive anomalies have their values above the average and negative anomalies have their values below the average. Trend lines were fitted on the anomaly graphs to show changes in rainfall over time. The rainfall standard deviation (SD) and CV were also calculated to show rainfall reliability. CV is calculated thus:

Table 3.2 RAI classification

RAI range	*Class description*
≥3.0	Extremely wet
2.00 to 2.99	Very wet
1.00 to 1.99	Moderately wet
0.50 to 0.99	Slightly wet
0.49 to −.49	Near normal
−0.50 to −0.99	Slightly dry
−1.00 to −1.99	Moderately dry
−2.00 to −2.99	Very dry
≤−3.00	Extremely dry

Source: Van Rooy, 1965

$$\sigma = \frac{\sqrt{\Sigma\left(Y - \bar{Y}\right)^2}}{N}, CV = \sigma * \frac{100}{\bar{Y}}$$

Where: Ȳ = mean, σ = Standard deviation, N = sample size (number of years of available rainfall data)

3.3 Results

3.3.1 Farmers' Perception of Rainfall Variability

Rainfall variability is perceived differently by different persons in different dimensions and localities. Since perception is a function of educational level, longevity in agricultural practice and gender in all the three agro-ecological zones of Jakiri, it was obtained that 65.9% of the sampled population have lived in the study area for more than 20 years, whereas 57.3% were male and 42.7% female (Table 3.3).

Rainfall variability means different things to indigenous farmers. Some attribute it to the prolonged dry season (91%), others to a shorter rainy season (70.6%), some perceived a decrease in rainfall (64.5%), while (79.1%) have a perception that there is an increase in rainfall (Table 3.4).

It should be noted that these responses on farmers' views about their perception of rainfall in all the three agro-ecological zones are a function of the socio-demographic correlates. Only 24.5% of the farmers in the different agro-ecological zones were neutral and had no idea of what is happening to rainfall, but they observed that there is a variation in rainfall. In an interview with a

Table 3.3 Agro-ecological zones, duration of stay and gender of the sampled population

Variables	*Frequency*	*Percentage*
Agro-ecological zone		
Wahsi-Ber-Mbokam	39	18.5
Jakiri-Nkar-Sop	77	36.5
Kisenjam-Vekovi	95	45.0
Total	211	100.0
Duration of stay		
6–10 years	19	9.0
16–20 years	53	25.1
20 years+	139	65.9
Total	211	100.0
Gender		
Male	121	57.3
Female	90	42.7
Total	211	100.0

Source: Fieldwork, July 2019

Table 3.4 Farmers' perception of rainfall variability

Perception	*Frequency*	*%*
Prolonged dryness and longer dry season	192	91
Decease rainfall	167	79.1
Increase rainfall	132	64.5
Shorter rainy season	149	70.6
Loss of crops seems to be on the increase	43	24.5

Source: Fieldwork, July 2019

59-year-old farmer (man) in Wahsi, he affirmed that "yes I believe there is rainfall variability, but I have no idea of what is causing it".

3.3.2 *Actual Rainfall Variability within the Jakiri Ecological Zones*

There exists variation in rainfall in the different agro-ecological zones of Jakiri Sub-Division. Different crops cultivated reflect the different agro-ecological zones of Jakiri as determined by rainfall and soil moisture. The high altitude agro-ecological zone (Vekovi, Nkarkui, Kisenjam) experiences more rainfall. The mean annual rainfall here is about 171.3 mm. Minimum rainfall occurs from December to March (dry season). Maximum rainfall is recorded from June to September. The lowland areas especially those on valley sides and the upper edge of the Ndop plain such as Ber, Wahsi, and Tan (part of the hollow frontier ecological zone) located on the south-eastern part of the Sub-Division have less rainfall comparable to the high-altitude zone. Here, the mean annual rainfall is 158 mm. This transitional ecological zone has a mean annual rainfall of 165.74 mm. Like the other ecological zones, the lowest amounts of rainfall are recorded from December to March, while high amounts are recorded from June to September. This variation reveals that rainfall decreases with an increase in altitude in the Jakiri agro-ecological zones

3.3.3 *Inter-annual Rainfall Variability*

To examine spatial variation of rainfall incidence in the three agro-ecological zones of Jakiri, the rainfall SD, CV and the RAI have been used. The lowland agro-ecological zones have rainfall characteristics similar to those of the Ndop plain. The inter-annual rainfall anomaly graph was used to show the trend of rainfall variation for a period of 28 years (1990–2018) (Figure 3.2).

Inter-annual rainfall in the lowland agro-ecological zone shows slight fluctuations. The rainfall trend for this lowland ecological zone depicts a slight increase with a coefficient of determination (R^2) of 0.0017 (0.17%). It is noted that the highest rainfall amount measured was in 2013 around (125 mm) and the lowest in 2001 (−45 mm). The rainfall SD here is 34.65, with a CV of

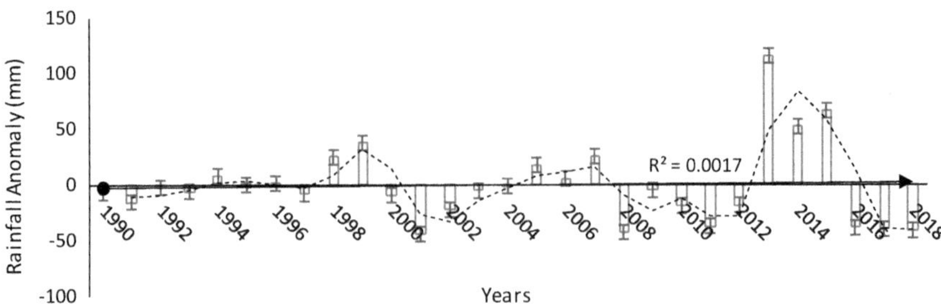

Figure 3.2 Inter-annual rainfall anomaly for lowland ecological zone

Data source: UNVDA, Ndop (2018)

21.93% (unreliable because it is more than the tropical threshold of 20%). The rainfall for the transitional or mid-latitude agro-ecological zone presents a different situation from the lowland ecological zone. The rainfall pattern is not evenly distributed in the two agro-ecological zones. Each zone is delimited depending on the rainfall characteristics of the zone. The rainfall data used in this zone ranges from 1961 to 2018 (Figure 3.3).

The results revealed that the R^2 is 0.029. This depicts a decreasing trend in the inter-annual rainfall of the mid-altitude agro-ecological zone. The years 1963, 1983, 2006, and 2010 recorded very low rainfall. This could be attributed to prolonged dry spells and the irregular onset of the first rains and the early cessation of the rains. The SD here stands at 22.9, with a CV of 13.81% (reliable). The highland ecological zone is situated on the highland extension of the Kilum-Ijim mountain forest. There has been a general notion that rainfall in this zone is heavier as compared to two previous agro-ecological zones. The rainfall data used in this zone range from 1986 to 2018 (Figure 3.4).

The inter-annual rainfall coefficient of determination for the highland ecological zone was recorded as R^2=0.3163. From the analysis, this zone experienced increasing rainfall from 1986 to 2005, but there has recently been a

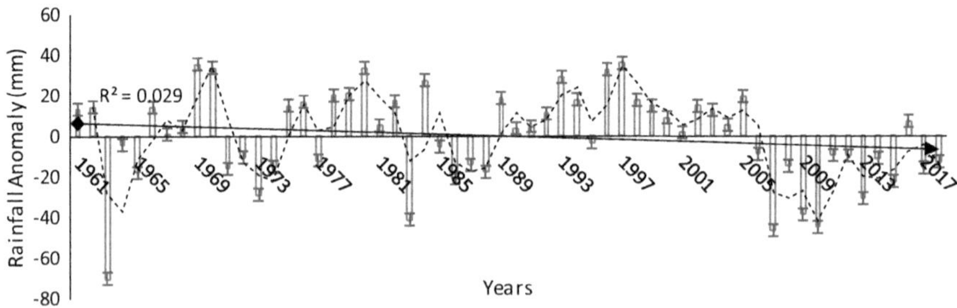

Figure 3.3 Inter-annual rainfall anomaly for mid-altitude ecological zone

Data source: National Meteorological Service (2018)

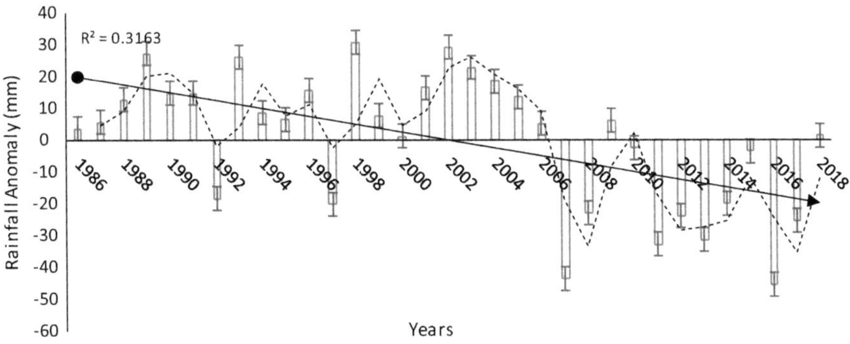

Figure 3.4 Inter-annual rainfall anomaly for highland ecological zone

Data source: Kilum Wildlife Sanctuary, 2018

drastic decline in rainfall for the past ten years (2008–2018). This means that high altitude being noted for heavy rainfall is gradually changing because the results we have show a negative trend. It is also noted that the amount of rainfall in mm is showing a negative with −55 mm in 2016. The SD here is 21.38 and a CV of 12.48%.

The rainfall SD and CV are also used to determine rainfall variability in the three agro-ecological zones. The mean annual rainfall as well as the maximum and minimum rainfall for the three stations representing each agro-ecological zone has been calculated to show more impetus in variation (Table 3.5).

The value of CV has limits of reliability and unreliability. When CV is ≤20 it means that rainfall variability is reliable but when it is ≥20 it indicates unreliable. Hence, rainfall is unreliable for the lowland ecological zone (21.93%), and reliable for the mid-latitude and highland ecological zones respectively with values of 13.81% and 12.48%. The data used to determine variation in the lowland ecological zone was obtained from the Ndop plain since it shares similar climatic characteristics with the hollow frontier agro-ecological zones. To analyse the RAI, the trend line in the graph determines if each agro-ecological zone is experiencing rainfall deficits or surplus (Figure 3.5).

From 1990 to 1998 the zone experienced dryness with the highest recorded between 1995 and 1996 with −0.5. This indicates a slight dryness from the RAI classification table (−0.5 to −0.99). In 1990 there was a slight increase in the

Table 3.5 Rainfall standard deviation and coefficient variation

Station	*MAR (mm)*	*Min (mm)*	*Max (mm)*	*SD*	*CV (%)*	*Remark*
Ndop	153.43	127.85	202.17	34.65	21.93	Unreliable
Jakiri	165.74	119.33	201.90	22.90	13.81	Reliable
Oku	171.30	112.60	269.47	21.38	12.48	Reliable

MAR: Mean Annual Rainfall, SD: Standard Deviation, CV: Coefficient of Variation

Source: Calculated from climatic data for the three ecological zones

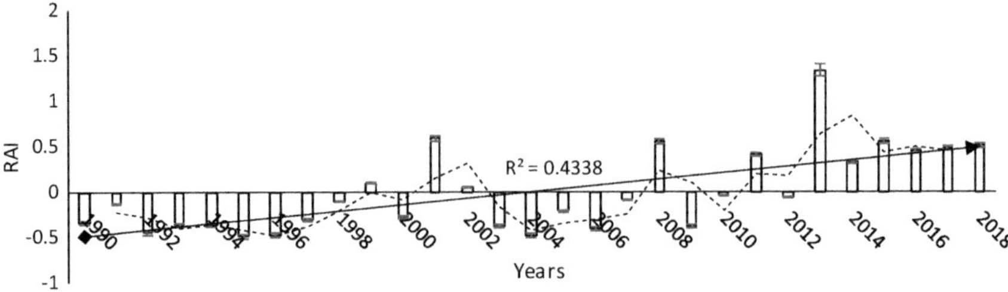

Figure 3.5 Rainfall Anomaly Index for Ndop (1990–2018)

RAI (0.1) but it dropped to −0.2 the following year. From 2002 to 2007 the RAI was still negative. The situation in 2008 shows a 0.5 RAI meaning the year was slightly wet. The rainfall anomaly index of the mid-attitude is agro-ecological zone is analysed using data collected from 1961 to 2017 (56 years) (Figure 3.6).

There is variation in rainfall throughout the area. These negatives values range between −0.1 and −0.5, which indicates slight dryness. Looking at the positive values, rainfall has not been evenly distributed annually. Some years recorded high rainfall than others. In 1963 after the dryness in the two previous years, a 1.3 RAI was recorded signifying a moderately wet year. This zone also experienced slightly wet years from 1970 to 1971. In 1974 and 1982 the trend was increasing up to 1986, as 1.7 was recorded – a moderately wet year. Moreover, in the years 2007, 2010, 2011, 2014, 2016 the zone has been slightly wet. Essentially the trend line determining the variation using RAI is just slightly above average or zero which is signifying slightly dryness. Data obtained for this analysis are records for the past 32 years analysing rainfall anomalies for the high-altitude zone (Figure 3.7).

The time series shows inter-annual rainfall fluctuation which is a usual phenomenon for all the zones. The fluctuation between the negatives and positives values of RAI throughout is between slightly wet or slightly dry meaning this zone is only liable to wetness rainfall variation which is not close to extremes events of water deficits drought and floods. The highest positive values of RAI (0.6) recorded were in 1986 and 2008, while the lowest

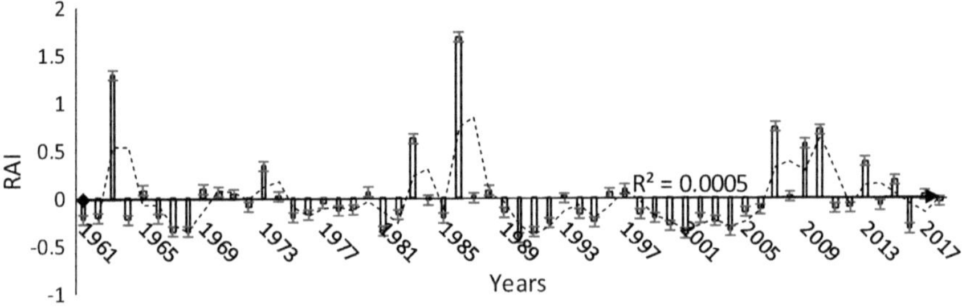

Figure 3.6 Rainfall Anomaly Index for Jakiri (1961–2018)

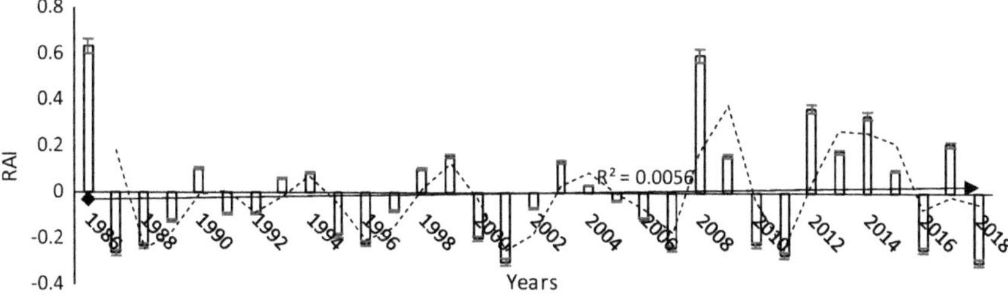

Figure 3.7 Rainfall Anomaly Index for Oku (1986–2018)

RAI of (0.6) was recorded in 1996 and 2018. With variations in rainfall conditions, RAI trends are increasing, suggesting that rainfall is gradually reducing in the ecological zones. It has been discovered that rainfall in the low altitudes zone is reducing more as compared to the mid and high-altitude zones (Table 3.6).

These characteristics reveal that there have been 63 incidents of slightly dry conditions, 23 near-normal episodes, 24 slightly wet incidents and 3 moderately

Table 3.6 Summary of decadal RAI for Jakiri Sub-Division

Decades	*RAI class descriptions*	*Number of incidents*
Lowland ecological zone		
1990–1999	Slightly dry	9
	Near normal	1
2000–2009	Slightly dry	7
	Slightly wet	2
	Near normal	1
2010–2018	Slightly dry	2
	Moderately wet	1
	Near normal	6
Mid-altitude ecological zone		
1961–1970	Slightly dry	6
	Slightly wet	3
	Moderately wet	1
1971–1980	Slightly dry	6
	Slightly wet	4
1981–1990	Slightly dry	6
	Slightly wet	3
	Moderately wet	1
1991–2000	Slightly dry	7
	Near normal	3
2001–2010	Slightly dry	6
	Slightly wet	3
	Near normal	1

(*Continued*)

Table 3.6 (Continued)

Decades	*RAI class descriptions*	*Number of incidents*
2011–2018	Slightly dry	3
	Slightly wet	1
	Near normal	4
Highland ecological zone		
1986–1996	Slightly dry	7
	Slightly wet	1
	Near normal	3
1997–2007	Slightly dry	7
	Near normal	4
2008–2018	Slightly dry	4
	Slightly wet	7

Source: Calculated from climatic data

dry instances. Slightly dry and moderately dry episodes are periods of meteorological droughts. From this data, the Jakiri agro-ecological zones have been prone to meteorological and agricultural droughts, which threaten food crop production and it is a threat to food security.

3.4 Effects of Rainfall Variability on Jakiri Agricultural Production

Variations in rainfall in the study area are being felt by farmers because of the impacts it has on productivity. It is these anomalies in the intensity and duration of rainfall that lead to extremes and slight extremes events of rainfall characteristics. The occurrences of floods, dry spells, and soil erosion reduction in water sources give us an impression that rainfall varies over time and space. The highest problem affecting productivity is pest and crop diseases (91%), followed by soil erosion (81%), and then decreasing and drying water bodies (79%) (Table 3.7).

Soil erosion is very common in all the zones but highest in the high-altitude zone because it's a steep slope and any intensive rainfall washes the soils away downward rendering the farms with fewer nutrients. On the other hand, the low-altitude zones experienced floods in the rainy periods since most streams convey around there. The water overflows into farmlands destroying crops. This is worse when crops are still at an early stage of growth. Prolonged dry seasons and dryness reduce the water table of most rivers and renders some streams dry. That is why irrigation farming is limited around Ber and Washi. Rainfall is the key determining factor of crop production in Jakiri Sub-Division since crop cultivation is rain-fed dependent. Less rainfall in terms of duration reduces the growing season. This in turn determines crops' ability to resist dry spells. Unpredictable annual oscillation in rainfall causes crop destruction. Less rainfall in terms of intensity contributes to failure in crops like maize and rice because they require much water for growth.

Table 3.7 Impacts of rainfall variability

Impacts	*Frequency*	%
Decrease and drying water sources	166	78.7
Decreased in crop output	124	58.8
Pests and crop diseases	192	91
Soil erosion	171	81
Flood occurrences	81	38.4
Livestock diseases have increased	161	76.3
Increased livestock mortality	103	48.8
Strange insects are increasing within the local vicinity	114	54

Source: Fieldwork, July 2019

3.4.1 Effects of Rainfall Variability on Crop Output

Generally, the effects of rainfall variability on crop output depend on either excess or deficit in rainfall; in other words, climatic extremes of flood and dryness affect crop productivity and lead to occurrences of other factors like a dry spell and soil erosion which contributes to low productivity and later leads to food shortage and insecurity in the study area (Table 3.8).

The farmers perceive that the main effects of rainfall variability in Jakiri Sub-Division are decreasing production (55.5%), emerging food insecurity (27%), floods (69.2), the prevalence of crop pests and diseases (84.8%), rising temperatures (74.4%), longer periods of dry spells after the onset of first rains (74.9%), high soil erosion and leaching (83.9%), drying up of streams and springs (73.5%), reduction in the volume of water in rivers and streams (90.5), and prevalence of cultivator-grazier conflicts (49.3%). The Wahsi and Ber plain is the only area where rice cultivation is carried out in Jakiri Sub-Division. The agro-ecological zone has climatic conditions like the Ndop plain and supports rice cultivation. Rice production in Wahsi-Ber plain has been increasing, despite variations in rainfall and environmental conditions. Data collected

Table 3.8 Effects of rainfall variability

Effects	*Frequency*	%
Decrease in production	117	55.5
Food insecurity	57	27
Floods	146	69.2
Prevalence crop pests and diseases	179	84.8
Rising temperatures	157	74.4
Long periods of a dry spell after the first rain	209	74.9
High soil erosion and leaching	158	83.9
Drying up of streams and springs	155	73.5
Reduction in the volume of water in rivers and streams	191	90.5
Farmer-graziers conflicts	104	49.3

Source: Fieldwork, July 2019

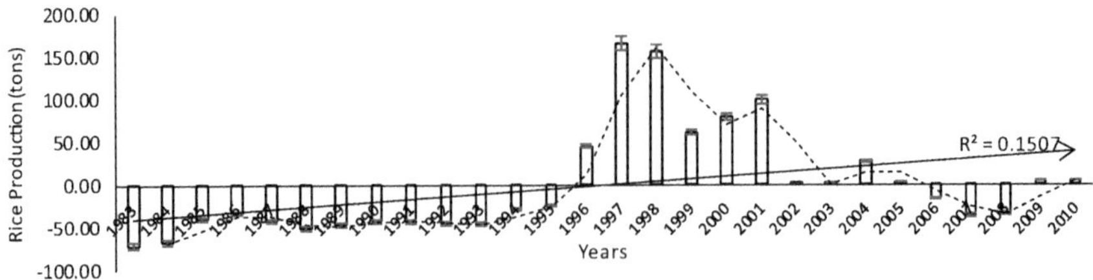

Figure 3.8 Rice production for Wahsi Plain

Data source: UNVDA, Ndop, 2018

from the UNVDA in Ndop for this production basin show an increasing trend of 15.07% from 1983 to 2010 (Figure 3.8).

Rice production is increasing. From 1983 to 1995, the production was a negative anomaly, due to a lack of interest in rice cultivation. From 1996 the trend line started increasing and with a positive anomaly till 2001. From 2002 to 2010 there was a positive fluctuation in rice productivity probably due to variation in rainfall.

3.4.2 Farmers' Perception of Changes in Crops Output

Farmers from the three agro-ecological zones have different perceptions about each variety of crop cultivated in each zone. There is an increase, decrease, no change, and not applicable for the cultivation of each crop species for the past years (Table 3.9).

Crops like beans, vegetables, and cocoyams show a positive change, which is probably because they are applicable in all the agro-ecological zones, but maize which is the most applicable in the three-zones is reducing in general production by 1%. The main food crops are cultivated in all the agro-ecological zones, and they include maize, beans, vegetables, solanum potatoes, plantain and egusi (pumpkin seeds). Beans, cocoyam, vegetables, plantain, and egusi are increasing with a positive percentage change, but a slight decrease in maize solanum potato, and yams gives a negative percentage change (Figure 3.9).

Cash crops are not cultivated in all zones. Rice and cassava are cultivated in the lower and transitional ecological zones, but rice shows a negative general percentage change because it is cultivated just in the lowland agro-ecological zone. The high-altitude zone is noted for market gardening and others. There is a general increase in cash crops in Jakiri Sub-Division. Most cash crops that are not applicable in all three agro-ecological zones represent the lowest negative anomaly; they include rice, soybeans, and pineapple, while cattle, vegetables, tomatoes and cocoyam have positive changes. Despite the perceived decrease in groundnuts, field observation proved the contrary. The farmers also perceived that cassava production is on the decrease, but field observations and some secondary data proved otherwise. The main cassava production

Table 3.9 Perceived changes in crops and animal production

Crops	*Increase*		*Decrease*		*No change*		*Not applicable*	
	F	*%*	*F*	*%*	*F*	*%*	*F*	*%*
Maize	95	45	97	46	19	9	0	0
Solanum potato	94	44.5	78	37	39	18.5	0	0
Beans	139	65.9	54	25.6	18	8.5	0	0
Groundnuts	60	28.4	52	24.6	36	17.1	63	29.9
Cassava	88	41.7	1	0.5	38	18	84	39.8
Cocoyam	156	73.9	36	17.1	19	9	0	0
Yams	91	43.1	47	22.3	73	34.6	0	0
Tomato	132	62.6	1	0.5	40	19	38	18
Onion	89	42.2	44	20.9	17	8.1	61	28.9
Soybeans	26	12.3	54	25.6	70	33.2	61	28.9
Coffee	59	28	125	59.2	27	12.8	0	0
Vegetables	171	81	15	7.1	25	11.8	0	0
Market gardening	102	48.3	1	0.5	19	9	89	42.2
Rice	22	10.4	20	9.2	8	3.8	161	76.3
Cowpeas	100	47.4	56	26.5	36	17.1	19	9
Pineapple	37	17.5	3	1.4	0	0	171	81
Plantain and egusi	115	54.5	34	46.1	62	29.4	0	0

Source: Fieldwork, July 2019

Notes: F: Frequency

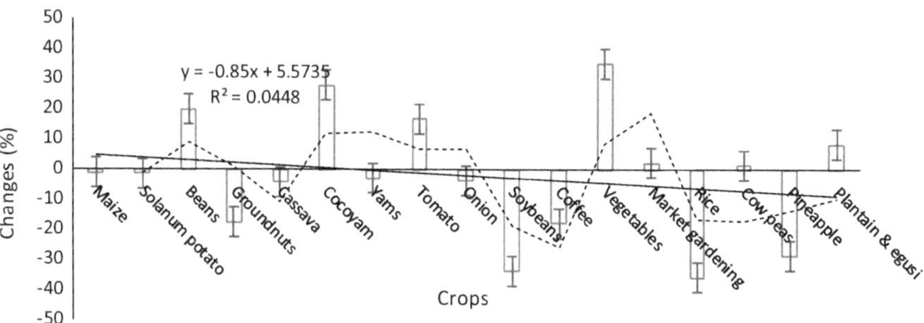

Figure 3.9 Perceived changes in food crop production in Jakir Sub-Division

Source: Fieldwork, July 2019

basins are in the lowland ecological zone. Data show increasing trends for Mbokam and Limbo (Figures 3.10 and 3.11).

Note should be taken that all cash crops show a general decrease in their output, but field observation confirms an increase in productivity since these crops are grown only in specific zones. Farmers in the Mbokam area have increasingly improved on cassava production in terms of areal coverage by extending farmlands only for cassava rather than the old way of planting at farm boundaries. Farmers have recently begun producing new kinds of

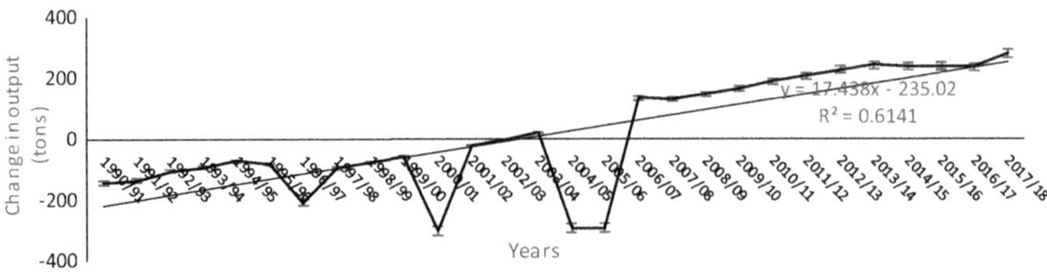

Figure 3.10 Cassava production trends for Mbokam

Data source: Mbokam cassava farmers' groups (2019)

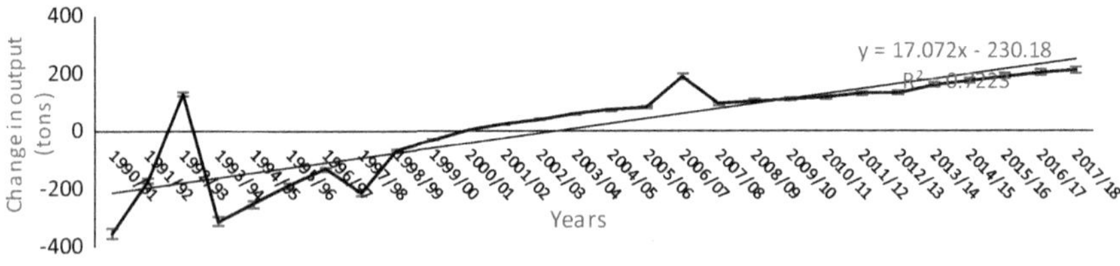

Figure 3.11 Cassava production trends for Limbo

Data source: Limbo cassava farmers' groups (2019)

cassava, which only take a year to mature, using new agricultural techniques. A mature cassava stand at fresh harvest can give approximately 8–15 kg. Cassava is transformed into several by-products, such as cassava flour, starch, and many others. All these by-products are meant for consumption and commercial purposes in local markets in Jakiri, Bui Division, and beyond. Increased productivity of cassava is normal for reliable rainfall conditions of the lowland agro-ecological zones. Adverse climatic conditions have physiological effects on crops and livestock. Higher temperatures reduce water available for crops by drying out air and soils, while they increase pest and disease pressures, directly stress livestock. The physical harm to food production and the consequences for people will be greatest in the most food-insecure regions, particularly in Sub-Saharan Africa where tens of millions of hectares of agricultural lands will become substantially drier. Populations in these regions (including Jakiri Sub-Division) already experience high rates of undernutrition, have limited or no social insurance, have less government support, and are most likely to experience direct health threats from extreme temperatures.

3.5 Discussion

Understanding and predicting temporal variations in African climate has become the major challenge facing African agriculture (Eriksen, 2005). While seasonal climate forecasts have taken great strides forward, rural farmers remain unsure of the ultimate causes of the lower frequency decadal and multi-decadal rainfall variability that affects some African climate regimes, especially in the African savannahs (where Jakiri Sub-Division is located) (Humle *et al.*, 2005). Many socio-economic activities in Africa heavily depend on climate and specifically on rainfall. The formal and informal economies of most African countries are strongly based on natural resources, especially agro-pastoralism (Obasi, 2005). Consequently, climatic variations alter the production systems of these activities have very high leverage on the local economy. It is estimated that on a global scale, about 75% of the natural disasters are associated with extreme weather and climate events such as floods and droughts, with devastating consequences on rain-fed agro-pastoral systems (like Jakiri Sub-Division) (Mairomi and Tume, 2021). Such extreme events increase the vulnerability of society to climate variability and change (Tume *et al.*, 2020).

Climate variability and change has already impacted and will further impact the agriculture sector and food production. Sensitivity to climate as well as several other driving forces, especially from the economic and societal domains, will determine the future evolution of the sector. But climate influence must be considered as a first-order factor in the context of the enormous challenge of providing food for about 8 billion people by 2,100 compared to the current 7.1 billion today (Zilberman, 2018). Assessments of climate influence on crop functionality should consider stimulation of photosynthesis by the elevation of CO_2 atmospheric concentration. The direct influence of climate variations on crop function involves temperature, the effects of which may be quite variable. Higher temperatures are generally favourable for growth in cold climates (except in extreme events) and generally unfavourable for warm areas. Further warming has increasingly negative impacts in all regions (Zilberman *et al.*, 2018).

On the other hand, rainfall variability seriously modulates the potential changes in plant growth resulting from the effects of increasing temperatures. Tendencies towards drier conditions in some areas such as the West and Southern Africa may cancel, at least partially, the positive potential impact due to higher CO_2 or milder temperatures (IPCC, 2019). Such combined climate influence leads to a variety of contrasting effects on crop production, depending upon the type, the geographical zone and the level of adaptation. Farmers' perceptions were recorded on changing crop patterns, with decreasing trends in food and cash crops (Tume *et al.*, 2020).

The global projected temperature rise of 1.4–5.8°C over the period 1990 to 2100 (Collier *et al.*, 2008) will result in large changes in the frequency of extreme events which can have severe impacts on agriculture. Increases in surface temperatures will increase soil temperatures which will, in turn, affect plant metabolism through the degradation of plant enzymes, limiting photosynthesis and affecting plant growth and yields (Sivakumar *et al.*, 2005). Increases in soil temperature will also increase potential evapotranspiration, which causes damage, especially to those crops with surface root systems that utilize mostly precipitation moisture (Collier *et al.*, 2008). It increases leaf-surface temperatures hence affecting crop metabolism and yields making crops more sensitive to moisture stress conditions. Such crops include groundnuts, soybeans, maize and fruit trees (all cultivated in the ecological zones of Jakiri Sub-Division) (Agba *et al.*, 2017).

Blanc (2012) revealed that crop yields change in 2100 will be near zero for cassava, –19% to +6% for maize, –38% to –13% for millet and –47% to –7% for sorghum under alternative climate change scenarios in SSA. Basic foodstuff on which people live daily has been declining in Jakiri Sub-Division and the Bui Plateau in general because food crop production systems are rain-fed (Tume and Fogwe, 2018; Tume and Tani, 2018). When there is, thus, hydrological, meteorological or agricultural drought, threats to food security and crop failure will be evident (Tume *et al.*, 2020). Cassava has proven to be weather-proof and resilient to changing environmental conditions. It is grown almost throughout the year in Jakiri Sub-Division. According to Ajiere and Weli (2018), the new improved cassava stems harvesting is done 6 to 18 months after planting making it preferable to the seasonal crops of yams, beans or solanum potato.

Shifts in seasonal patterns, recurrent meteorological droughts, and unpredictable precipitation affect crop yields and make some crops and varieties inviable, while extreme weather events, floods, and other disasters destroy crops (Tall *et al.*, 2021). Climate variability and change also increases the incidence of pests and diseases. In the face of these environmental adversaries, switching to crop varieties that are resistant to heat, drought and/or floods, diversification of crops, irrigation, sustainable water management practices, climate-smart agriculture and regenerative farming techniques and practices, off-farm activities, and crop insurance will go a long way to enhance farmers' resilience and adaptive capacity (Tume, 2019; Tume *et al.*, 2020). In addition to resilience, these endeavours can also create new business opportunities and equity funds to invest in agro-production and develop lending products tailored to smallholder and larger-scale agri-producers to invest in climate-resilient farming practices (Tume *et al.*, 2018).

3.6 Conclusion

There is still a slight increase in the inter-annual rainfall for the lowland agro-ecological zone, while mid and high-altitude areas are already witnessing significant drops in rainfall. The rainfall CV revealed that rainfall is still reliable in the low and mid-agro-ecological zones while the high zone is getting unreliable. The values are 21.93%, 13.81%, and 12.48% for the lowland, mid-altitude, and highland zones respectively. The main indices used to determine rainfall variability in Jakiri Sub-Division is the Rainfall Anomaly Index (RAI). The RAI is increasing for all the ecological zones. An increasing RAI, show that dry spells are recurrent. Recurrent dry spells mean that rainfall is generally decreasing. Climate variation poses some problems which hinder agricultural production in Jakiri Sub-Division. Some of these problems are highly felt while others are not considered as a threat to agriculture by farmers maybe because their manifestation is not felt. The greatest problem faced by farmers is water scarcity during the dry season, declining soil fertility, deforestation, inadequate grazing land, while early onsets rain and pest and insect infestations is what farmers consider to be a rare problem. The drivers that cause more problems to agricultural production in Jakiri Sub-Division as perceived by farmers are the late onset of the first rains, plant diseases, floods, soil infertility, lack of pesticides, wind hazards, deforestation, competition of arable land with graziers, and water scarcity. Minor drivers are early onset of the first rains, persisting droughts after the first rains, insect pests, soil erosion, inadequate farming land, lack of fertilizers, and bush fires. Other control factors such as temperature as relief and anthropogenic factors come into play. This explains the reason why there is just a slight increase in both food and cash crops in the entire Sub-Division. A crop like cassava has proven to be resilient to changing climatic and environmental conditions in Jakiri Sub-Division and its production should be encouraged. The three agro-ecological zones of production should embark on proactive and reactive adaptive measures to further increase general crop outputs or the trend might easily be negative in the nearest future if a quick measure to improve on agricultural activity with variation in rainfall patterns is not taken. Cultivating more drought-resistant food crops can help to help sustain rural populations. If these measures are taken, then Jakiri Sub-Division will be on the right path to achieving food security in line with SDG-2 and proactive to combat climate change and its impacts at the local level (SDG-13).

Bibliography

Agba, D.Z., Adewara, S.O., Adama, J.I., Adzer, K.T., & Atoyebi, G.O., (2017): Analysis of the effects of climate change on crop output in Nigeria. *American Journal of Climate Change*, 6, 554–571.

Ajiere, S.I., & Weli, V.E., (2018): Assessing the impact of climate change on maize (*Zea mays*) and cassava (*Manihot esculenta*) yields in Rivers State, Nigeria. *Atmospheric and Climate Sciences*, 8, 274–285.

Blanc, E., (2012): The impact of climate change on crop yields in Sub-Saharan Africa. *American Journal of Climate Change*, 1, 1–13.

Campbell, B., Mann, W., Meléndez-Ortiz, R., Streck, C., Tennigkeit, T., Christophe Bellmann, C., Meijer, E., Wilkes, A., & Vermeulen, S., (2011): A*griculture and climate change: A scoping report*. Meridian Institute, Bangkok, 98.

Collier, P., Conway, G., & Venables, T., (2008): Climate change and Africa. *Oxford Review of Economic Policy*, 24, 337–353.

Eriksen, S., (2005): The role of indigenous plants in household adaptation to climate change: The Kenyan experience. In Low, P.S., (2005): *Climate change and Africa*. Cambridge University Press, 248–259.

Feenstra, J.F., Burton, I., Smith, J.B., & Tol, R.S.J., (1998): *Handbook on methods for climate change impact assessment and adaptation strategies*. UNEP & Vrije Universiteit Amsterdam, 2-13–2-15.

Hulme, M., Doherty, R., Ngara, T., & New, M., (2005): Global warming and African climate: A reassessment. In Low, P.S., (2005): *Climate Change and Africa*. Cambridge University Press, 29–40.

Jakiri Council (2012): Jakiri council development plan. Available: www.pndp.org/documents/27_CDP_Jakiri.pdf

Intergovernmental Panel on Climate Change-IPCC, (2007): Climate Change 2007: Impacts, Adaptation and Vulnerability. Contribution of Working Group II to the Fourth Assessment Report of the Intergovernmental Panel on Climate Change. Cambridge University Press, Cambridge UK, 976.

Intergovernmental Panel on Climate Change-IPCC, (2019): Special Report on Climate Change, Desertification, Land Degradation, Sustainable Land Management, Food Security, and Greenhouse Gas fluxes in Terrestrial Ecosystems. IPCC Secretariat, Bonn, 1–35.

Mairomi, H.W., & Tume, S.J.P., (2021): Standardized precipitation index valuation of seasonal transitions and adaptation of pastoralist to climate variability in rangelands of the Bamenda Highlands of Cameroon. *Journal of Ecology & Natural Resources*, 5(1), 000229, 1–20.

McSweeney, C., New, M., & Lizcano, G., (2012): *UNDP climate change country, Profiles: Cameroon. School of Geography and Environment*. University of Oxford, 7–8.

National Meteorological Service (2018): https://cdsp.imdpune.gov.in/

Obasi, G.O.P., (2005): The Impacts of ENSO in Africa. In Low, P.S., (2005): *Climate change and Africa*. Cambridge University Press, 226.

Shames, S., Scherr, S.J., De Pinto, A., Ringler, C., Cenacchi, N., Thornton, P., Loboguerrero, A.M., Campbell, B., Shackleton, S., Fergus, S., Wezel, de Pinto, Mbow, C., Susan, C., Robiglio, V., & Harrison, R., (2019): *Food security and livelihoods of small-scale producers: Adapt Now: A global call for leadership on climate resilience*. Global Centre on Adaptation and World Resources Institute, 23–29.

Sivakumar, M., Das, H., & Brunini, O., (2005): Impacts of present and future climate variability and change on agriculture and forestry in the arid and semi-arid tropics. *Climatic Change*, 70, 31–72.

Tall, A., Lynagh, S., Vecchi, C.B., Bardouille, P., Pino, F.M., Shabahat, E., Stenek, V., Stewart, F., Power, S., Paladines, C., Neves, P., & Kerr, L., (2021): *Enabling private investment in climate adaptation and resilience: Current status, Barriers to Investments and Blueprint for Action*. World Bank and Global Facility for Risk Reduction and Recovery, 70 p.

Tani, B.V., & Tume, S.J.P., (2019): The role of municipal councils in climate change mitigation the northwest region of Cameroon. *International Journal of Resource and Environmental Management (JOREM)*, 4(2), 3–17.

Toulmin, C., & Huq, S., (2006): *Sustainable development: Africa and climate change*. International Institute for Environment and Development (IIED), 1–2.

Tume, S.J.P., & Fogwe, Z.N., (2018): Standardised precipitation index valuation of crop production responses to climate variability on the Bui Plateau, northwest region of Cameroon. *Journal of Arts and Humanities (JAH)*, 1(2), 21–38.

Tume, S.J.P., & Tani, B.V., (2018): Stakeholders' signature to climate change adaptation in the agrarian Sector of Bui Plateau, Northwest Cameroon. *Journal of Environmental Issues and Agriculture in Developing Countries*, 10(3), 140–156.

Tume, S.J.P., (2019): Standardised precipitation valuation of water resources vulnerability to climate variability on the Bui Plateau, Northwest Cameroon. *Environment and Ecology Research*, 7(2), 83–92.

Tume, S.J.P., Kimengsi, J.N. & Fogwe, Z.N., (2019): Indigenous knowledge and farmer perceptions of climate and ecological changes in the Bamenda Highlands of Cameroon: Insights from the Bui Plateau. *Climate*, 7, 138, 1–18.

Tume, S.J.P., Zetem, C.C., Nulah, S.M., Ateh, E.N., Mbuh, B.K., Nyuyfoni, S.R., Ahfembombi, L.L., & Kwei, J., (2020): Climate change and food security in the Bamenda Highlands of Cameroon. In Squires, V.R., Gaur, M.K., (2020): *Food security and land use change under conditions of climate variability: A multidimensional perspective*. Springer, 107–124.

Tume, S.J.P., Kongnso, M.E., Nyukighan, M.B., Dindze, N.E., & Njodzeka, G.N. (2018): Stakeholders in climate change communication in the northwest region of Cameroon. In: Tume, S.J.P., & Tanyanyiwa, V.I., (2018): *Climate Change perception and changing agents in Africa & South Asia*. Vernon Press. First Edition, 97–116.

UNVDA – Upper Nun Valley Development Authority Ndop. (2018). *Production statistics*, UNVDA Ndop Cameroon.

Van Rooy, M.P. (1965): A rainfall anomaly index independent of time and space. *Notos*, 14, 43–48.

World Meteorological Organization (WMO) and Global Water Partnership (GWP), (2016): Handbook of Drought Indicators and Indices. Integrated Drought Management Programme (IDMP), Integrated Drought Management Tools and Guidelines Series 2. Geneva, 45.

Zilberman, D., (2018): Conclusion and policy implication to climate-smart agriculture: Building resilience to climate change. In Lipper, L., McCarthy, N., Zilberman, D., Asfaw, S., & Branca, G., (2018): *Climate-smart agriculture: Building resilience to climate change*. FAO and Springer, 421–426.

Zilberman, D., Lipper, L., McCarthy, N., & Gordon, B., (2018): Innovation in response to climate change. In Lipper, L., McCarthy, N., Zilberman, D., Asfaw, S., & Branca, G., (2018): *Climate-smart agriculture: Building resilience to climate change*. FAO and Springer, 49–76.

4 Vulnerability Assessment of Soil Erosion

A Geographical Study of the Upper Catchment Area of the Tista River, India

Asraful Alam, Lakshminarayan Satpati, and Arijit Ghosh

4.1 Introduction

Soil is an important natural resource that provides significant physical, social, and economic roles, but, on the other hand, due to the high demand of goods and services generated from soils, much pressure has been exerted, principally from underdeveloped and developing countries, which largely depends on primary sectors (Wessels et al., 2007). Soil erosion is one of the major causal factors of environmental degradation as it deteriorates the quality of top soil by removing the soil particles including the nutrients and water supply, and thus depriving the plants of their basic life support systems. As much as 175 million ha (about 53%) of land in India suffers from soil erosion. It has been estimated that about 16.4 tons/ha of soil is detached annually in the country (Singh, 2000). During the last 40 years, almost one-third of the world's arable land has been reduced due to erosion, with loss ongoing at the rate of more than 10 million ha per year (Pimental et al., 1995). To concentrate on this issue World Soil Charter define that 'Soils are fundamental to life on Earth but human pressures on soil resources are reaching critical limits'. Taking into account the importance of protecting the soil resource, it is progressively recognized by the world community (Lal, 2001). Sustainable management of soil received strong support at the Rio summit in 1992 and its Agenda 21 (UNCED, 1992). Careful soil management is a fundamental element of sustainable agriculture which provides an important switch for climate regulation and a pathway for protecting ecosystem services and biodiversity (FAO and ITPS, 2015). The main factors responsible for soil erosion in India include rainfall, soil type, vegetation, and topographic and morphological characteristics of the river basin (Jain and Kothiyari, 2000). Remote Sensing (RS) and Geographical Information System (GIS) have already been proved to play a very important role in analysing soil erosion and sediment yield, as evident from the recent studies conducted in the Indian Peninsula and other parts of the world (Jain and Kothiyari, 2000; Nooka Ratnam et al., 2005; Kusre et al., 2010; Sharma et al., 2013). Soil erosion creates different types of environmental problems because it removes rich

DOI: 10.4324/9781003275916-5

nutrients from soil and increases sedimentation in the rivers and reservoirs reducing their storage capacity (Pandey et al., 2007). In the present study, digital analysis of remotely sensed data has been carried out to assess the soil erosion. GIS is widely employed for assessment of spatial phenomena through amalgamation of various land surface data (Saraf and Choudhary, 1998). Topographic, soil, and morphometric indices have been generated under the GIS platform by using topographic information (contour and drainage). All of the indices have been combined to categorize the soil erosion vulnerability of the different sub-watersheds in the study area. Topography is one of the important factors that directly affect soil erosion (Sun et al., 2014). The properties mostly influencing soil erosion include soil structure and texture, organic matter content, moisture content, density, shear strength as well as chemical and biological characteristics of the concerned soil (Sharma and Bhatt, 1990). Soil erosion has been regarded as a serious problem arising from agricultural intensification, land degradation and possibly due to global climatic change also (Yang et al., 2003). Accelerated soil erosion has adverse economic and environmental impacts. Worldwide, each year about 75 billion tons of soil is eroded from the agricultural lands, a rate that is about 13–40 times as fast as the natural rate of erosion. Asia has the highest soil erosion rate of 30 tons/ha per year; and the Asian rivers contribute about 80% of the total sediments delivered to the world oceans, and amongst these Himalayan rivers contribute up to 50% of the world's total river sediment flux. In India about 16.4 tons/ha of soil is detached annually, about 29% is carried away by the rivers into the sea, and 10% is deposited in reservoirs resulting in considerable loss of the storage capacity. NRSA and NBSS & LUP have estimated that the extent of area under water erosion in the country is 23.62 Mha. (Saini et al., 2015; Nagarajan et al., 2020). Erosion assessment techniques are helpful for the evaluation of agricultural impacts on soil and water resources. The assemblage of landforms in the Teesta/Tista river basin including its innumerable tributaries is the result of climatic processes, tectonic deformation, denudation, and deposition (Mukhopadhyay, 1982). The Tista is one of the most important rivers of the eastern Himalayan region. Earlier it was the principal water source of the Karatoya, Atrai, and Jamuneshwari rivers. Various anthropogenic activities like construction of dams, river bridges, irrigation, and hydroelectric projects are used to reduce the morphological stability of the river. The main objective of this chapter is to assess the vulnerability of soil erosion which is based on multi-criteria evaluation (MCE), the technique applied in the upper catchment area of the Tista river. Long-term average soil loss can be predicted by several established empirical models such as 'rank sum methods'. In this study 'rank sum method' has been used to calculate the weights of the factors influencing soil erosion vulnerability. The Tista river is characterized by a complex hydrological regime as it is fed by precipitation in the form of orographic rains, and also by melt water from glaciers and snow as well as ground water from the upper catchments (Wiejaczka et al., 2014). The dominant factors influencing soil erosion, place emphasis on the roles of precipitation, soil moisture, soil porosity, slope steepness and length, vegetation, and soil organisms (Holz et al., 2015). GIS is vital in the development of computerized methods for quantifying the spatial variability of hazard and erosion related problems, and it is extensively used for modelling and hazard analysis (Rahman et al., 2009). Remote sensing data and GIS technique can form the information base on which sound planning for soil management can be made to achieve the desired results (Franklin et al., 2000). This type of soil erosion model can be an effective tool to predict excessive soil loss and to help in implementation of soil erosion control strategy (Ritchie et al., 2003).

4.2 The Study Area

The Teesta or Tista is originated from Pauhunri Glacier, and the main sources of the water are from glacial meltwater and precipitation (Mukhopadhyay, 1982) but more visibly as the Teesta Kangse from Khangchung Lake (27.590 N and 38.480 E, at an altitude of 2173 m) above mean sea level (Hanif, 1995) within the Eastern Himalayas of Sikkim in India. The river passes through a hilly region before entering Sivoke in the alluvial plains of North Bengal, India. From there, it travels 80 km in a braided route to join the Jamuna (Brahmaputra) river in Bangladesh before entering the Bay of Bengal. The Teesta River, is a source of fresh water for the aquatic ecosystem that is home to the activities of millions of people along its route (Khuman et al., 2018). However, this study area covers 2039 sq. km. of the upper catchment of the river, mainly located (Figure 4.1) in

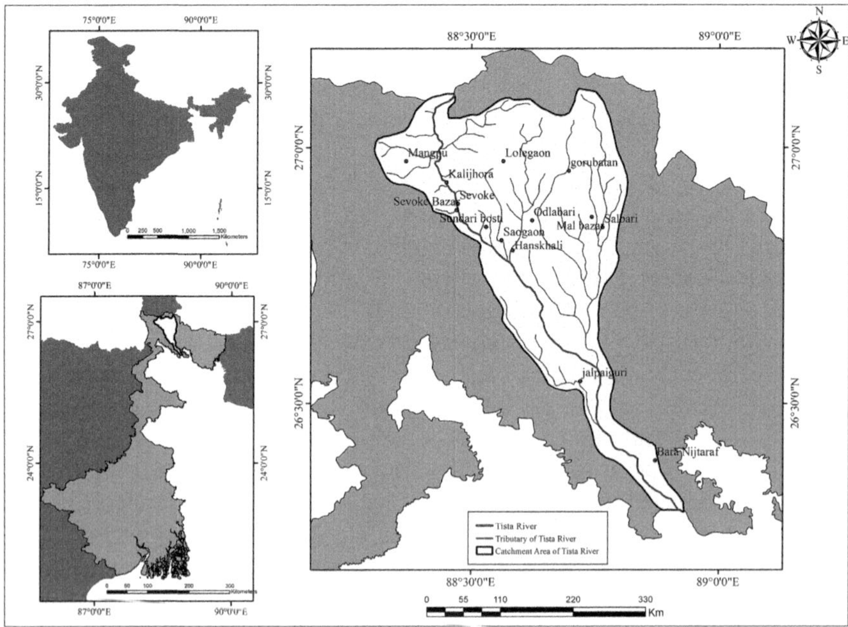

Figure 4.1 Location of the study area

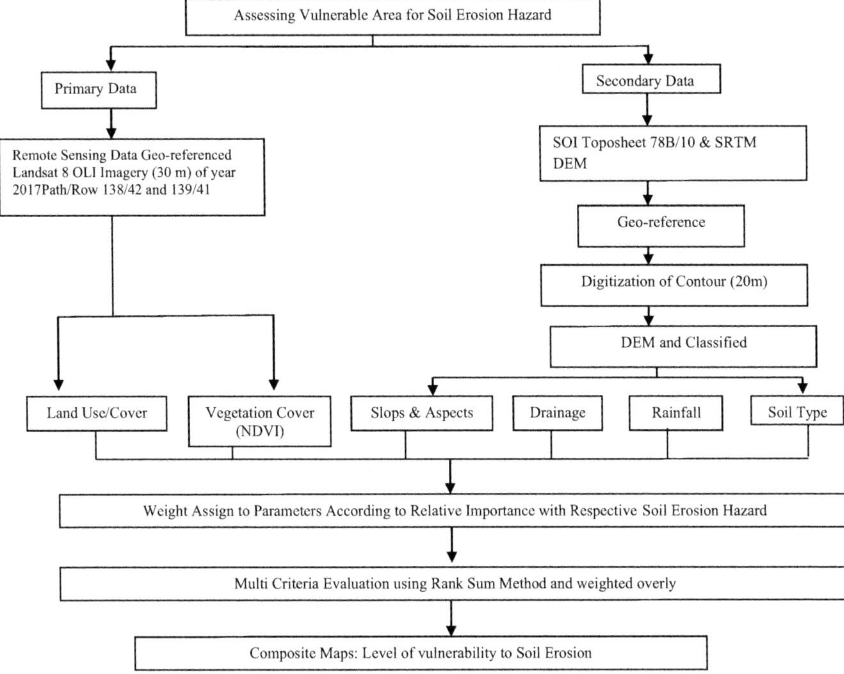

Figure 4.2 Flow chart of the research methodology and methods adopted in the study

the Darjiling Himalayas. The study sites in the Tista basin area were chosen for disaster-prone characteristics to enable investigation of the norms and practices contributing to these (Ferdous and Mallick, 2019). Agriculture is the major source of income of the local people and their economic activity depends on the Tista river (Figure 4.2).

4.3 Materials and Methods

The present chapter is mainly based on secondary sources of data which have been collected from various government and other sources. India Metrological Department (IMD) and local tea garden rain gauge stations are the primary sources of rainfall data. Magnitude and aspects of slopes, and drainage network have been extracted from Digital Elevation Model (DEM) prepared on the basis of SRTM data. Re-projection of DEM and the slope of the area have been calculated using ERDAS 2014. Similarly, aspects of the slopes and vegetation covers have been calculated on the basis of satellite images using Landsat 8 OLI (30 m) for calculation of the Normalized Difference Vegetation Index (NDVI). With ArcGIS 9.2 a Land Use and Land Cover (LULC) map has been prepared based on the training samples in different locations of the study area. Mainly supervised classification methods have been applied to understand the patterns of LULC. Field verification has been done to validate the LULC and other aspects of Kappa Statistics applied for accuracy assessment of the classified image. In order to estimate the importance of each criterion for the model of aggregation, we have applied the Analytic Hierarchy Process (AHP) to take out standard weights. The AHP method is well recognized as a multi-criteria technique (Saini et al., 2015) which is suitable for GIS based 'rank sum' weighted overlay analysis to find out the area or region from where soil erosion is high to low. Initially seven spatial indicators have been identified and given weightage and categorized into different layers. All output vector layer maps have been reclassified and converted into raster grids. For the fulfilment of this research work a five-point rating scale has been used in which 1, 2, 3, 4, and 5 represent very low, low, medium, high, and very high respectively. The weightage is given on the basis of field knowledge and experience of each parameter and their sub-categories (Table 4.1). Multi-criteria analysis using AHP have been applied for the prioritization of erosion prone areas. All the thematic layers have been converted into raster form and each theme is reclassified (Table 4.2). Weightage is given in the percentage scale according to the level of erosion proneness. Topography is a key factor in controlling soil processes and level of erosion under various conditions and, on the other hand, rainfall is one of the important factors causing soil erosion (Li and Huang, 2009). Slope length and steepness, rainfall intensity, and soil type are the most significant factors influencing soil loss (Fang et al., 2014). Drainage density also plays a vital role for soil erosion in the study area and the drainage density decreases with the intensity of precipitation (Moglen, 1998). These factors cause a wide variety of impacts on runoff generation and soil loss. Layers thus obtained are multiplied by the respective weightage of the parameters. Determination of the degree of erosion hazard for each individual factor is an important process for the assessment. To assign a hazard score for each factor, the 'rank sum

Table 4.1 Calculation of the weights by rank sum method

Sl. No.	*Parameters of soil erosion*	*Straight rank (rj)*	*Weight (n − rj + 1)*	*Normalized weight (wj)*	*Standard weight*
1	Rainfall	1	7	0.25	25.00
2	Vegetation	2	6	0.21	21.40
3	Slope	3	5	0.17	17.90
4	Soil	4	4	0.14	14.30
5	Drainage density	5	3	0.10	10.70
6	Land use and land cover	6	2	0.07	7.10
7	Aspects of slope	7	1	0.03	3.60
N = 7	Sum		28	1.00	100

Table 4.2 Spatial layers used in the weighted overlay analysis with sub-parameters' weightage and share of the of parameters in the weightage

Sl. No.	*Parameters of soil erosion*	*Sub-parameters*	*Sub-parameters' weightage*	*Share of parameters 'weightage*
1	Rainfall	Very high > 3800 mm	4	25.0
		High 3500 mm–3800 mm	3	
		Moderate 3300 mm–3500 mm	2	
		Low < 3300 mm	1	
2	Vegetation	Water bodies/others (−0.1 to −0.93)	5	21.4
		Sandy land (01– to −0.1)	4	
		Sparse vegetation (0.20–0.50)	3	
		Moderate vegetation (0.50–0.70)	2	
		Dense vegetation (0.70–0.96)	1	
3	Slope	Very steep (63.94°–81.94°)	5	17.9
		Steep (49.80°–63.94°)	4	
		Moderate (28.27°–49.80°)	3	
		Gentle (<9.31°–28.27°)	2	
		Very gentle (<9.3°)	1	
4	Soil	Loamy skeletal shallow	8	14.3
		Loamy moderately skeletal shallow	7	
		Fine loamy poorly drained	6	
		Fine loamy imperfectly drained	5	
		Fine loamy moderately drained	4	
		Coarse loamy very poorly drained	3	
		Coarse loamy poorly drained	2	
		Coarse loamy moderately shallow	1	
5	Drainage Density	High (1.17–1.56)	4	10.7
		Moderate (0.78–1.17)	3	
		Low (0.39–0.78)	2	
		Very low (0.00–0.39)	1	
6	Land use and land cover	Agricultural area	6	7.1
		Bare/sandy land	5	
		Water bodies	4	
		Vegetation cover	3	
		Tea garden	2	
		Built-up area	1	
7	Aspects of slope	East (67.5–112.5)	5	3.6
		West (247.5–292.5)	5	
		Southeast (112.5–157.5)	4	
		Northwest (292.5–337.5)	4	
		Northeast (22.5–67.5)	3	
		Southwest (202.5–247.5)	3	
		North (337.5–360)	2	
		North (0–22.5)	2	
		South (157.5–202.5)	1	

method' (Janssen, 1994) has been used, where 'wj' is the normalized weight for the 'jth' criterion, 'n' is the number of criteria (j = 1, 2…n) under consideration, and 'rj' is the rank position of the criterion. Each criterion is weighted as 'n − rj + 1' and then normalized by the sum of weights, that is Σ(n − rj + 1) (Saini et al., 2015).

Assessing the soil erosion rate is essential for the development of adequate erosion prevention measures for sustainable management of land and water resources. Raster layers of all the seven parameters have been integrated using weighted overlay in raster calculator under spatial analyst extension in ArcGIS 9.2. The model has generated a composite map with calculated scaling value ranging from 0 to 20.896, where a lower value represents less vulnerability and a high value represents high vulnerability to soil erosion (Samsom, 2017). Saini and Karale (1986) arranged erosion-assessment maps for a part of the Chambal catchment in Rajasthan to predict relative soil loss from different watersheds. Recently the National Bureau of Soil Survey & Land Use Planning (NBSS & LUP) successfully delineated and mapped three levels of soil erosion in part of granitic landscape of Andhra Pradesh using digital analysis of satellite data. In the Sikkim- Darjeeling Himalaya, Tista Watershed is a probable flash flood occurrence zone in India (Mandal and Chakrabarty, 2016). Relief is important since it may affect flash flood occurrence in particular catchments by a combination of two main mechanisms, i.e. orographic effects augmenting precipitation and anchoring convection, and relief promoting the rapid concentration of stream flow (Marchi et al., 2010). Finally a composite soil erosion map has been reclassified into four sub-groups depending on the level of vulnerability to soil erosion in the study area.

4.4 Results and Discussion

Based on the model and the calculated values of the composite map, a 138 sq km area was recorded under very high, 789 sq km under high, 107 sq km under moderate and 1005 sq km under low soil erosion categories. The area under very high, high and moderate vulnerability to soil erosion collectively covers more than about 50% of the study area which may be considered as a serious problem from the point of view of controlling the erosion. Soil erosion plays an important role on degradation of water quality downstream due to sediment deposition (Welde, 2016; Singh and Panda, 2017). In this study area, 821 sq km is under forest cover. DEM has been derived using contour information from the topographical map, and SRTM based DEM for estimation of slope (in degrees). The nature of slopes within a watershed influences the topographical conditions and drainage network. Slope also plays a key role in runoff and stream discharge. Slope is considered as an important criterion for selecting an erosion-prone area (Hlaing et al., 2008; Bandyopadhyay et al., 2009). The high values of drainage density, stream frequency, and drainage texture indicate that underlain by impermeable rocks, which are responsible for high runoff (Umrikar, 2017), these are the result of high soil erosion in the study area and other impacts of surface runoff and soil erosion were identified including eutrophication of water body, sedimentation of rivers and reservoirs, and muddy flooding of roads and residential areas (Boardman et al., 2009). In this research after the overlay analysis and identification of the specific areas, the very high soil erosion affected regions are found to be Saogaon, Sundarbasti, and the lower portion of Sevoke bazaar within the limits of the higher part of upper catchment of the Tista River, and also in the opposite side of Jalpaiguri, Uttar Sisuabari in the middle portion of the upper catchment of the river. The lower part of the catchment area near Bara Nijtaraf, which is occupied by agricultural land use, is also highly vulnerable to soil erosion. One of the important reasons for soil erosion is loss of soil fertility and productivity because it removes organic matter and important nutrients and prevents vegetation growth, which negatively affects overall biodiversity (Scherr, 2000). Soil degradation causes decline in soil feature and productivity (Panagos et al., 2018). On the other hand, floods also play an important role in soil erosion in the study area, and it is relatively common in the Tista valley. The river transports huge amount of silt load as its catchment lies in the tectonically active Himalayas to witness frequent landslides. Besides, the river and its adjoining catchment areas are marked by variegated geomorphic process and forms, unstable flow, bank erosion shifting of the channel, siltation etc. (Shaikh and Shetkar 2018). The loss of soil from land surfaces by erosion is widespread and reduces the productivity of all natural ecosystems as well as agricultural, forest, and pasture ecosystems (Troeh et al., 2004).

4.5 Conclusion

In this research, the basic objective is to determine the critical soil erosion-prone areas along with the spatial pattern of soil loss, using AHP methods. Soil loss in the different vulnerable areas of the upper Tista basin matches with the AHP based soil loss model. This study has also demonstrated the effectiveness of remote sensing and GIS and its usefulness in accepting AHP methods for modelling and evaluating the soil erosion status of the study area. Surface lowering rate is the maximum in the extremely vulnerable soil erosion areas and relatively low in less vulnerable areas and areas with varying and irrational agricultural practices, deforestation, settlement construction, together makes the area more vulnerable to soil erosion and associated environmental degradation. The model based integrated maps can help us better understand the causes, make predictions, and plan how to implement preventative and restorative strategies to prioritize the area according to severity of erosion. In soil erosion prone areas, hazard and risk assessment are key requirements for any planning and management, and soil erosion hazard mapping also can be a starting point of any regional intervention policy for erosion control and conservation (Rahman et al., 2015). Sub-watersheds with different soil erosion potential have been assessed here with a view to adopting soil conservation measures. The major aspects that put in to soil erosion potential in these sub-watersheds have been analysed so that conservation measures can be adopted. Therefore, the people of this region could be made aware and the loss of resources can be minimized. This research based on satellite data products analysed with the help of geospatial technology will also facilitate advancement of future research on this issue (Figure 4.3–4.6).

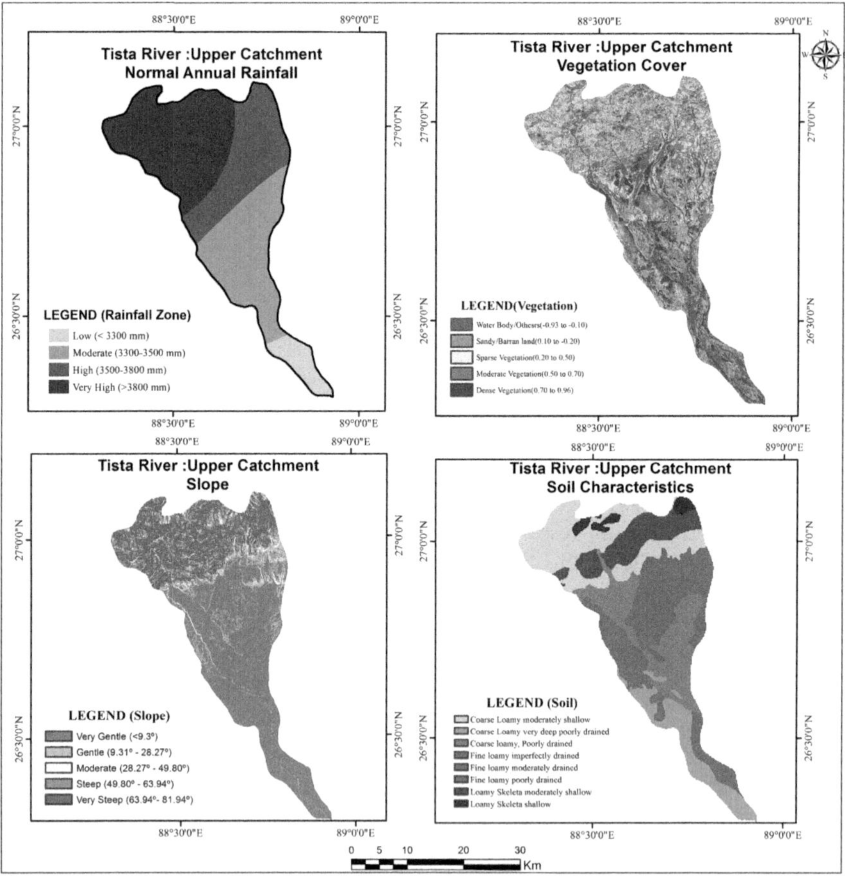

Figure 4.3 Maps of rainfall, vegetation, slope and soil characteristics

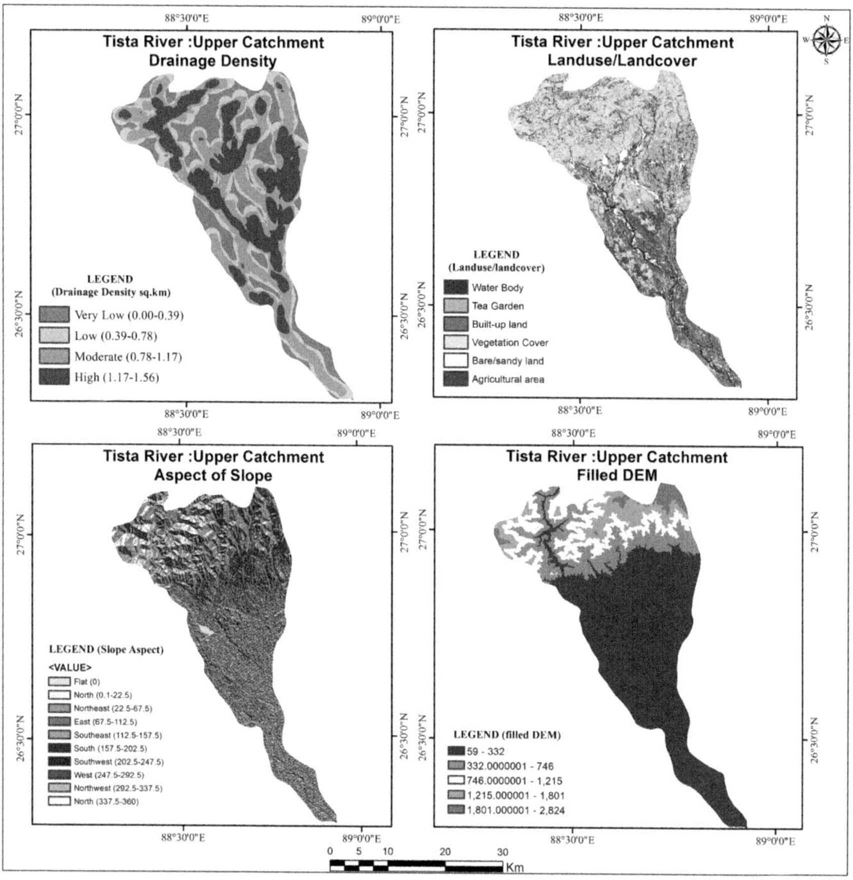

Figure 4.4 Maps of drainage, LULC, aspects of slope and filled DEM

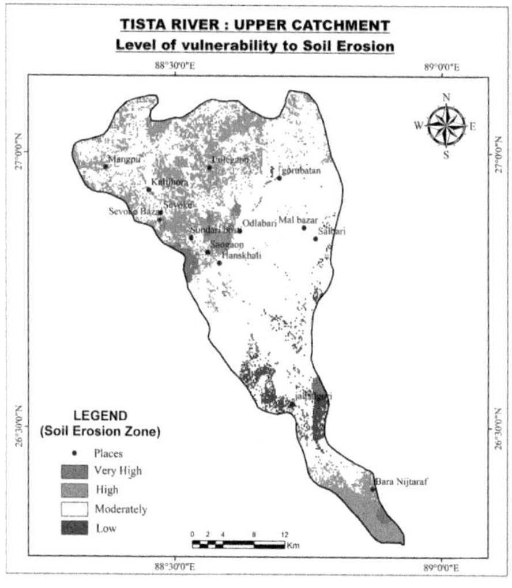

Figure 4.5 Map of the level of vulnerability to soil erosion

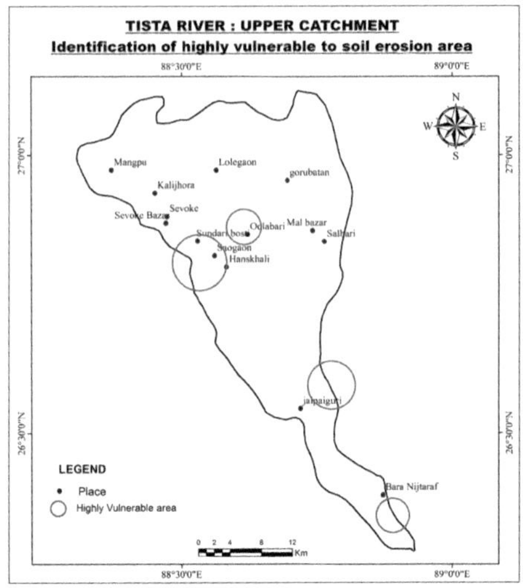

Figure 4.6 Map of highly vulnerable to soil erosion areas

Bibliography

Ahmad, N. and Pandey, P. (2018) 'Assessment and monitoring of land degradation using geospatial technology in Bathinda district, Punjab, India', *Solid Earth*, 9(1), 75–90. DOI: 10.5194/se-9-75-2018

Ahmed, I. (2012) 'Teesta, Tipaimukh and river linking: Danger to Bangladesh–India relations', *Economic & Political Weekly*, 47(16), 51.

Al-Shalabi, M. A. et al. (2006) 'GIS based multicriteria approaches to housing site suitability assessment', In XXIII FIG congress, shaping the change, Munich, Germany, pp 1–17, (8–13 October). https://citeseerx.ist.psu.edu/viewdoc/download?doi=10.1.1.124.9679&rep=rep1&type=pdf

Al-Soufi, R. (2004) 'Soil erosion and sediment transport in the Mekong basin', *Proc. of 2nd APHW conference, Singapore*, 4. Available at: http://rwes.dpri.kyoto-u.ac.jp/~tanaka/APHW/APHW2004/proceedings/OHS/56-OHS-A339/56-OHS-A339_resubmit.pdf

Balica, S. F. et al. (2013) 'Parametric and physically based modelling techniques for flood risk and vulnerability assessment: A comparison', *Environmental Modelling and Software*, 41, 84–92. DOI: 10.1016/j.envsoft.2012.11.002

Bandyopadhyay, S., Jaiswal, R. K., Hegde, V. S. and Jayaraman, V. 2009. 'Assessment of land suitability potentials for agriculture using a remote sensing and GIS based approach', *International Journal of Remote Sensing*, 30(4), 879–895. DOI: 10.1080/01431160802395235

Bera, A., Mukhopadhyay, B. and Das, D. (2018) 'Morphometric analysis of Adula river basin in Maharashtra, India using GIS and remote sensing techniques', *Geo-spatial Data in Natural Resources*, (July), 13–35. DOI: 10.21523/gcb5.1702

Boardman, J., Shepheard, M. L., Walker, E. and Foster, I. D. L., (2009) 'Soil erosion and risk-assessment for on- and off-farm impacts: A test case using the Midhurst area, West Sussex, UK', *Journal of Environmental Management*, 90(8), 2578–2588.

Borah, D. K. et al. (2013) 'Sediment and nutrient modeling for Tmdl development and implementation', *Transactions of the ASABE*, 49(4), 967–986. DOI: 10.13031/2013.21742

Clarke, J. (1966) *Morphometry from maps: Essays in geomorphology*. New York: Elsevier Publishing Company.

Dabral, P. P., Baithuri, N. and Pandey, A. (2008) 'Soil erosion assessment in a hilly catchment of North Eastern India using USLE, GIS and remote sensing', *Water Resources Management*, 22(12), 1783–1798. DOI: 10.1007/s11269-008-9253-9

Decoursey, D. G. et al. (n.d.) 'Integrated quantity/quality modeling: Receiving waters', In: Bowles, D.S., O'Connell, P.E. (eds) *Recent advances in the modeling of hydrologic systems*. NATO ASI Series, vol 345. Dordrecht: Springer. https://doi.org/10.1007/978-94-011-3480-4_15

Djendoel, D. (1973) 'Los limites del crecimiento', *Actualidades Biológicas*, 2(3), 18–20. http://matematicas.udea.edu.co/~actubiol/actualidadesbiologicas/raba1973v2n3art4.pdf

DPR. (2005) *Carrying capacity study of Teesta basin in Sikkim. Phase II report, ICISMHE, University of Delhi, Delhi.*

Dutta, S. (2016) 'Quantification and mapping of morphometric parameters of Subarnarekha river basin in Eastern India using geo-spatial techniques', (November).

Erosion, R. (n.d.) 'Effects of riverbank erosion on livelihood', pp. 1–39.

Fang, H., Sun, L., and Tang, Z. (2014). 'Effects of rainfall and slope on runoff, soil erosion and rill development: An experimental study using two loess soils', *Hydrological Processes*, 29(11), 2649–2658. DOI: 10.1002/hyp.10392

FAO and ITPS. (2015) Status of the world's soil resources (SWSR) – main report. Food and Agriculture Organization of the United Nations and Intergovernmental Technical Panel on Soils, Rome, Italy, Viewed 20 March 2018. www.bergen.kommune.no/bk/multimedia/archive/00316/FAO_status_jordress_316788a.pdf

Ferdous, Jannatul and Mallick, Dwijen, 2019 Norms, practices, and gendered vulnerabilities in the lower Teesta Basin, Bangladesh, *Environmental Development*. DOI: 10.1016/j.envdev.2018.10.003

Franklin, J., Woodcock, C. E. and Warbington, R., (2000) 'Digital vegetation maps of forest lands in California: Integrating satellite imagery, GIS modeling, and field data in support of resource management', *Photogrammetric Engineering and Remote Sensing*, 66, 1209–1217.

Gobin, A. et al. (n.d.) *Assessment and reporting on soil erosion.*

Hajkowicz, S. A., McDonald, G. T. and Smith, P. N. (2000) 'An evaluation of multiple objective decision support weighting techniques in natural resource management', *Journal of Environmental Planning and Management*, 43(4), 505–518. DOI: 10.1080/713676575

Hanif, S. (1995) *Hydro-geomorphic characteristics of the Teesta floodplain, Bangladesh*, Unpublished PhD thesis. Rajshahi: Department of Geography, University of Rajshahi.

Hlaing, K. T., Haruyama, S. and Aye, M. M. (2008) 'Using GIS-based distributed soil loss modeling and morphometric analysis to prioritize watershed for soil conservation in Bago river basin of Lower Myanmar', *Frontiers of Earth Science in China*, 2(4), 465–478. DOI: 10.1007/s11707-008-0048-3

Holz, D. J., Williard, K. W. J., Edwards, P. J. and Schoonover, J. E. (2015) 'Soil erosion in humid regions: A review', *Journal of Contemporary Water Research & Education*, 154(1), 48–59.

Huq, S., Rahman, A. A. and Mallick, D. (1998) 'Population and environment in Bangladesh', *Workshop on Population and Environment in Bangladesh*, (November). http://docs.ims.ids.ac.uk/migr/upload/fulltext/bcaspop.pdf

Islam, F. and Higano, Y. (2011) 'Equitable sharing of bilateral international water: A policy measure for optimal utilization of the Teesta River', *Studies in Regional Science*, 32(1), 17–32. DOI: 10.2457/srs.32.17

Jain, M. K. and Kothiyari, U. (2000) 'Estimation of soil erosion and sediment yield using GIS', *Hydrological Sciences Journal*, 45(5), 771–786.

Janssen, R. V. H. (1994) *Multi-objective decision support for environmental management +DEFINITE Decisions on an FINITE set of alternatives: Demonstration disks and instruction*. 232.

Karthick, P., Lakshumanan, C. and Ramki, P. (2017) 'Estimation of soil erosion vulnerability in Perambalur Taluk, Tamilnadu using revised universal soil loss equation model (RUSLE) and geo information technology', 5(8), 8–14.

Khuman, S. N., Chakraborty, P., Cincinelli, A., Snow, D., & Kumar, B. (2018). Polycyclic aromatic hydrocarbons in surface waters and riverine sediments of the Hooghly and Brahmaputra Rivers in the Eastern and Northeastern India. *Science of the Total Environment 636*, 751–760. DOI: 10.1016/j.scitotenv.2018.04.109

Kirschbaum, D. B. et al. (2010) 'A global landslide catalog for hazard applications: Method, results, and limitations', *Natural Hazards*, 52(3), 561–575. DOI: 10.1007/s11069-009-9401-4

Kusre, B. C. et al. (2010) 'Assessment of hydropower potential using GIS and hydrological modeling technique in Kopili River basin in Assam (India)', *Applied Energy*, 87(1), 298–309. DOI: 10.1016/j.apenergy.2009.07.019

Lal, R. (2001) 'Soil degradation by erosion', *Land Degradation & Development*, 12, 519–539.

Lal, R. (2010) 'Critical reviews in plant sciences soil erosion impact on agronomic productivity and environment quality soil erosion impact on agronomic productivity and environment quality', 2689. DOI: 10.1080/07352689891304249

Lee, S. (2004) 'Soil erosion assessment and its verification using the universal soil loss equation and geographic information system: A case study at Boun, Korea', *Environmental Geology*, 45(4), 457–465. DOI: 10.1007/s00254-003-0897-8

Li, A. et al. (2006) 'Eco-environmental vulnerability evaluation in mountainous region using remote sensing and GIS – A case study in the upper reaches of Minjiang River, China', *Ecological Modelling*, 192(1–2), 175–187. DOI: 10.1016/j.ecolmodel.2005.07.005

Li, G. and Huang, G. B. (2009) Effects of rainfall intensity and land use on soil and water loss in loess Hilly Region. *Journal of Agricultural Engineering*, 25(11): 85–90.

Mandal, S. P. and Chakrabarty, A. (2016) 'Flash flood risk assessment for upper Teesta river basin: Using the hydrological modeling system (HEC-HMS) software', *Modeling Earth Systems and Environment*, 2(2), 1–10. DOI: 10.1007/s40808-016-0110-1

Marchi, L., Borga, M., Preciso, E., & Gaume, E. (2010). Characterisation of selected extreme flash floods in Europe and implications for flood risk management. *Journal of Hydrology*, *394*(1-2), 118–133.

Mishra, S. and Nagarajan, R. (2010) 'Morphometric analysis and prioritization of sub watersheds using GIS and Remote Sensing techniques: A case study of Odisha, India', *International Journal of Geomatics and Geosciences*, 1(3), 501–510.

Moglen, Glenn E. (1998) 'On the sensitivity of drainage density to climate change', *Water Resources Research*, 34(4), 855–862.

Moss, T. (2004) 'The governance of land use in river basins: Prospects for overcoming problems of institutional interplay with the EU Water Framework Directive', 21, 85–94. DOI: 10.1016/j.landusepol.2003.10.001

Mukhopadhyay, S. C., (1982) *The Tista basin: A study in fluvial geomorphology*. Calcutta: K.P. Bagchi and Company, 308pp.

Nagarajan, D., Lee, D., Chen, C., & Chang, J. (2020). Resource recovery from wastewaters using microalgae-based approaches: A circular bioeconomy perspective. *Bioresource Technology*, *302*, 122817. doi:10.1016/j.biortech.2020.122817

Nooka Ratnam, K., Srivastava, Y. K., Venkateswara Rao, V. et al. (2005) Check dam positioning by prioritization of micro-watersheds using SYI model and morphometric analysis: Remote sensing and GIS perspective. *Journal of the Indian Society of Remote Sensing 33*, 25–38. https://doi.org/10.1007/BF02989988

Pal, S. (2016) 'Identification of soil erosion vulnerable areas in Chandrabhaga river basin: A multi-criteria decision approach', *Modeling Earth Systems and Environment*, 2(1), 1–11. DOI: 10.1007/s40808-015-0052-z

Panagos, P., Standardi, G., Borrelli, P., Lugato, E., Montanarella, L. and Bosello, F. (2018) Cost of agricultural productivity loss due to soil erosion in the European Union: From direct cost evaluation approaches to the use of macroeconomic models. *Land Degradation & Development*, 29, 471–484.

Pandey, A., Chowdary, V. M. and Mal, B. C. (2007) 'Identification of critical erosion prone area in the small agricultural watershed using USLE, GIS and remote sensing', *Water Resources Management*, 21, 729–746.

Patil, R. J., Sharma, S. K. and Tignath, S. (2015) 'Remote sensing and GIS based soil erosion assessment from an agricultural watershed', *Arabian Journal of Geosciences*, 8(9), 6967–6984. DOI: 10.1007/s12517-014-1718-y

Pimental, D., Harvey, C. and Resosudarmo, P. (1995) Environmental and economic costs of soil erosion and conservation benefits. *Science*, 267, 111741123.

Rahman, M. R., Shi, Z. H. and Cai, C. (2009) Soil erosion hazard evaluation-an integrated use of remote sensing, GIS and statistical approaches with biophysical parameters towards management strategies. *Ecological Modelling*, 220(13–14), 1724–1734.

Rahman, M. R., Shi, Z. H., Chongfa, C. and Dun, Z. (2015). Assessing soil erosion hazard – A raster based GIS approach with spatial principal component analysis (SPCA). *Earth Science Informatics*, 8(4), 853–865. DOI: 10.1007/s12145-015-0219-1

Ritchie, J. C., Walling, D. E. and Peters, J. (2003) Application of geographic information systems and remote sensing for quantifying patterns of erosion and water quality. *Hydrological Processes*, 17(5), 885–886.

Saini, K. M., & Karale, R. L. (1986). Application of Remote Sensing for catchment characterization and assessment of run-off potential in upper Ganga Catchment, UP (UNDP/FAO Project Tech. Report 02). *Remote Sensing Centre, All India Soil & Land Use Survey, New Del hi.*

Saini, S. S., Jangra, R. and Kaushik, S. P. (2015) 'Vulnerability assessment of soil erosion using geospatial techniques – A pilot study of upper catchment of markanda river', *International Journal of Advancement in Remote Sensing, GIS and Geography*, 3(1), 9–21.

Samsom, S. A., Auanlade, A., Alabi, O., Alaga, A. T., Oloko-Oba, M. O., Ogunyemi, S. A., & Badru, R. A. (2017). Soil erosion vulnerability mapping and implication on vegetation in parts of Oshun River Basin, Nigeria. *International Journal of Scientific Research in Science, Engineering and Technology*, *3*(1), 82–91.

Saraf, A. K. and Choudhary, P. R. (1998). 'Integrated remote sensing and GIS for groundwater exploration and identification of artificial recharge sites', *International Journal of Remote Sensing*, 19(10), 1825–1841.

Scherr, S. J. (2000). A downward spiral? Research evidence on the relationship between poverty and natural resource degradation. *Food policy*, *25*(4), 479–498.

Shaikh, M. S., & Shetkar, R. V. (2018). Soil Erosion Estimation Modelling by Revised Universal Soil Loss Equation and Soil and Water Assessment Tool on Geographic Information System Platform. *Journal of Water Resource Engineering and Management*, 5(2).

Sharma, S. A. and Bhatt, H. P. (1990) 'Generation of brightness and greenness transformations for IRS-LISS II data', *Journal of the Indian Society of Remote Sensing*, 18(3), 25–31.

Sharma, S. B., Sayyed, R. Z., Trivedi, M. H. et al. (2013) Phosphate solubilizing microbes: sustainable approach for managing phosphorus deficiency in agricultural soils. *Springerplus*, *2*, 587. https://doi.org/10.1186/2193-1801-2-587

Singh, G. and Panda, R. K. (2017) Grid-cell based assessment of soil erosion potential for identification of critical erosion prone areas using USLE, GIS and remote sensing: A case study in the Kapgari watershed, India. *International Soil and Water Conservation Research*, 5(3), 202–211. DOI: 10.1016/ j.iswcr.2017.05.006

Singh, R. (2000) *Watershed planning and management*. Yash Publishing House.

Sinha Roy, S. (1980) 'Terrace systems in the Tista valley of Sikkim Darjeeling Himalayas and the adjoining Piedmont region', *Indian Journal of Earth Sciences*, 7, 146–161.

Solaimani, K., Modallaldoust, S. and Lotfi, S. (2009) 'Investigation of land use changes on soil erosion process using geographical information system', 6(3), 415–424.

Sun, W., Shao, Q., Liu, J., and Zhai, J. (2014) 'Assessing the effects of land use and topography on soil erosion on the Loess Plateau in China', *Catena*, 121, 151–163. DOI: 10.1016/j.catena.2014.05.009

Troeh, F. R., Hobbs, A. H. and Donahue, R. L. (2004) *Soil and water conservation: For productivity and environmental protection*. Prentice Hall.

Umrikar, B. N. (2017) 'Morphometric analysis of Andhale watershed, Taluka Mulshi, District Pune, India'. *Applied Water Science*, 7(5), 2231–2243. DOI: 10.1007/ s13201-016-0390-7

UNCED. (1992) UNCED Agenda 21: Programme of action for sustainable development, Rio declaration on environment and development, statement of principles. In *Final Text of Agreement Negotiated by Governments at the United Nations Conference on Environment and Development (UNCED)*; United Nations Development Programme (UNDP): New York, NY, pp. 3–14.

Hoenselaar, T. Van, Verkerk-Bakkels, B. and Janssen, R. (1995) 'Species identification of cyst and root-knot nematodes from potato by electrophoresis of individual females', pp. 105–109.

Vogiatzakis, I. N. (2003) 'GIS-based modelling and ecology: A review of tools and methods', *Geographical Paper No. 170*, 170, 1–34.

Welde, K. (2016) 'Identification and prioritization of subwatersheds for land and water management in Tekeze dam watershed, Northern Ethiopia', *International Soil and Water Conservation Research*, 4(1), 30–38. DOI: 10.1016/j.iswcr.2016.02.006

Wessels, K. J., Prince, S. D., Malherbe, J., Small, J., Frost, P. E., and VanZyl, D.(1991) 'Soil erosion assessment and mapping of the algar river watershed (uttar pradesh) using remote sensing technique', *Journal of the Indian Society of Remote Sensing*, 19(2), pp. 67–76. DOI: 10.1007/BF03008122

Wessels, K. J., Prince, S. D., Malherbe, J., Small, J., Frost, P. E. and VanZyl, D. (2007). Can human-induced land degradation be distinguished from the effects of rainfall variability? A case study in South Africa. *Journal of Arid Environments*, 68(2), 271–297. DOI: 10.1016/j.jaridenv.2006.05.015

Wiejaczka, L., Bucała, A. and Sarkar, S. (2014) 'Human role in shaping the hydromorphology of Himalayan rivers: Study of the Tista River in Darjeeling Himalaya', *Current Science*, 106(5), 717–724.

Yang, D., Kanae, S., Oki, T., Koike, T., & Musiake, K. (2003). Global potential soil erosion with reference to land use and climate changes. *Hydrological Processes*, 17(14), 2913–2928.

Young, R. A. (1987) *AGNPS: An agricultural non point source pollution model*, Washington, DC: US Dept. Agric. Res. Services.

5 Assessing the Dynamics of Agriculture under the Changing Climate in Gosaba Block of Indian Sundarban

Biraj Kanti Mondal

5.1 Introduction

The majority of each block of the Indian Sundarban are dependent on primary activities, and more than 80% of the populace is solely associated with agriculture. Most of the blocks have mono-crop land, small land holdings, traditional means of agriculture, and less irrigation opportunity, and therefore, the productivity is much lower than the state average. The Gosaba block amongst the 19 blocks of Sundarban, is seriously facing such problems, which are aggravated during a hazard or disaster situation. The situation becomes more adverse when there is any sort of hazard, like a cyclone hitting the area, because of embankment collapse, saline water inundation, water stagnation, crop damage etc. Thus, low productivity, low capital for investment, mono crop nature etc. have pushed the stakeholders towards wage labour and frequently they choose to migrate from fishing and the collection of prawn seeds, and collection of honey, particularly during agricultural lean season. Such negative impact on the agricultural sector is increasing day by day due to the increasing frequency and intensity of cyclones in the area, enhancing the risk for sustaining livelihood and often creates climate-induced migration.

Some of the earlier studies analysed the linkages between climate change and crop yield and migration in other geographical areas and in dissimilar contexts (Ahsan & Warner, 2014; Dumenu & Obeng, 2016; Feng et al., 2010; Gornall et al., 2010; Iqbal & Roy, 2014; Kumar & Viswanathan, 2015; Pound et al., 2018); but there is very limited research on Sundarban, especially on Gosaba (Ghosh, 2012; Debnath, 2013a; Halder & Debnath, 2014; Dutta et al., 2019; Mondal, 2018; Ghosh et al., 2015, 2018; Danda, 2010, 2019; Pramanik, 2015; Sahana et al., 2019a; Sahana et al., 2020) and on agricultural productivity and allied issues (Mandal et al., 2011, 2013; Chand et al., 2012a; Jana et al., 2012; Mondal, 2012; Debnath, 2013b; Hazra et al., 2014; Hajra et al., 2016; Roy & Guha, 2017; Laha, 2019).

The Gosaba block is probably the most backward block of Sundarban and, therefore, weaker people with higher poverty have the least capacity to respond to hazards like a super cyclone, flood, etc., and they often face higher risk and greater burdens. The islanders of the block constantly have to

DOI: 10.4324/9781003275916-6

tolerate both detectable and not-so-visible uncertainties to uphold their livelihood especially associated with agriculture. The agriculture-dependent inhabitants often live I danger of climatic hazards and related effects like sea level rise, erosion of earthen embankments, flooded land, and salinity incursion in land and water. The in-depth analysis of the agricultural patterns and its dynamics in the Gosaba block and its adjustment to the climate change-induced hazards has portrayed some important issues that have been empirically conferred in the current effort. Thus, the story of the inhabitants has always been a sad one, and their livelihood faces troubles, and they have constantly been coping with every vulnerable situation and trying to adapt. The risk and vulnerability status due to changing climate-related issues and hazards were established as the pillar of the current study. Henceforth, this submission seeks to delve deep into the agricultural issues of Gosaba block with a serious prominence of climate change-induced hazards, vulnerability and livelihood. This chapter endeavours to tease out data from field surveys, interviews, and minute observation along with secondary data sources and employing GIS mapping.

5.2 Study Area

Indian Sundarban is distributed across the North and South 24 Parganas districts of West Bengal (2016) and is designated as a World Heritage Site and Ramsar Site. The present study is concentrated on the Gosaba block of Sundarban, which is considered as one of the least accessible blocks of Sundarban in terms of connectivity and remotely located island position (Figure 5.1). According to the climatic vulnerability and the risk zonation map, the Gosaba block is a highly vulnerable or very risky zone with respect to any climatic events. The block is characterized entirely as an rural area: 15 Gram Panchayat; 51 Mouzas; moderate population density (825 person/sq. km.); moderate decadal population growth rate (8.83%). Moreover, it is overly dependent on primary economic activities like agriculture (70%) and fishing (20%); there is a high concentration of poor households. Most of the households are thatched and kutcha (80%), they are vulnerable; less than 2% of households have electricity; telecommunication facilities and connection levels are not satisfactory. Consequently, inhabitants here are isolated as it interlinked by the crisscross drainage network and livelihood opportunities are very much restricted; thereafter the condition of climate change affected stockholders are extremely susceptible not only in terms of hazards but also socio-economic position and empowerment.

As the block is entirely surrounded by rivers and earthen embankments, the coastal exposure is very high; intended to amplify the challenges during any climate-induced hazards. Many of such embankments and roads were destroyed during cyclone Aila leading to prolonged flooding, saline water stagnation, damaging agricultural land along with the existing issues of man-animal conflicts due to close location to forest. Thus, the block is frequently

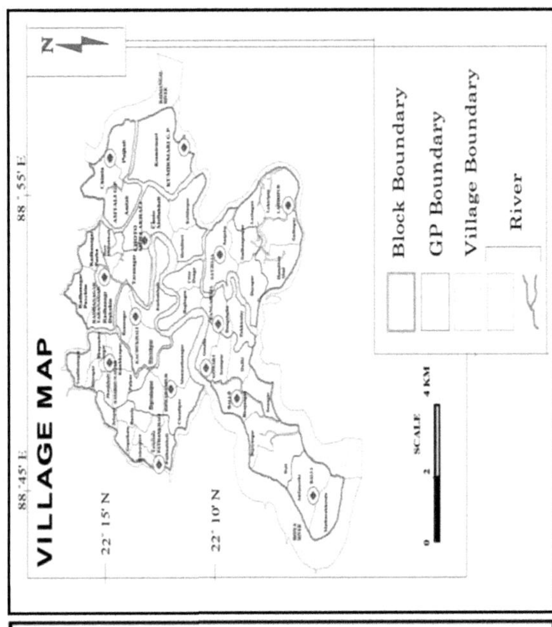

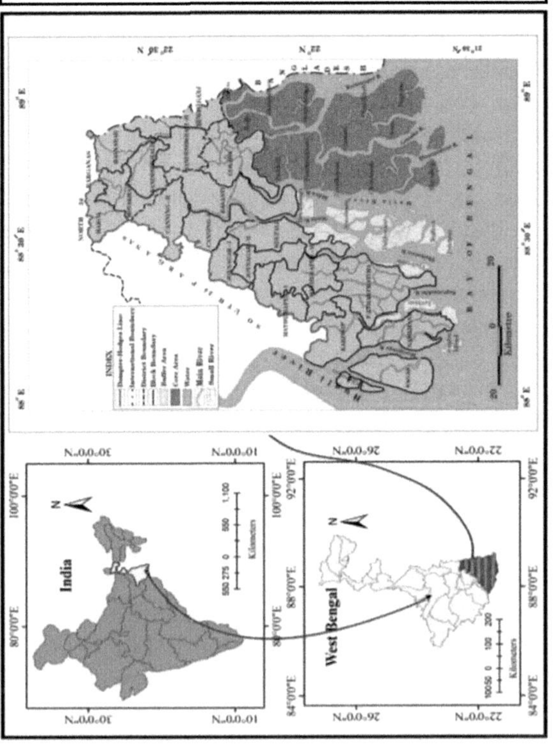

Figure 5.1 Location map of Gosaba block with villages

exposed to all sorts of hazards and the entire area remains vulnerable and at risky as it is experiencing cyclones, sea level rise, coastal erosion, changing of the river course, saline water inclusion, and many more.

5.3 Database and Methodology

The current research followed suitable methodological steps (Figure 5.2) with a fusion of primary and secondary data. The primary data were collected through a primary survey, field work, observation, and direct interviews and discussions with households and individuals were conducted following a two-stage cluster purposive random sampling method. At the first stage, three villages were chosen randomly from dissimilar Gram Panchayats positioned in a diverse island of Gosaba block [1. Rangabelia G.P: Pakhiralay Mouza (22.142079° N, 88.829524° E); 2. Amtali G.P.: Puijhali Mouza (22.223345°N, 88.83164°E); 3. Radhanagar Taranagar G.P.: Boramollakhali Mouza (22.2318461°N, 88.85417474°E)]; while in the second stage, households within these villages were selected randomly for the survey. Direct interviews was carried out during the survey, choosing 50 households from each village covering a total of 150 households in the year 2018–19. The questionnaire scheduled was designed by the questions about occupation, major economic engagement, agricultural land, family size, land use change etc. Some other questions were asked concerning the nature of climate change and its consequences, like losses of agricultural land, saline water intrusion, reduction of agricultural production, reduction of the vegetables production, embankment collapse, impact on life and livelihood, temperament of vulnerability, and adaptation challenges faced. Whereas, the secondary data has collected from the Census of India, District Statistical Handbook, District Human Development Report, Report of State Disaster Management Authority, and published reports/literatures. Land use land cover (LULC) classification of the Gosaba block of 2020 was also prepared based on the Landsat 8 OLI satellite images by using Arc GIS 10 software. In due course, to analyse all the data, statistical techniques, quantitative methods, and GIS mapping were carried out using geoinformatics as a key tool.

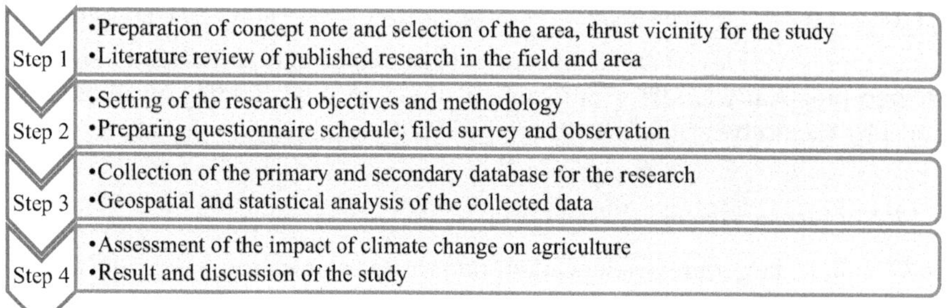

Figure 5.2 Structural framework of the study

5.4 Results and Discussion

5.4.1 Proviso of Agriculture in Gosaba Block

Agriculture is considered as the backbone of the economy of Gosaba as more than 80% people are diversely associated with the largely important economic sector. Agriculture is operated in the traditional way in the reclaimed land which consumes a large proportion of poor people and the agricultural land is surrounded by embankments to protect the inflow of tidal saline water of the river. The area is penetrated by limited irrigation facilities, leading to mono-cropping agriculture, which again is hamstrung by salinity through both ground water and surface water. Moreover, the majority of the farmers have small land holding, marginal land holding and landless in nature. Hence, the required capital investment in agriculture is not taking place. Currently, due to the negative impact of climate change-induced hazards, especially in the economic sector, people are engaging themselves in fishing and the collection of prawn seeds, collection of honey, risking their lives from man-eating tigers and crocodiles particularly during the agricultural lean season.

5.4.1.1 Agriculture, Farmers and Land Holdings

Agriculture is the mainstay of the economy of the Gosaba despite the geophysical adversity along with the fact that 55% of the population is landless. The land holding on the islands were very little, each family is like a marginal farming family with only 0.63 hectare in some small islands of Gosaba, like Mollakhali according to the 2001 census. In West Bengal, farmers with landholding below 2.5 acres (1 hectare) are defined as marginal farmers and the ones with landholding between 2.5 and 5 acres (2 hectares) are deemed as small farmers. However, the concept of marginal and small farmers as perceived by the peasantry of the Gosaba block is a few notches below that defined by the state, a farmer with a holding of between 0.13 and 0.27 hectares (2 *bigha*) is considered marginal while a farmer with agricultural land between 0.27 and 0.67 hectares (2 to 5 *bigha*) is perceived as small. About 54.21% have been recorded as landless labourers, 85.22% are small and marginal farmers with an average holding of 0.82 hectare per family (District Statistical Handbook). Households with less than 0.13 hectares (1 *bigha*) of agricultural land are considered practically landless, but some of them have no holding at all but still consider themselves as farmers.

5.4.1.2 Cropping Pattern and Intensity

In Gosaba, a larger area is dependent on rain-fed single crop, i.e. paddy (aman paddy); and other crops are aus and boro paddy, wheat, potato, jute, pulses, oilseed, and some vegetables in other seasons. The season-wise cropping pattern is important as most of the land is mono-crop and mostly mono-usable, thus the harvesting seasons with the respective crops are tabulated (Table 5.1).

Table 5.1 Crops grown in different seasons

Seasons	Crops grown
Autumn *(Bhadoi)*	Aus paddy, vegetables
Winter *(Haimantik)*	Aman paddy, pulses, vegetables, chilli, potato
Spring *(Rabi)*	Pulses, sugarcane, oilseeds, tomato
Summer *(Baisakhi)*	Boro paddy, vegetables, fruits and others
Vegetables (yearly)	Tomato, brinjal, chili, bhindri, cucumber, cabbage, cauliflower, pumpkin, bitter gourd, potato, green gram

Source: District Statistical Handbook

5.4.1.3 Agricultural Productivity and Efficiency

Agricultural productivity is the overall productivity per unit area and it is low in the Gosaba block as it suffers from lack of irrigation and other difficulties. The agricultural labourers occupy a greater percentage (55%) than the cultivator and there are a good number of landless labourers. The landless labourers work in terms of money or personal benefit as an agricultural worker or marginal workers in those agricultural fields. The land owners distribute their lands for cultivation to those labourers in terms of some money and they are engaging themselves in some other money earning process. The percentage of agricultural labourers is high in Gosaba amongst the Sundarban blocks. The proportion of the cultivator signifies that the population are mostly engaged in the cultivation of paddy, food grains, or vegetables with a greater percentage.

5.4.2 An Investigation of Climate Change in Gosaba Block of Sundarban

The study region is often devastated by super-cyclones and floods, but what is even more alarming is the fact that as a consequence of these hazards and climate change, this area is facing tremendous pressure due to sea level rise that is culminating into loss of habitat and cultivable land, compelling people to dislocate. The occurrence of cyclones, floods, sea level rise, river bank erosion, and storm surges has a strong impact on reducing agricultural potentialities, inundation of land by saline water, damage embankment, and diminish ecosystem services, mass displacement and relocation of people. The occurrence of various climate change events and effects on various segments in Gosaba as well as Sundarban are accumulating various literature (Hazra et al., 2002; Mitra et al., 2009; WWF, 2010; Nandy & Bandyopadhyay, 2011; Banerjee, 2013; Raha et al., 2012; Raha, 2014; Chakraborty, 2015; Mondal, 2018; Pramanik, 2015; Kusche et al., 2016; Sahana et al., 2019b), which is tabulated below (Table 5.2).

It is revealed that here climate change and extreme weather events, like a severe cyclone, are integrally related to increasing vulnerability and resulting agricultural loss and ultimately poor socio-economic conditions of the already

Table 5.2 Issues and consequences of climate change in Sundarban

Climate Change	*Facts*	*Impact*	*Facts*
Temperature	Increased 0.5°C	*Flood hazard*	Increased
Rainfall amount	Decreases	*Embankment damaged*	Increased
Frequency of cyclone	Increased	*Soil erosion*	Increased
Intensity of cyclone	Changed	*Saline water inclusion*	Increased
Super cyclone	Greater than before	*Salinity of soil*	Increased
Damages by cyclone	Increased	*Fresh water*	Less available
Rainy days	Fewer	*Surface water*	Decreased
Rainy season	Shorter	*Ground water*	Decreased
Winter	Late coming and warmer	*Agricultural area*	Decreased
Summer	Coming early and Longer	*Inundation of land*	Increased
Overall weather	Changed	*Land degradation*	Increased
Sea level rise	Increased (3.14 mm / year)	*Agricultural potentialities*	Decreased
Coastal areas	Destroyed	*Crop production*	Decreased

Source: Tabulated by the author (based on literature review)

marginalized group. Most of this block is coastal and riverine and thus it is a high risk zone in terms of cyclones (Figure 5.3) and other atmospheric hazards and therefore, natural calamities become a part and parcel of the larger community who are depending on agriculture. It is also disclosed that the most

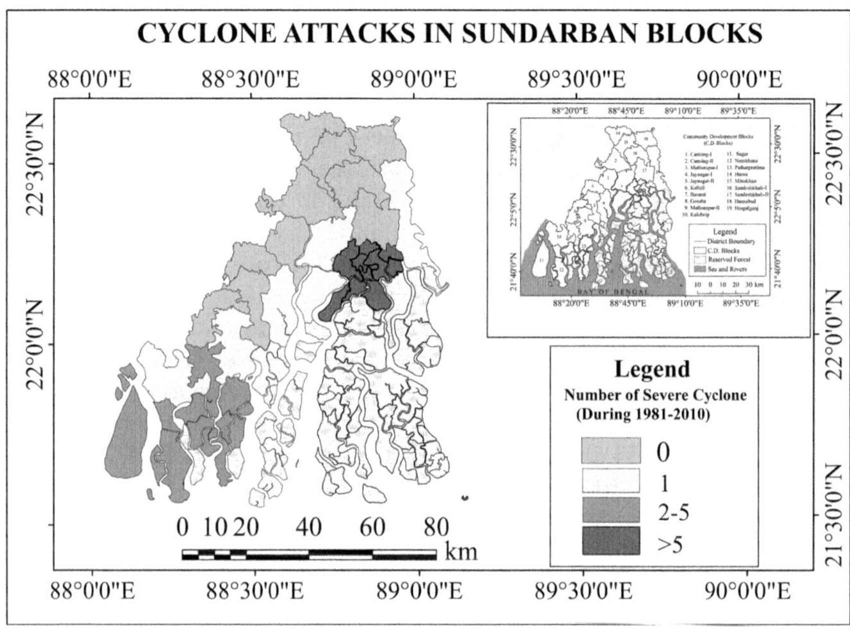

Figure 5.3 Cyclone attacks in Gosaba amongst Sundarban blocks

vulnerable embankments are associated largely with Gosaba block (Ghosh & Mistri, 2020). People recover land for agriculture or convert it to aquaculture that is appropriate for them and the circumstances, whenever a severe cyclonic event occurs such as coastal erosion, flooding, etc. Henceforth, implementation of any kind of policies or developmental schemes is very critical in the block mostly due to physical isolation and ignorance of people about different issues also plays a vital role in this regard.

After the cyclonic attack in 1999, Gosaba block was again dreadfully affected in 2009, due to the attack of cyclone Aila on 25 May, and the inhabitants of Gosaba totally became helpless as most of the embankment of the block was damaged (Ghosh & Mistri, 2020; Kundu, 2014) and the saline water encroached land (Ghosh & Mukhopadhyay, 2016; Mondal, 2015), and there was damage to agricultural crops, shortage of drinking water etc. Consequently, almost all the inhabitants were stuck in a waterlogged situation for numerous days, and the majority of the thatched houses were smashed. The people didn't have the land for crop production as most of agricultural land became salt ridden and remained non-productive for a couple of years after the event (Mukherjee et al., 2012; Mukherjee et al., 2019). Again cyclone *Bulbul* in 2019 threatened the inhabitants with damages and loss and harked back to their state of panic during the period of Aila. The further attack of cyclone Amphan (May 2020) completely devastated the entire region and severe damages were recorded. The breakage and damages of embankments; land erosion, shoreline erosion and sediment movement in sea surface and estuaries; damage of agricultural lands; coastal flood, storm surge; salt water intrusion; inundation of land; soil nutrient loss; paddy and vegetables affected; road damages; coastal drainage system becoming problematic; culvert, bridge, drinking water problem; houses were damaged and trees were uprooted due its very high velocity (>180 km/hour) and the consequences of such are still continuing. Some of these problems occurred again in May 2021 because of the *Yaas* cyclone. The damaged embankments were not recovered properly and people suffered a distressing situation, which became acute due to the lockdown during the outbreak of COVID-19; consequently, these situations made the region susceptible to physical, biological, social, and health disasters.

5.4.3 Impact of Climate Change on Agriculture in Gosaba

As the study region is a physically feeble region where most of the inhabitants depend on natural resources, especially on the mangrove forest and primary activities like agriculture and fishing; it is therefore crucial to understand how the issues of climate change-induced hazards amplify vulnerability, diminish agriculture capacity, and increase uncertainty by reducing crop variety and adaptability. Climate change makes people more risk prone and thereafter their capacity in response to vulnerable situation was judged with the concentration of agricultural labourers and cultivators, concentration of crops and yield, reduction of agricultural efficiency. The climate induced physical vulnerability boosts demographic and socio-economic vulnerability and makes people more sensitive, makes them more exposed to risk, and reduces their adaptive capacity which was judged from micro-level investigation in the entire Gosaba block.

In 2012–13, the production of Boro paddy was reduced to 20,800 kg/hectares from 34,671 kg/hectares of 2008–09 (i.e. before the cyclone Aila attack) and the highest reduction was recorded in Rangabelia mouza (Figure 5.4). While the Aman paddy was also reduced to 14,425 kg/hectares from 28,000 kg/hectares during the same period and the highest reduction of that was noticed in Hentalbari mouza of Gosaba (Debnath, 2013a). It is revealed from another study (Figure 5.5) that the yield of Aman paddy was drastically reduced from 360–480 kg/bigha (i.e. 0.13 hectares) to 180–240 kg/bigha and further recovered to 330–440 kg/bigha in the pre-Aila period (2005–08), immediate after Aila (2009–13), and the post-Aila period (2014–18) respectively (Ghosh & Mistri, 2020). Moreover, the yield of Boro paddy was drastically reduced from 720–900 kg/bigha to 240–420 kg/bigha and further recovered to 680–860 kg/bigha in the pre-Aila (2005–08), immediate after Aila (2009–13) and post-Aila period (2014–18) period respectively. The average salinity of the region during winter increased to 2–2.6‰ from 1.4–1.6‰ in the post-Aila season, which reduced to 1.8–2‰ in 2018–19 (Tagore Society of Rural Development). However, a salt-tolerant variety, like Nonabokra, Matla, Hamilton, Patnai 23 Ghetu, Lunishhree etc. was cultivated in the region and that gave comparatively higher production (2–2.5 tons/hectares).

The situation of agriculture and related livelihood was often distorted after any super-cyclone, like the attack of cyclone Aila in the area as found from several studies (Das, 2016; Debnath, 2018; Kar & Bandyopadhyay, 2015, 2016). It was revealed from such studies that most of the islands of the Gosaba block were affected and a notable LULC change was occurring. The LULC of 2009 just after cyclone Aila was drastically changed as the salt water inundated the land for several days (Das, 2016; Debnath, 2018; Ghosh & Mistri, 2020; Kar & Bandyopadhyay, 2015, 2016; Kundu, 2014). Most of the areas of the entire block have a below 10 metre altitude, and there are several earthen embankments besides the crisscross drainage network, which are easily damaged and collapsed due cyclone Aila and the saline and turbid water encroached in the agricultural field (Kar & Bandyopadhyay, 2015, 2016). During the post-Aila period for a long time, the north-western part of the block had some cropped area and fallow land with grass and the rest of the entire block area was inundated by less turbid water and highly turbid water behind the embankment parts. The entire block area was inundated from 25 May almost up to October 2009, which resulted in less land available for crops thereafter, a small percentage of land remaining dry, and a large wet region that made matters worse even after October 2009 and for around two months in 2010 (Kar & Bandyopadhyay, 2015, 2016). This situation even aggravated some islands of the block and thus the overall subsequent

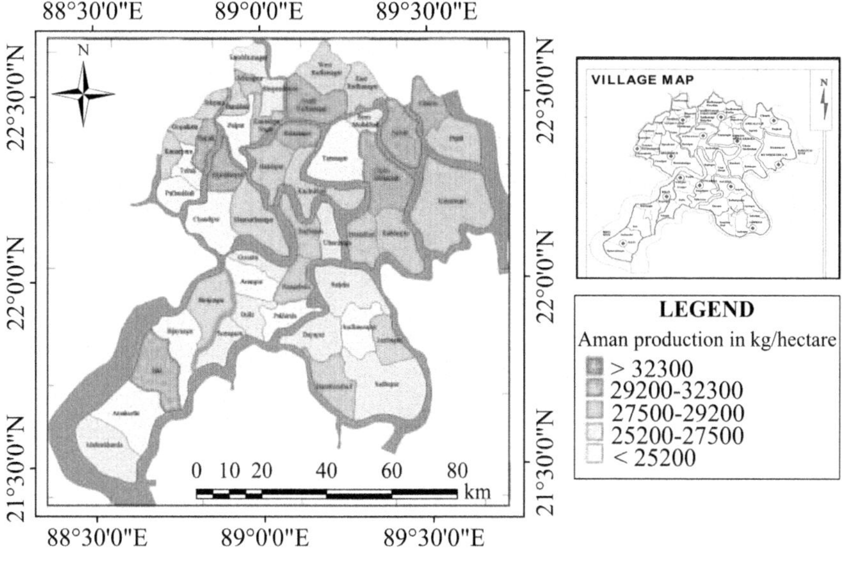

(a)

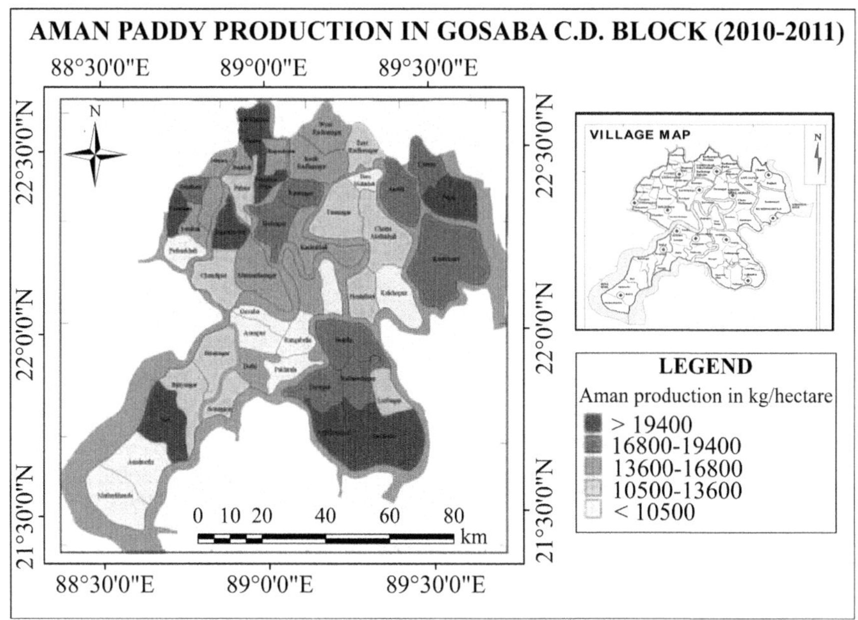

(b)

Figure 5.4 Boro and Aman paddy production of Gosaba (2008 and 2011)

Source: Debnath, 2013b

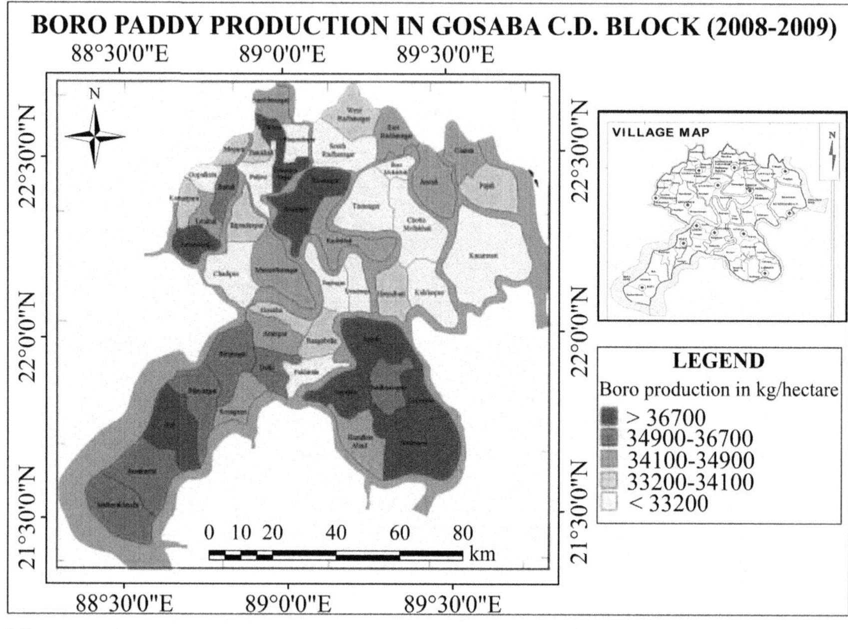

(c)

BORO PADDY PRODUCTION IN GOSABA C.D. BLOCK (2012-2013)

88°30'0"E 89°0'0"E 89°30'0"E

22°30'0"N 22°0'0"N 21°30'0"N

N

0 10 20 40 60 80 km

VILLAGE MAP

LEGEND

Boro production in kg/hectare

> 24300
23000-24300
20400-23000
17700-20400
< 17700

(d)

Figure 5.4 (Continued)

recovery from this situation were quite challenging and as a consequence, rice production was reduced to 32–40 quintal from 64–80 quintal per 1.6 hectares (Debnath, 2018). The recent (July 2020) LULC classification of the Gosaba block has been prepared (Figure 5.6), which also revealed the continuation of the same LULC changes as previously and found a major proportion is covered by agricultural land. It is also found that agricultural land has further been reduced as the settlement area capturing agriculture land.

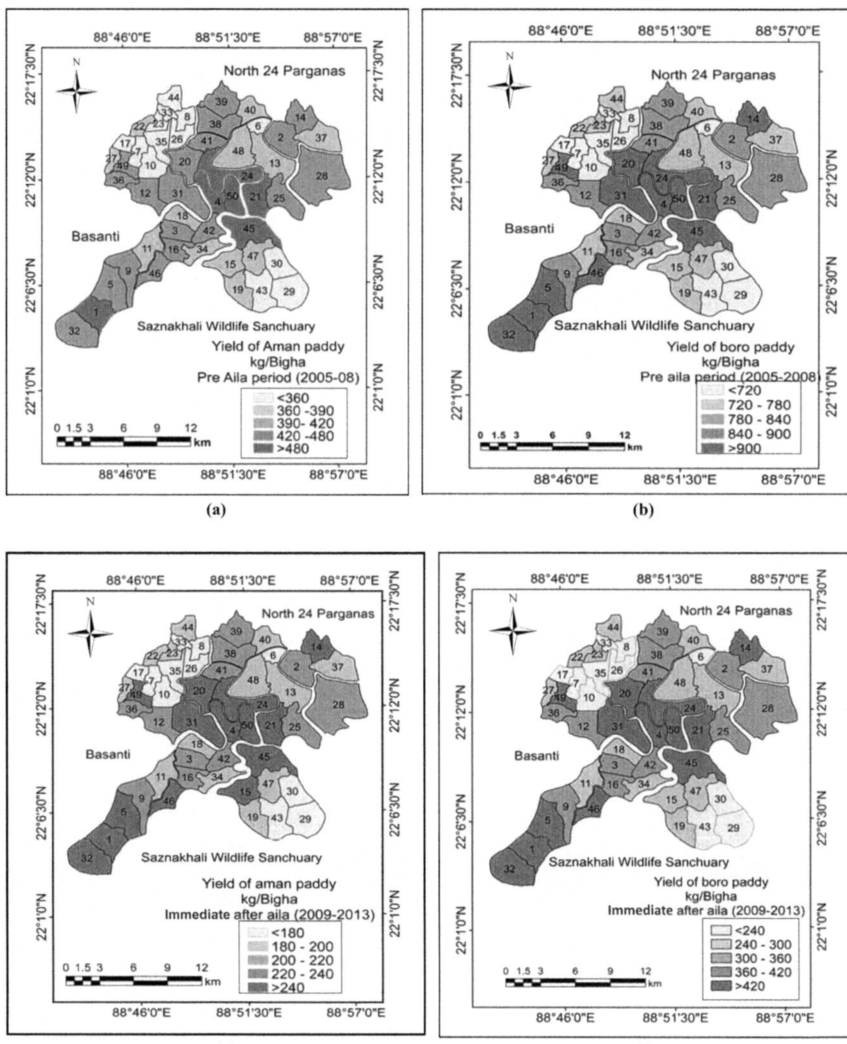

Figure 5.5 Production of Aman and Boro paddy

Source: Ghosh & Mistri, 2020

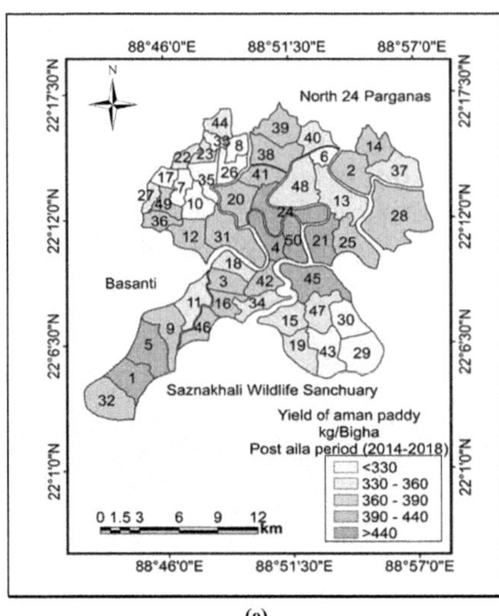

(e)

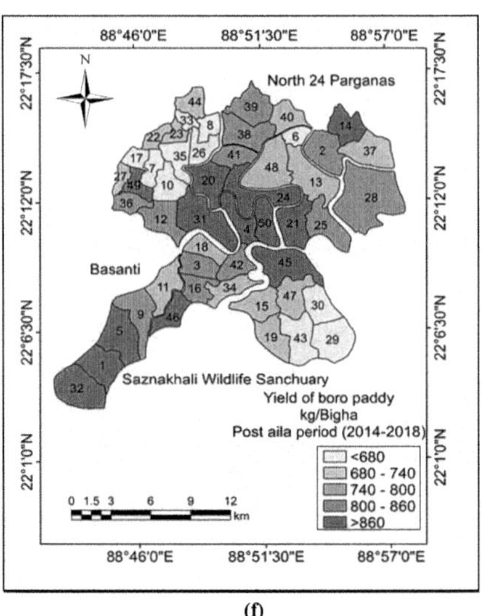

(f)

Figure 5.5 (Continued)

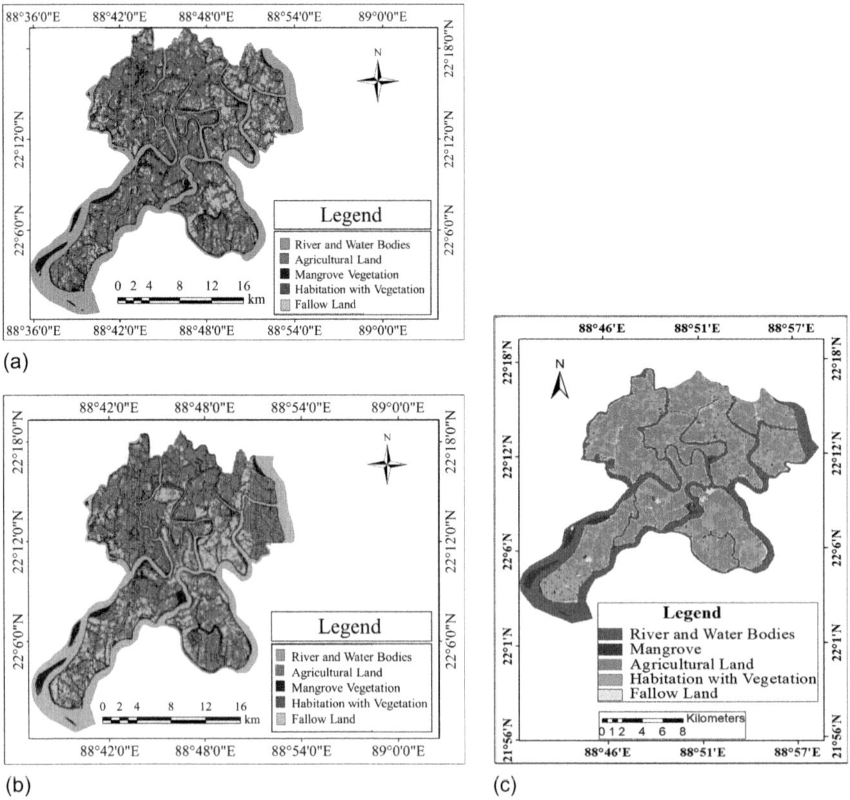

(a)

(b)

(c)

Figure 5.6 Land Use Land Cover of Gosaba, 2001; 2010

Source: Debnath, 2018; 2020 (Author)

The close location of the mangrove forest makes the inhabitants inclined towards the forest resources and probably they are preparing land for settlement, or aquaculture as per their requirement. The respondents of the study villages also reaffirmed the LULC changes as the consequences of the cyclone and their pathetic situation and tremendous challenges for adaptation, especially since the attack of cyclone Aila continued up to Bulbul, Amphan, ad Yaas.

5.4.4 Reaffirmation of Adaptation Challenges by Micro Level Scrutiny

After the serious emphasis on the climate change related issues, the status of agricultural workers, labourers was trying to understand by analysing their socio-economic conditions and thereafter, the livelihood of such inhabitants was reaffirmed by the primary survey in order to identify the challenges and uncertainties in the tremendous environmental state of affairs. This primary survey was conducted in the year 2019 and after that it was not possible to conduct further field investigation after cyclones Amphan (2020) and Yaas (2021) in the Gosaba block, due to the outbreak of COVID-19 pandemic situation; but a timely telephonic survey was carried out with some of the previous key respondents to gather information and photographs related to the adverse effect of such two major cyclonic events. Most of the respondents strongly agreed (>50%) that the temperature increased, rainfall decreased, and there was a warm winter; further about 20% agreed with the occurrence of a longer summer, early summer, late winter, warm winter, shorter rainy season, decrease of rainfall and increase of temperature due to the impact of climate change.

The adaptation challenges that are faced by the people associated with agriculture in the Gosaba block was reaffirmed by in-depth micro-level primary survey and observation, focus group discussion. The status of the working population of the Gosaba block is alarming and the graph reveals (Figure 5.7) that here working groups are main workers, cultivators, agricultural labourers, marginal and other workers, household, industrial workers, and a large proportion remain non-working among which females are double in number than their counterparts. Here the household, industrial workers are very insignificant and the proportion of agricultural labourers is more than cultivators, which signifies the landlessness of the population of the block. In all study villages, more than 70% population are landed peasants and more or less only 5% people have their own land in spite of having a four to five members in a family (Figure 5.8). That signifies a stress figure of man–land ratio. The man–land ratios of the three study villages are 5:3 in Pakhiralay, 4:2.5 in Puijhali, and 5:2 in Boramollakhali.

The majority of the inhabitants in the area spend their time working in agricultural land, riverside, fishing, going to the forest, and in some of the unpaid household activities (Figure 5.9). Over 30% partake in agricultural activities, followed by fishing and prawn seed collection (>20%), followed by livestock in all the three study mouzas; while a smaller proportion were engaged in daily wage, business, and jobs including tourism.

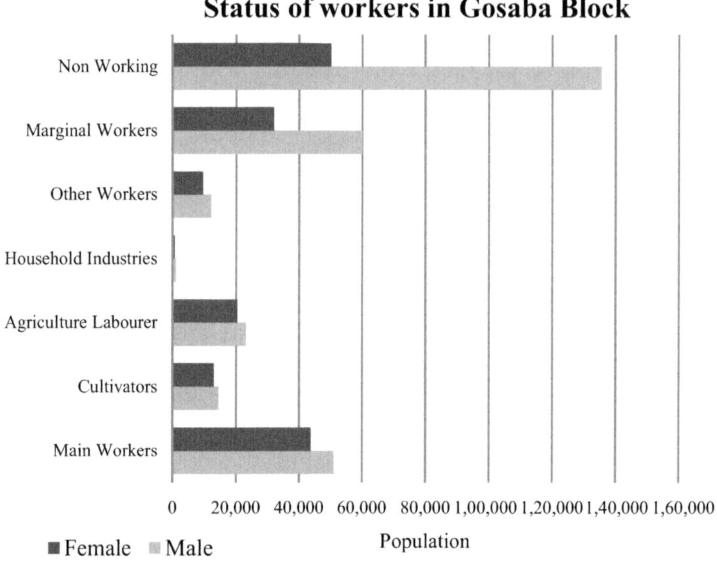

Figure 5.7 Status of workers of Gosaba block

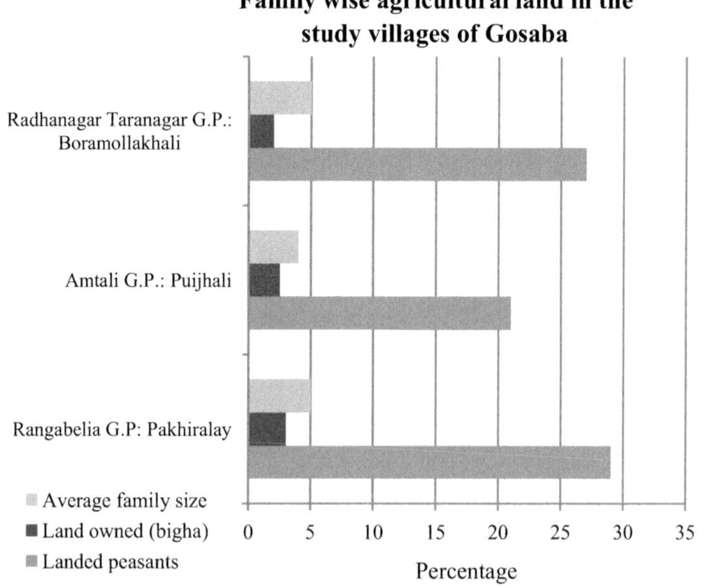

Figure 5.8 Agriculture of Gosaba villages

Data Source: Census of India, 2011 and Primary Survey, 2019

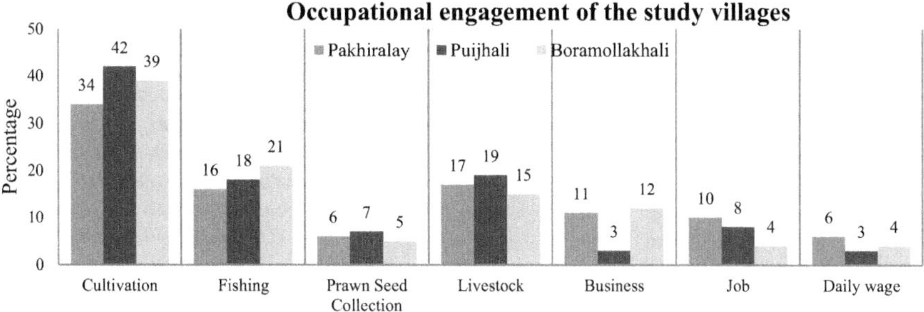

Figure 5.9 Occupational engagement

Data Source: Primary Survey, 2019

The outcomes of the perceptual study during primary survey about the impact of climate change on agriculture and allied sectors is tabulated (Table 5.3), which is essential to recognize the affective livelihood issues. The natures of adaptations of the stakeholders are tabulated (Table 5.4) which are self-explanatory.

Table 5.3 Livelihood issues affected by climate change-induced condition

Livelihood issues	*Cyclone*	*Flood*	*River bank erosion*	*Embankment damages*	*Salinity*	*Sea level rise*	*Soil erosion*
Agriculture	Yes	Yes	Yes	Yes	Yes	Yes	Yes
Livestock	Yes	Yes	Yes	Yes	Yes	No	Yes
Fishing	Yes	Yes	Yes	Yes	Yes	Yes	Yes
Shrimp culture	Yes	Yes	Yes	Yes	Yes	Yes	No
Honey production	Yes	Yes	Yes	Yes	No	Yes	No

Source: Primary survey, 2019

Table 5.4 Nature and degree of adaptations of agriculture

Nature and degree	*Cyclone and related hazards*
Impact	Inundation of agricultural land, reduction of land for cropping, crop loss, cross disease, increases soil salinity,
Level of impact	Very severe to severe, increasing helplessness, Food crisis, Malnutrition, disease, increasing poverty
Coping style	Cultivate salinity resistant crops, trying to collect salt tolerance paddy seeds & HYV seeds, cultivate vegetables, migrated, became wage labour, trying for an alternate income generation

Source: Primary Survey, 2019

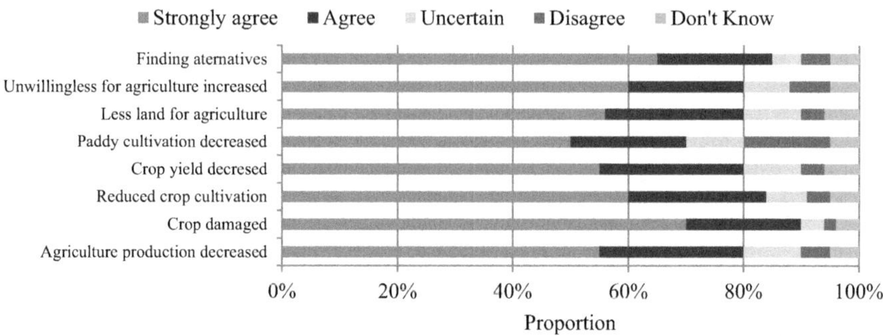

Figure 5.10 Impact of climate-induced hazard on agriculture

Data Source: Primary Survey, 2019

While tracing the impact of climate change-induced hazards on agriculture through the reaffirmation from the respondents, significant outcomes are observed and revealed. Here, more than 65% respondents strongly agreed and agree with the decrease of paddy and crop cultivation, lessening of crop yield and agricultural production due to the impact of hazards (Figure 5.10). Furthermore, crop damages and a decrease in arable land are also present; these issues arise mostly when dangers of any kind render people susceptible, leading them to seek other means of subsistence farming. A negligible percentage of respondents disagreed with these facts, whereas the majority of respondents could not connect climate change to agriculture and consequently answered "don't know" or "uncertainly."

Further, to abridge the inspection of how the various climate change-induced hazards and vulnerability affect economic stress, life and livelihood, the respondents' perceptions were surveyed and graphed (Figure 5.11) to understand the status of suffering of the people in Gosaba block. It reveals that around 60% of the respondents have severely suffered, about 20% or more have moderately suffered and the remaining respondents suffered least from various events like occurrence of flood hazard, river bank erosion, changes in river flow, embankment collapse, decrease of mangrove, decrease in fish and honey collection, lessening of crop production, salinization of agricultural land, land inundation and degradation, soil erosion, damages to house; only some of the respondents replied that they are not affected by some of the events directly but still suffered a lot, and their lives were always been at risk. The inundation of land, land degradation, salinization of agricultural land of Sundarban is decreasing the agricultural production, which has reached its daily minimum required for sustainment.

5.4.4.1 *Adaptive Agriculture and Land Use Pattern of Recent Times*

In the Gosaba block the total cultivable area is 78.11 hectares, and the cereal yield is 2831.80 kg/hectares, whereas 10.81% of the total area has the opportunity of net pisciculture; the net area available for pisciculture is 4282 hectares

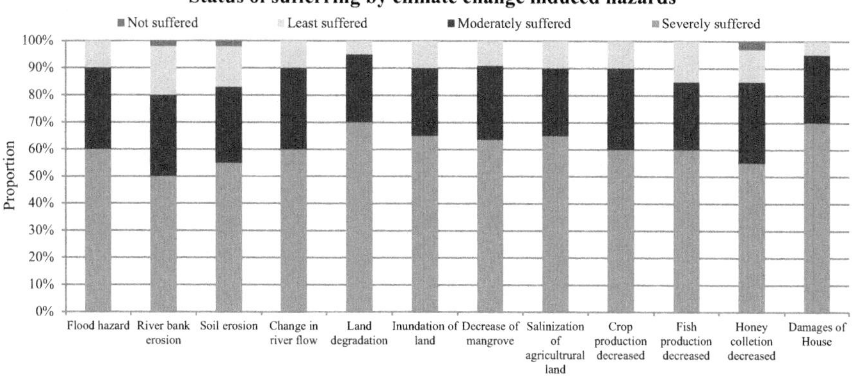

Figure 5.11 Status of suffering by climate-induced hazards

Data Source: Primary Survey, 2019

and the net area under effective pisciculture is 3208 hectares; thus the values signify that this block is dependent on agriculture and fishing and specifically more inclined to pisciculture in recent times. Agriculture in the islands of the block is quite a hard task because of the high salinity of soil, heavy rain, and storm incidences. Thus, the amount of cultivable area is gradually decreasing mostly due to the division of land, property after breaking of joint families and the establishment of new human settlements. Unpredictable floods is a frequent cause of crop failure and often intensifies human sufferings. There are no irrigation facilities; the mono crop paddy cultivation is totally dependent on rain and thus the productivity of paddies (aman) is negatively related with rainfall as the climate change and related excess rainfall reduces its productivity. The winter boro paddy is not responsive to rainfall but shows a decline with increasing temperature. Moreover, there is inadequate drainage during monsoons; saline water inclusion and water logging very frequently create problems for agriculture. Thus it reveals that the most trusted productivity and subsistence of the aman and boro paddy is becoming under serious threat in response to the changing climate, especially the increase of temperature and rainfall. The cropping variety, especially the paddy cultivation, has drastically changed due to such prolonged consequences. Initially Patnai, Banskati, Kalomota, Morishali, Kanakchur, Dhudeswar, Nona bokra, Talmagur, Lal getu, Sada setu, Annada and Hamilton, etc. are the major paddy variety and these were known for a long time (80–100 years) in the area but currently lost. The cultivation of paddy in the acceptable salinity limit is 4–6 ppt (parts per thousand) but following cyclone Aila the salinity levels are up to 20 ppt in the region (CSSRI, 2009; Chand et al., 2012b; Das et al., 2013).

In this block, the Rupshal, Boyerbat, Chinekamini, Dudheswar, Gheus, Gobindobhog, Gopalbhog, Kalomota, Khejurchori, Lilabati, Narasinghajatta, Malabati, Lal dhan, Paloi, Patnai are some of the traditional varieties; whereas CR 1017, Pratikha, Pankaj, Ranjit, Barsha, BN 20, Niranjan,

Santoshi, Swarna masuri, Super-Shyamali, Maharaj, Dharitri are the recently preferable HYV (high-yielding variety) rice seeds that are cultivated, but the production has drastically reduced. In order to safeguard the best yield of paddies, farmers use, often overuse, different types of pesticides. Pesticides are easily available, even in grocery shops. Pesticides are related to deliberate self-harm and suicide is a major health issue. There have been at least 15 cases (12 female and 3 male) of deliberate self-harm either by pesticide ingestion or yellow seed ingestion in these villages during the last two years. Many farmers have pesticide-related illnesses. Pesticide contamination is also causing great damage to the environment.

Very recently to recover from the cyclone Yaas damages, in the Amtali GP area, the farmers have adopted a very new technique to cultivate, which is also adopted by 500 males and females over the Sundarban. In this process, first the bamboo scaffolding was made on plastic drums on the ponds and a special type of soil was then placed in a sack (grow bag) on the floating scaffolding. The soils are a mix of different ingredients including earthworm manure, coconut husk, mustard husk, neem husk and bone powder. Thereafter, cultivation starts by putting seeds in the grow bag and thus various seasonal vegetables, including eggplant, tomato, gourd, and papaya were cultivated. Even rice seedbeds, like Charmarmani, Banshkathi, Dudheswhar, can be made in this way and apart from this, the farmers are able to cultivate fish and crabs in the pond with equal rhythm. An NGO is helping to employ this new method and after their successful initiation, they have trained several groups of farmers in Deulbari, Kumirmari, Chhotomollakhali in the Gosaba block on how to cultivate in this way.

5.5 Conclusion

This chapter attempts to record extensive scrutiny on various agricultural issues and its distortion or deviation of climate change-related shocks and stresses which are adding pressure to the already precarious livelihoods of marginalized people especially associated with agriculture in the Gosaba block of Sundarban. The climatic hazards enhance the vulnerability to the inhabitants who are already suffering from poverty, massive illiteracy, health hazards, dependent on natural resources, often get displaced, migrate for better wages, and so on and so forth. The increasing economic stress due to the climate change-induced physical vulnerability, which is strong enough to boost other sorts of vulnerability is adding to the adverse state of affairs and creating pressure to the marginalized, coastal, and remotely positioned populace of Gosaba. Therefore, agricultural sectors necessitate instantaneous attention and an effective strategy process for the holistic progress in Gosaba and strengthening the public transport system with provision of monetary opening in economic sector is essential for shrinking the effect of adverse climate change. The strengths and opportunities of the agricultural sectors should be improved; more salt tolerance crop variety; better crop combination and alternative cash crop cultivation should be introduced. Moreover, adaptation of new technology; creation of good market of the agricultural outputs; diminishing the threats and weakness of agricultural sectors; increasing irrigation facilities for crop diversification; rainwater harvesting; strengthening of agricultural cooperatives are the some pillars to be prioritized for the betterment of the people associated with agricultural sectors of Gosaba. In this regards intra-block level variations of the agricultural scenario and its dynamics as found in such research outcomes may ameliorate the planning process to prioritize the adaptation programmes multiplicity to address the current issue of long-term sustainability for the agricultural sector (Figure 5.12, Figure 5.13).

Figure 5.12 Preparing bamboo scaffolding over pond

Figure 5.13 Cultivation in grow bags on floating scaffolding

Acknowledgement

The author would like to thank to Sri Subhas Acharya, former Joint Director, Sundarban Development Board for his valuable comments and Sri Subrata Sarkar, without whom the field work in Gosaba would have been a hundred times difficult. This effort is a tribute to the inhabitants of Sundarban, especially the farmers, for their spontaneous response during discussion.

Consent to Participate

All the respondents were informed about the nature of the study before proceeding with the interview and survey, and thereafter participation in the research was voluntary. The participants were assured that the data would be used for academic purposes only.

Disclosure Statement

The author declares that there are no potential conflict of interest.

Funding

This research did not receive any specific grant from any funding agency.

Bibliography

Ahsan, M. N., & Warner, J. (2014). The socioeconomic vulnerability index: A pragmatic approach for assessing climate change led risks – A case study in the south–western coastal Bangladesh. *International Journal of Disaster Risk Reduction*, *8*, 32–49.

Banerjee, K. (2013). Decadal change in the surface water salinity profile of Indian Sundarbans: A potential indicator of climate change. *Journal of Marine Science: Research & Development*. https://doi.org/10.4172/2155-9910.S11-002

Census of India. (1991, 2001, 2011). Provisional Population Totals. Registrar General and Census Commissioner of India, Ministry of Home Affairs, New Delhi, India. https://censusindia.gov.in/2011census/dchb/DCHB_A/19/1917_PART_A_DCHB_SOUTH%20TWENTY%20FOUR%20PARGANAS.pdf Accessed 02 March 2020

Central Soil Salinity Research Institute (CSSRI). (2009). *Project on 'Strategies for Sustainable Management of Degraded Coastal Land and Water for Enhancing Livelihood Security of Farming Communities', Project Document Submitted to GEF through NAIP*. Karnal, Haryana: Central Soil Salinity Research Institute.

Chakraborty, S. (2015). Investigating the impact of severe cyclone Aila and the role of disaster management department: A study of Kultali block of Sundarban. *American Journal of Theoretical and Applied Business*, *1*(1), 6–13.

Chand, B. K., Trivedi, R. K., Biswas, A., Dubey, S. K., & Beg, M. M. (2012a). Study on impact of saline water inundation on freshwater aquaculture in Sundarban using risk analysis tools. *Exploratory Animal and Medical Research*, *2*, 170–178.

Chand, B. K., Trivedi R. K., Dubey, S. K., & Beg, M. M. (2012b). *Aquaculture in changing climate of sundarban: Survey report on climate change vulnerabilities, aquaculture practices & coping measures in sagar and basanti blocks of Indian Sundarban*. Kolkata: West Bengal University of Animal & Fishery Sciences.

Danda, A. (2010). *Sundarbans: Future imperfect climate adaptation report*. New Delhi: World Wide Fund for Nature. WWF (World Wildlife Fund). www.eldis.org/document/A59893. Accessed 02 March 2020.

Danda, A. (2019). *Environmental security in the Sundarban in the current climate change era: Strenghening India–Bangladesh cooperation*. New Delhi: Observer Research Foundation. https://www.orfonline.org/research/environmental-security-in-the-sundarban-in-the-current-climate-change-era-strengthening-india-bangladesh-cooperation-57191/ Accessed 02 February 2020.

Das, K. (2016). Livelihood dynamics as a response to natural hazards: A case study of selected places of Basanti and Gosaba Blocks, West Bengal. *Earth Science*, *5*(1), 13. https://doi.org/10.11648/j.earth.20160501.12

Das, S., RoyChoudhury, M., Das, S., & Nagarajan, M. (2013). Integrated geospatial technologies for soil salinity assessment over South 24 PGS, West Bengal. *The International Journal of Geoscience, Engineering and Technology 1*(2), 41–85. ISSN: 2321-2144.

Debnath, A. (2013a). Condition of agricultural productivity of Gosaba C.D. Block, South 24 Parganas, West Bengal, India after severe cyclone Aila. *International Journal of Scientific and Research Publications*, *3*(7), ISSN 2250-3153.

Debnath, A. (2013b). Impact of cyclone Aila on paddy cultivation in Gosaba island of the Indian Sundarban region. *The Indian Journal of Spatial Science*, *5*(1), 7–13.

Debnath, A. (2018). Land use and land cover change detection of Gosaba island of the Indian Sundarban region by using multitemporal satellite image. *International Journal of Humanities and Social Science*, *7*(1), 209–217. www.thecho.in/files/26_uf3v344k.-Dr.-Ajay-Debnath.pdf Accessed 2 February 2020

District Statistical Handbook of South and North 24 Parganas (2016). Development & Planning Department, Government of West Bengal. 1–20. www.wbpspm.gov.in/publications/District%20Statistical%20Handbook Accesseded on 12 March 2020

Dumenu, W. K., & Obeng, E. A. (2016). Climate change and rural communities in Ghana: Social vulnerability, impacts, adaptations and policy implications. *Environmental Science & Policy*, *55*, 208–217.

Dutta, S., Maiti, S., Garai, S., Bhakat, M., & Mandal, S. (2019). Socio economic scenario of the farming community living in climate sensitive Indian Sundarbans, *International Journal of Current Microbiology and Applied Sciences*, *8*(02), 3156–3164. https://doi.org/10.20546/ijcmas.2019.802.369

Feng, S., Krueger, A. B., & Oppenheimer, M. (2010). Linkages among climate change, crop yields and Mexico–US cross-border migration. *Proceedings of the National Academy of Science*, *107*(32), 14257–14262. https://doi.org/10.1073/pnas.1002632107

Ghosh, A. (2012). *Living with changing climate: Impact, vulnerability, and adaptation challenges in Indian Sundarbans*. Delhi, India: Centre for Science and Environment.

Ghosh, A., & Mukhopadhyay, S. (2016). Bank erosion and its management: Case study in Muriganga-Saptamukhi interfluves Sundarban, India. *Geographical Review of India*, *78*(2), 146–161. https://doi.org/10.1007/s40808-016-0130-x

Ghosh, A., Schmidt, S., Fickert, T., & Nüsser, M. (2015). The Indian Sundarban mangrove forests: History, utilization, conservation strategies and local perception. *Diversity*, *7*(2), 1–13.

Ghosh, S., Chakraborty, D., Dash, P., Patra, S., Nandy, P., & Mondal, P.P. (2018). Climate risks adaptation strategies for Indian Sundarbans. https://doi.org/10.7287/peerj.preprints.26963v2

Ghosh, S., & Mistri, B. (2020). Geo-historical appraisal of embankment breaching and its management on active tidal land of Sundarban: A case study in Gosaba Island, South 24 Parganas, West Bengal. *Space and Culture, India*, *7*(4), 166–180. https://doi.org/10.20896/SACI.V7I4.587

Gornall, J., Betts, R., Burke, E., Clark, R., Camp, J., Willett, K., & Wiltshire, A. (2010). Implications of climate change for agricultural productivity in the early twenty-first century. *Philosophical Transactions of the Royal Society of London B: Biological Sciences*, 365(1554), 2973–2989. https://doi.org/10.1098/rstb.2010.0158

Hajra, R., Ghosh, A., & Ghosh, T. (2016). Comparative assessment of morphological and landuse/landcover change pattern of Sagar, Ghoramara, and Mousani island of Indian Sundarban Delta through remote sensing. In Hazra, et al. (Eds.), *Environment and Earth observation: Case studies in India*. Springer. ISBN: 978-3-319-46008-6. https://doi.org/10.1007/978-3-319-46010-9_11

Halder A., & Debnath, A. (2014). Assessment of climate induced soil salinity conditions of Gosaba Island, West Bengal and its influence on local livelihood. In M. Singh et al. (eds.), *Climate Change and Biodiversity: Proceedings of IGU Rohtak Conference*, Vol. 1, Advances in Geographical and Environmental Sciences. https://doi.org/10.1007/978-4-431-54838-6_3

Hazra, S., Das, I., Samanta, K., & Bhadra, T. (2014). Impact of climate change in Sundarban Area West Bengal, India. School of Oceanographic Studies. *Earth Science and Climate Book*. 9326/17.02.00. Report submitted to Caritus India, SCiAF.

Hazra, S., Ghosh, T., DasGupta, R., & Sen, G. (2002). Sea Level and associated changes in the Sundarbans. *Science and Culture*, *68*(9–12): 309–321. ISSN: 0036-8156.

Iqbal, K., & Roy, P. K. (2014). Examining the impact of climate change on migration through the agricultural channel: Evidence from district level panel data from Bangladesh. SANDEE Working Papers, ISSN: 1893-1891. WP 84–14. ISBN: 978-9937-596-14-5.

Jana, A., Sheena, S., & Biswas, A. (2012). Morphological change study of Ghoramara Island, Eastern India using multi temporal satellite data. *Research Journal of Recent Sciences*, *1*(10), 72–81.

Kar, N. S., & Bandyopadhyay, S. (2015). Tropical storm Aila in Gosaba block of Indian Sundarban: Remote sensing based assessment of impact and recovery. *Geographical Review of India*, *77*, 40–54.

Kar, N. S., & Bandyopadhyay, S. (2016). Tropical storm Aila in Gosaba block of Indian Sundarban: Remote sensing based assessment of impact and recovery. Accessed 10 March 2020. www.academia.edu/27512407/

Kumar, K. S., & Viswanathan, B. (2015). Weather variability, agriculture and rural migration: Evidence from state and district level migration in India. SANDEE Working Papers, ISSN 1893-1891; WP 83–14. ISBN: 978-9937-596-13-8. https://efaidnbmnnnibpcajpcglclefindmkaj/www.sandeeonline.org/uploads/documents/publication/1026_PUB_Final_Working_Paper_83_14_WEB.pdf Accessed 20 March, 2020.

Kundu, A. K. (2014). Embankment in Sundarban and reconstruction of damaged embankment in Aila. *Sechpatra*, Waterways and Irrigation Department, Government of West Bengal, 19–25.

Kusche, J., Uebbing, B., Rietbroek, R., Shum, C. K., & Khan, Z. H. (2016). Sea level budget in the Bay of Bengal (2002–2014) from GRACE and altimetry. *Journal of Geophysical Research: Oceans*, *121*(2), 1194–1217.

Laha, A. (2019). Mitigating climate change in Sundarbans role of social and solidarity economy in mangrove conservation and livelihood generation implementing the sustainable development goals: What role for social and solidarity economy? http://unsse.org/wp-content/uploads/2019/07/258_Laha_Mitigating-Climate-Change-in-Sundarbans_En.pdf Accessed 10 March 2020

Mandal, S., Bandyopadhyay, B. K., Burman, D., Sarangi, S. K., & Mahanta, K. K. (2011). Baseline Report of the NAIP project on 'strategies for sustainable management of degraded coastal land and water for enhancing livelihood security of farming Communities', Regional Research Station, Central Soil Salinity Research Institute, Canning Town, South 24 parganas, West Bengal.

Mandal, S., Sarangi, S. K., Burman, D., Bandyopadhyay, B. K., Maji, B., Mandal, U. K., & Sharma, D.K. (2013). Land shaping models for enhancing agricultural productivity in salt affected coastal areas of West Bengal – An economic analysis. *Indian Journal of Agricultural Economics*, *68*, 389–401.

Mitra, A., Banerjee, K., Sengupta, K., & Gangopadhyay, A. (2009). Pulse of climate change in Indian Sundarbans: A myth or reality? *National Academy Science Letters (India)*, *32*(1), 19.

Mondal, B. K. (2015). Nature of propensity of Indian Sundarban. *International Journal of Applied Research and Studies*, *4* (1), 1–17.

Mondal, B. K. (2018). Assessment of effects of global warming and climate change on the vulnerability of Indian Sundarban. In P. S. Shukla (Ed.), *Sustainable development: Dynamic perspective* (pp. 63–74). Anjan Publisher.

Mondal, T.K. (2012). Environmental hazards and community responses in Sagar Island, India. Ocean and Coastal Management. https://doi.org/10.1016/j. ocecoaman. 2012.10.001

Mukherjee, N. et al. (2019). Climate change and livelihood vulnerability of the local population on Sagar island, India. *Chinese Geographical Science*, *29*(3), 417–436. https://doi.org/10.1007/s11769-019-1042-2

Mukherjee, S., Chaudhuri, A., Sen, S., & Homechaudhuri, S. (2012). Effect of cyclone Aila on estuarine fish assemblages in the Matla river of the Indian Sundarbans. *Journal of Tropical Ecology*, 28(4), 405–415.

Nandy, S., & Bandyopadhyay, S. (2011). Trend of sea level change in the Hugli estuary, India. *IJMS*, *40*(6), 802–812. http://hdl.handle.net/123456789/13266

Pound, B. Lamboll, R., Croxton, S., & Gupta, N. (2018). Climate-resilient agriculture in South Asia: An analytical framework and insights from practice. https://reliefweb.int/sites/reliefweb.int/files/resources/OPM_Agriculture_Pr2Final_WEB.pdf Accessed 20 March, 2020

Pramanik, M. K. (2015). Assessment of the impacts of sea level rise on mangrove dynamics in the Indian part of Sundarbans using geospatial techniques. *Journal of Biodiversity, Bioprospecting and Development*, *3*, 155. https://doi.org/10.4172/2376-0214.1000155

Raha, A., Das, S., Banerjee, K., & Mitra, A. (2012). Climate change impacts on Indian Sunderbans: A time series analysis (1924–2008). *Biodiversity and Conservation*, *21*(5), 1289–1307. https://doi.org/10.1007/s10531-012-0260-z

Raha, A. K. (2014). Sea level rise and submergence of Sundarban Islands: A time series study of estuarine dynamics. *Journal of Ecology and Environmental Sciences*, 5(1), 114–123. ISSN 0976-9900.

Roy, C., & Guha, I. (2017). Economics of climate change in the Indian Sundarbans. *Global Business Review*, *18*(2), 493–508.

Sahana, M., Hong, H., Ahmed, R. Patel, P.P., Bhakat, P., & Sajjad, H. (2019a). Assessing coastal island vulnerability in the Sundarban Biosphere Reserve, India, using geospatial technology. *Environmental Earth Sciences*. *78*(10), 1–22. https://doi.org/10.1007/s12665-019-8293-1

Sahana, M., Rehman, S., Ahmed, R., & Sajjad, H. (2020). Analyzing climate variability and its effects in Sundarban Biosphere Reserve, India: Reaffirmation from local communities. *Environment, Development and Sustainability*. https://doi.org/10.1007/s10668-020-00682-5

Sahana, M., Rehman, S., Paul, A. K., & Sajjad, H. (2019b). Assessing socio-economic vulnerability to climate change-induced disasters: Evidence from Sundarban Biosphere Reserve, India. *Geology, Ecology, and Landscapes*, 1–13. https://doi.org/10.1080/24749508.2019.1700670

World Wide Fund for Nature (WWF) Report. (2010). Sundarban: Future Imperfect Climate Adaptation Report Edited By Anurag Danda, pp. 1–2. http://awsassets.wwfindia.org/downloads/sundarbans_future_imperfect__climate_adaptation_report_1.pdf Accessed 7 March, 2020.

6 Dynamics of Vegetation and Land-Use Land-cover Assessments of Kolkata District

Najib Ansari and Rukhsana

6.1 Introduction

The fast-increasing population in different countries of the world is a major challenge, especially for developing nations. The United Nations (2014) projected that with this augmenting population, it will be 68% by 2050 compared with the 55% of the total population that currently lives in urban areas in the world which will accelerate the spatial expansion of urban areas beyond the administrative limit (Mosammam et al., 2017; Hasnin and Rukhsana 2020). In developing countries like India, the population grew very rapidly from 68.33 million in 1981 to 121.01 million in 2011, and it is predicted that India will be the largest populated country in the world by the year 2028 (Census of India, 2011; World Population, 2002; Hasnin and Rukhsana, 2020).

Land use and land cover (hereinafter referred to as LULC) dynamics provide important data for policy makers to harmonize forest management practice and crop cultivation in the agro-forestry landscape (Pareta and Pareta, 2011; National Climatic Data Center, 2013). In the broad sense of the ecosystem, land refers to landforms, climate, edaphic features, plants, and water resources. Changes in the LULC date back to prehistoric times and are a direct and indirect result of human activities on the integrated elements of these resources (US Environmental Protection Agency, 1999; Wulder et al., 2016). Negative change contributes significantly to biodiversity loss and adverse climate change, mainly due to conversion of forest cover to agricultural land, human settlement, and infrastructure (FC, 2011; Duguma et al., 2019).The dynamics of LULC with an increasing human population is affecting global atmospheric concentrations of greenhouse gases (GHGs) in different ways (Albrecht & Kandji, 2003; Bălteanu et al., 2013).

Understanding the status of LULCs is critical to the selection of possibilities for optimal use of land-use types to meet the increasing demands of basic human needs in harmony with environmental protection (Pareta and Pareta, 2011; Rawat and Kumar, 2015). The combination of crops and cultivated trees in agroforestry land-use type generally provides countless results in mitigating climate change and in ecological, economic, and social services (FAO, 2011a;

DOI: 10.4324/9781003275916-7

FAO, 2011b; Meragiaw, 2017; Mbow et al., 2014; Smith & Wollenberg, 2012; Helen, Jarzebski & Gasparatos, 2019).

The main objectives of the present research work to assess the suburb migration of people from the Kolkata megacity of India and to understand the changing pattern of built-up areas, and the encroachment of vegetation by urban growth.

6.2 Study Area

Kolkata, located between 22°001900 N to 23°000100 N latitude and 88°000400 E to 88°003300 E longitude, is an urban agglomeration in the Indian state of West Bengal (Figure 6.1), which has been selected for this study. It is the third most populous metropolitan area in India after Delhi and Mumbai. The region is administered by the Kolkata Metropolitan Development Authority (KMDA), which consists of 37 municipalities and four municipal corporations and comprises the six main districts of West Bengal, namely Kolkata District, North 24 Parganas, South 24 Parganas, Nadia, Howrah, and Hooghly. The Kolkata urban agglomeration has a linear urban prototype corresponding to both the eastern and western banks of the Hooghly River, one of the lifelines of southern Bengal. The Kolkata urban agglomeration with a population of over 10 million is one of the 30 largest megacities in the world (UN, 2014). As a growing metropolitan city in a developing country, Kolkata has considerable urban pollution in terms of air-water and noise, traffic congestion, poverty, overpopulation as well as several socio-economic nuisances (Bhatta, 2009).

6.3 Database and Methodology

This study used satellite images derived from the United States Geological Survey (http://glovis.usgs.gov). Table 6.1 shows the details of the images used. For the analysis of migration flow, district data has been collected from the Census of India (2011). The study has incorporated the software like Erdas Imagine 2014, Arc GIS 10.7.1, to fulfil the objectives of the study. Cloud free digital data were selected for LULC assessment five-year group data from 2006, 2011, 2016 (Table 6.1), as the built-up and non-built-up area can be easily assessed during this period. Initially, standard image processing techniques have been applied for the analysis of satellite data such as correction, enhancement, band extraction, restoration and classification. The hybrid image classification technique was employed for image classification using ERDAS imaging software (Jaybhaye and Mundhe, 2013). All layer's stack in GIS was applied to generate FCCs and the supervised classification method. The maximum likelihood algorithm was executed to find out the land use/cover including five land use/cover classes such as (a) Built-up, (b) Vegetation, (c) Water Body, (d) Vacant Land, and (e) Agricultural land (Mosammam et al. 2017).

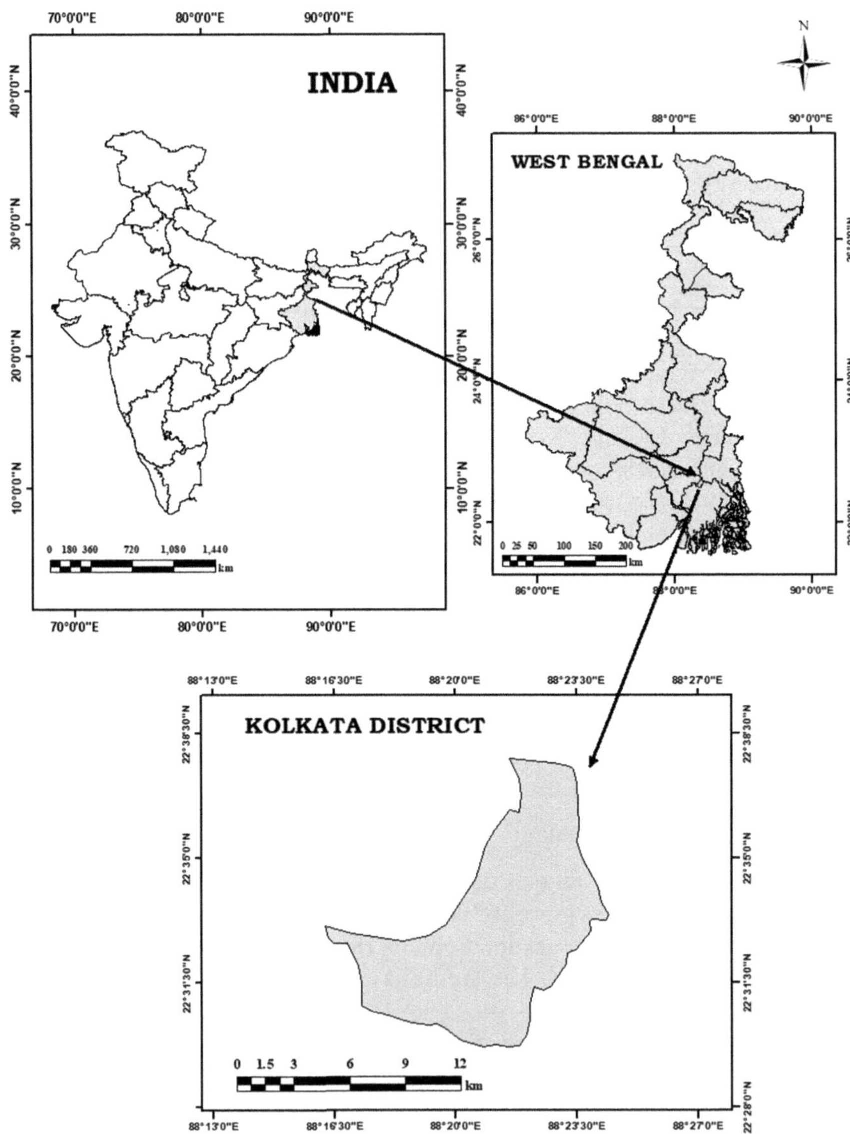

Figure 6.1 Study area map of Kolkata district

The LULC classification of six categories is mention below:

- **Built-up area**: Residences, commercial centres, industrial zones, railways, highways, expressways and others. It is an area of human habitation developed due to non-agricultural use and that has a cover of buildings, transport and communication, utilities in association with water, vegetation, and vacant lands. The web LULC map consists of three classes under built-up viz., urban, rural and mining

Table 6.1 List of multi-sensor data collected from Landsat

Satellite	*Spatial resolution (m)*	*Date of acquisition*	*Map projection*	*No. of band*	*Path & row*
Landsat 4–5	30	2006 June	UTM – WGS 84 Polar Stereographic for the continent of Antarctica.	7	138/44
Landsat 4–5	30	2011 November	UTM – WGS 84 Polar Stereographic for the continent of Antarctica.	7	138/44
Landsat 8	30	2016 October	UTM – WGS 84 Polar Stereographic for the continent of Antarctica.	9	138/44

Note: TM: Thematic Mapper, UTM: Universal Transverse Mercator, WGS: World Geodetic System

- **Vegetation cover**: Natural forest land, shrub lands, and other agricultural land. Agricultural land defined as arable agricultural land and others. These are the lands primarily used for farming and for production of food, fibre, and other commercial and horticultural crops. These lands consist of cropland, plantation, fallow land, and current shifting cultivation areas.
- **Water body**: Rivers, ponds, lakes, reservoirs, permanent and seasonal wetlands, inland natural (ox-bow lake, cut off meander, waterlogged etc.), and inland manmade (water logged, saltpans etc.)
- **Sandy area**: Sandy areas or sandbars can occur in coastal, riverine, or inland areas. Desertic sands are characterized by the accumulation of sand developed in situ or transported by Aeolian processes. Coastal sands are the sands that are accumulated as a strip along the sea-coast. Riverine sands are those that are seen as accumulations in the flood plain as sheets which are the resultant phenomena of river flooding.

There are also some criteria for inclusion and exclusion of neighbouring out-migration districts. We take some threshold populations for inclusion of neighbouring outmigration districts. In the case of Kolkata the minimum population is 10,000, which means that from the source of origin district to the source of the destination district the minimum out-migrated population should be 10,000 then those districts are selected for showing a higher out-migration flow. Below that population we are not including for showing out-migration flow of the district.

The method includes compilation, calculation, and presentation of collected data. After getting the secondary database, series of analyses have been done with the help of MS Excel and MS Word, etc. Various quantitative, statistical and cartographic methods and techniques are also used for data compilation, tabulation, analysis, and representation.

Analysis and assessment of LU/LC changes can provide a tool to assess ecosystem changes and their environmental impacts at various temporal and spatial scales (Gregorio, 2005; Lambin, 1997). Current technologies such as geographic information systems (GIS) and remote sensing provide a cost-effective and accurate alternative to understanding landscape dynamics. GIS and remote sensing are related to land and are therefore very useful in formulation implementation and monitoring of urban development in moving towards a sustainable development strategy (Yeh and Li, 1998). Remote sensing and GIS tools have been applied in many urban studies to detect, monitor, and simulate urban land use changes which have great potential for the acquisition of detailed and accurate surface information to manage and plan the urban areas (Herold et al., 2002). In addition, demographic data (population size, density and migration) are collected from the national data source Census of India and the Mumbai Metropolitan Regional Development Authority (MMRDA). The base map and other reference maps are collected from the MMRDA website. Basic GIS algorithms were used for slope and altitude; Euclidean distances were aggregated for water bodies, roads, rails and rivers.

6.4 Results and Discussions

Figure 6.2 shows that the nearer district like North 24 Paraganas has the highest outmigration flow and after that South 24 Paraganas, Hugli, Haora. People decided to move to this district because of the highly congested land on the Kolkata district and the population pressure is also high. The city has a very high population density of 24,000 people per square kilometre or 63,000 per square mile. This is one of the world's highest densities. The city itself qualifies as a megacity and covers an incredible amount of surface area that comes to a total of 205.00 square kilometres (79.150 square miles), according to World Population Review (2021).

On the other hand, Kolkata district is the capital city of West Bengal and it has been a lifeline of this state. The city has high connectivity in terms of roadways and railways. People who live in this district connect daily with this main city for their work purpose. Moreover, this district has an advanced information technology hub which is the centre of some of the notable IT/ITES Indian and multinational companies. Approximately 1,500 companies have their offices in Sector V. The majority of the corporate offices are situated in Sector V and Sector III. Around 3.5 Lakh (2017) people are employed in Salt Lake City.

6.5 Land Use Classification of Kolkata District (Area in Square Miles)

Figure 6.3 and Table 6.2 depict LULC change detection in different LULC classes between 2006, 2011, 2016 identifying the vegetation, waterbody, and

Flow Map of Outmigration From Kolkata District to Neighbor Districts in 2011

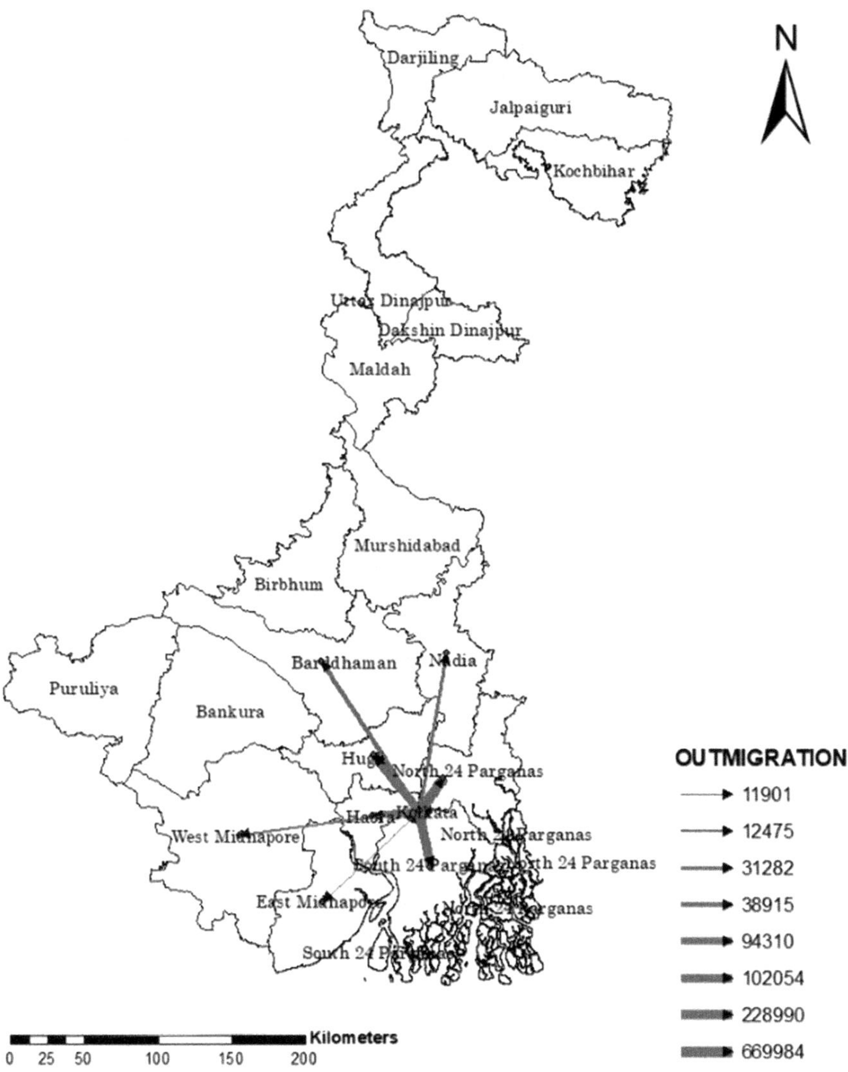

Figure 6.2 Flow map of outmigration from core district to outskirts district of KMA

wetland were converted into a built-up area. From these different years of district map, the built-up area continuously increased; in 2006 it was 51.90 per square miles, it reached 67.08 per square miles in 2011, and in 2016 it was 70.46 per square miles. Besides that, the waterbody, wetland, and vegetation land continuously decreased throughout these periods.

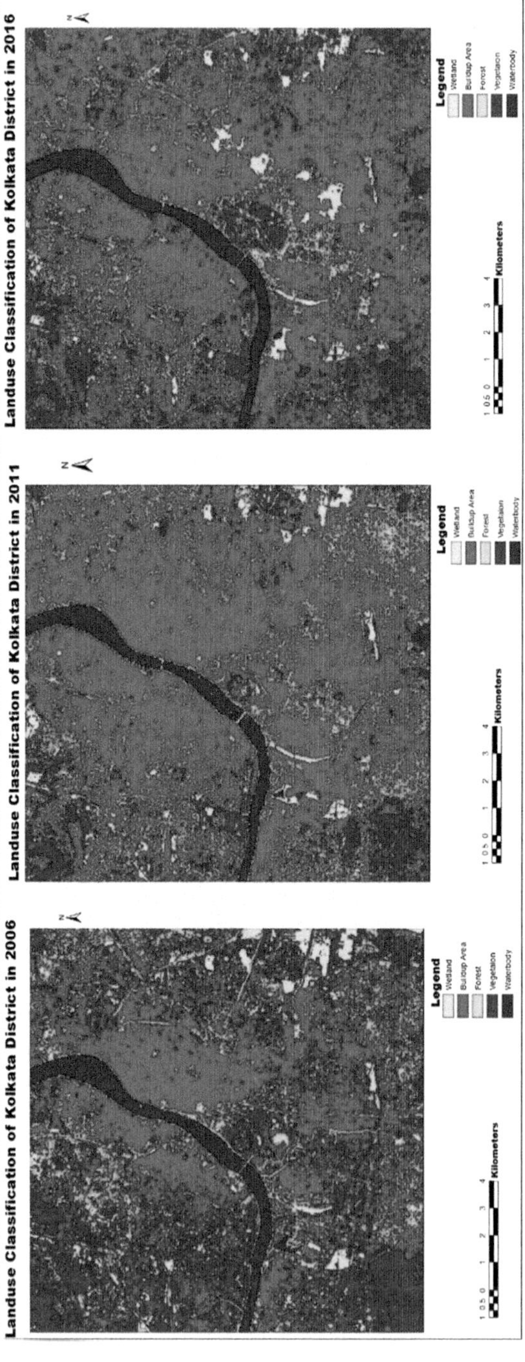

Figure 6.3 Land use Land cover of Kolkata district during 2006–2016

Table 6.2 Dynamics of land use land cover (LULC) change

Land Class	*Area 2006 (sq. mile)*	*Area 2011 (sq. mile)*	*Area 2016 (sq. mile)*	*Change in area (sq. mile) 2006–2016*	*Change in percentage (2006–2016)*
Built-up area	51.90	67.08	70.46	18.56	35.77
Vegetation	27.85	15.66	13.58	−14.27	−51.23
Waterbody	12.22	10.07	3.79	−8.43	−68.99
Wetland	4.33	4.11	2.74	−1.58	−36.62
Forest	3.70	3.08	9.42	5.72	154.31

It also seen from the Figure 6.4, that the built-up area in 2006 is 52%, and it is increased to 67% in 2011 and to 70% in 2016. On the opposite side vegetation, waterbody, and wetland continuously decreased because of high construction of the built-up area. Due to those types of land use and land cover change, this area is facing a huge amount of land degradation, soil erosion, land subsidence, and also unavailability of groundwater (Halder et al., 2021).

Table 6.3 illustrates that vegetation cover has been constantly decreased whereas the built-up area was dominated with a replacement of vegetation cover. It is evident from the study that out of these five- land-use and land-cover classes vegetation has been mostly changed. The percentage of negative change from 2006 to 2016 is 51.23, which means that vegetation has faced a greater impact for built-up area expansion. This directly creates a lot of problems like, urban pollution, traffic congestion, land degradation, soil erosion, land subsidence, and also unavailability of groundwater.

It may be noted from Table 6.3 and Figure 6.5 that vegetation has been changed constantly through this ten-year period. The area under vegetation was recorded to decrease from 2006–2011 to 2011–2016, which was reported as 43.77 and 2.08% respectively. The most interesting part of Table 6.3 is that the highest transition of vegetation cover is found from 2006 to 2016, about 51.23 percentage. And it has created a very worst situation for these study areas through creating a lot of problems.

In Ethiopia, the rapid expansion of agricultural land into a steeper slope has aggravated soil erosion and land degradation (Tegene, 2002). The change in land use and land cover class significantly affected surface runoff, soil erosion, sedimentation, land degradation, flood, drought, migration, biodiversity change, and decrease of agricultural productivity (Abbas et al., 2013). In this study area, massive land use and land cover change due to metro expansion were mostly affected in the overall environment (Halder et al., 2021).

6.6 Conclusion

The lack of metro train in the southern part of Kolkata has increased urban convenience such as transport access. The development of urban facilities has also led to a rapid increase in population. From the survey, it was found that

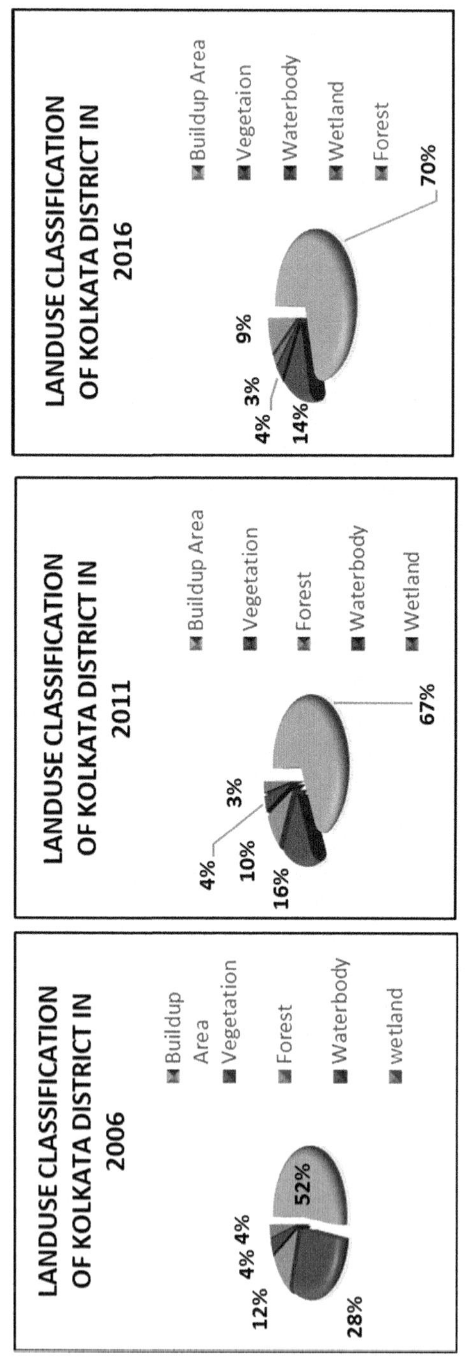

Figure 6.4 Land use land cover percentage of Kolkata district during 2006–2016

Table 6.3 Changing pattern of vegetation in Kolkata

Land class	*Area 2006 (sq. mile)*	*Area 2011 (sq. mile)*	*Area 2016 (sq. mile)*	*Change in area (sq. mile) from 2006–2011*	*Change in area from 2006–2011 in percentage*	*Change in area (sq. mile) from 2011–2016*	*Change in area from 2006–2016 in percentage*	*Change in area (sq. mile) from 2006–2016*	*Change in area from 2006–2016 in percentage*
Vegetation	27.85	15.66	13.58	−12.19	−43.77	−2.08	−13.26	−14.27	−51.23

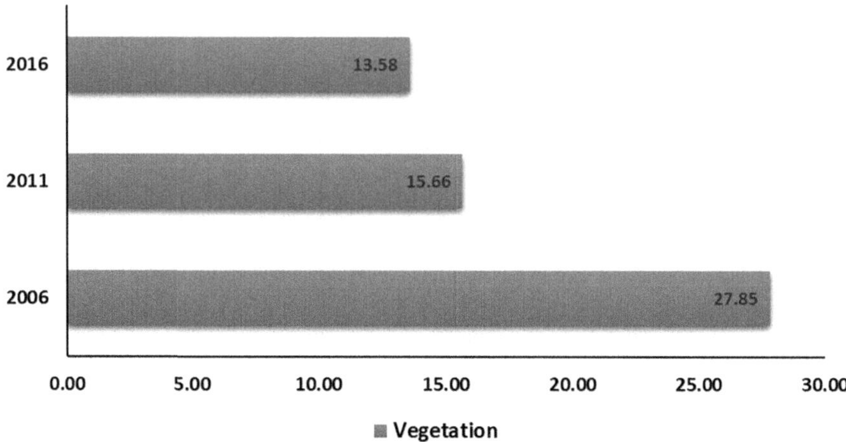

Figure 6.5 Vegetation pattern of Kolkata district during 2006–2016

the southern wards of Kolkata are highly affected by the expansion of the Kolkata Metro (Halder et al., 2021). The current trend of urban expansion has the most obvious environmental impacts on neighbouring ecosystems, land resources, the structure and pattern of the urban area, and therefore the quality of life. This chapter mainly focused on the detection of urban change from 2006 to 2016. The land-use and land-cover of the city of Kolkata has changed a lot during these ten years. In this study the vegetation has mostly changed with the replacement of built-up area construction. Similarly, waterways and wetlands have been changed due to population pressure, high construction area, road construction, flyover construction, car parking plot, parking ground etc. In 2006 the vegetation land was found to cover 27.86 sq. miles, which reduced to 13.56 sq. miles in 2016. Meanwhile, the built-up area expanded from 51.90 sq. miles in 2006 to 70.46 sq. miles in 2016. Besides, the wetland and waterbody have also been reduced throughout these periods.

Bibliography

Abbas, M., Ebeling, A., Oelmann, Y., Ptacnik, R., Roscher, C., Weigelt, A. et al (2013). Biodiversity effects on plant stoichiometry. *PLoS One* 8(3). DOI 10.1371/2Fjournal.pone.0058179

Albrecht, A., & Kandji, S.T. (2003). Carbon sequestration in tropical agroforestry systems. *Ecosystems and Environment* 99:15–27. DOI 10.1016/S0167-8809(03)00138-5

Bălteanu, D., Dragotă, C.-S., Popovici, A.M., Dumitrascu, A., Kucsicsa, G., & Grigorescu, I. 2013. Land use and crop dynamics related to climate change signals. *Proceedings of the Romanian Academy Series A* 15(3):265–278.

Bhatta, B. (2009). Analysis of urban growth pattern using remote sensing and GIS: A case study of Kolkata, India. *International Journal of Remote Sensing* 30 (18:20):4733–4746.

Census of India (2011). *Primary census abstract, census of India*. The government of India, Registrar General and Census Commissioner of India, Ministry of Home Affairs, New Delhi, India

Duguma, L. A., Atela, J., Minang, P. A., Ayana, A. N., Gizachew, B., Nzyoka, J. M., & Bernard, F. (2019). Deforestation and forest degradation as an environmental behavior: Unpacking realities shaping community actions. *Land* 8:26 DOI 10.3390/land8020026

FAO. (2011a). *Tees outside forests: Towards rural and urban integrated resources management*. Rome: FAO.

FAO. (2011b). *State of the world's forests*. Rome: FAO.

Forestry Commission (FC). (2011) *Forests and climate change: UK forestry standard guidelines*. Edinburgh: Forestry Commission.

Gregorio Di Antonio, (2005) *Land cover classification system: classification concepts and user manual: LCCS*. Vol. 2. Food & Agriculture Org.

Halder, B., Banik, P., & Bandyopadhyay, J. (2021). Mapping and monitoring land dynamic due to urban expansion using geospatial techniques on South Kolkata. *Safety in Extreme Environments* 3, 27–42. DOI 10.1007/s42797-021-00032-2

Hasnine, M., & Rukhsana (2020). an analysis of urban sprawl and prediction of future urban town in urban area of developing nation: Case Study in India. *Journal of the Indian Society of Remote Sensing* 48:909–920. DOI 10.1007/s12524-020-01123-6

Helen, J. M. P., & Gasparatos, A. (2019). Land-use change, carbon stocks, and tree species diversity in green spaces of a secondary city in Myanmar, Pyin Oo Lwin. *PLoS One* 14(11):e0225331. DOI 10.1371/journal.pone.0225331

Herold, M., Scepan, J., & Clarke, K. C. (2002). The use of remote sensing and landscape metrics to describe structures and changes in urban land uses. *Environment and Planning A* 34(8):1443–1458.

Jaybhaye, R. G., & Mundhe, N. N. (2013). Hybrid image classification technique for spatio-temporal analysis of Pune City. *Transactions of the Institute of Indian Geographers* 35(2):210–223.

Lambin, E. F. (1997). Modeling and monitoring land-cover change processes in tropical regions. *Progress in Physical Geography* 21(3):375–393.

Mbow, C., Smith, P., Skole, D., Dugumal, L., Bustamante, M. (2014). Achieving mitigation and adaptation to climate change through sustainable agroforestry practices in Africa. In: Sustainability challenges. *Current Opinion in Environmental Sustainability* 6:8–14. DOI 10.1016/j.cosust.2013.09.002

Meragiaw, M. (2017). Role of agroforestry and plantation on climate change mitigation and carbon sequestration in Ethiopia. *Journal of Tree Sciences* 36(1):1–15. DOI 10.5958/2455-7129.2017.00001.2

Mosammam, H. M., Nia, J. T., Khani, H., Teymouri, A. & Kazemi, M. (2017). Monitoring land use change and measuring urban sprawl based on its spatial forms: The case of Qom city, *Egyptian Journal*20(1):103–1016, 10.1016/j.ejrs.2016.08.002

National Climatic Data Center. (2013). *Historical land-cover change and land use conversions global dataset*. Asheville, Washington, DC: National Climatic Data Center.

Pareta, K., & Pareta, U. (2011). Forest carbon management using satellite remote sensing techniques a case study of Sagar district (MP). *E-International Scientific Research Journal* 3(4):335–348.

Rawat, J.S., & Kumar, M. (2015). Monitoring land use/cover change using remote sensing and GIS techniques: A case study of Hawalbagh block, district Almora, Uttarakhand. India. *The Egyptian Journal of Remote Sensing and Space Sciences* 18:77–84.

Smith, P., & Wollenberg, E.. (2012). Achieving mitigation through synergies with adaptation. In: E. Wollenberg, A. Nihart, M.-L. Tapio-Boström, M. Grieg-Gran, eds. *Climate change mitigation and agriculture*. London and New York: ICRAF-CIAT, 50–57.

Tegene, B. (2002). Land-cover/land-use changes in the derekolli catchment of the south Welo zone of Amhara region, Ethiopia. *Eastern Africa Social Science Research Review* 18(1):1–20. DOI: 10.1353/eas.2002.0005

U.S. Environmental Protection Agency. (1999). In: Loveland, T.R., ed. *Land cover trends: Rates, causes, and consequences of late-twentieth century*. Washington, DC: U.S. Environmental Protection Agency.

United Nations (2014). World urbanization prospects. The 2014 revision department of economic and social affairs population division New York.

World Population Review (2021). https://worldpopulationreview.com/continents/sub--saharan-africa-population

Wulder, M.A., White, J.C., Loveland, T.R., Woodcock, C.E., Belward, A.S., Cohen, W.B., Fosnight, E.A., Shaw, J., Masek, J.G., & Roy, D.P. (2016). The global Landsat archive: Status, consolidation, and direction. *Remote Sensing of Environment* 185:271–283. DOI: 10.1016/j.rse.2015.11.032

Yeh, A.G., & Li, X. (1998). Sustainable land development model for rapid growth areas using GIS. *International Journal of Geographical Information Science* 12:169–189.

Part II

Disaster Risk, Population and Livelihood

7 Self-Reported Health Problems Due to Disasters in India

Evidence from the Longitudinal Ageing Study in India Survey

Margubur Rahaman, Amiya Saha, Kailash Chandra Das, and Md. Juel Rana

7.1 Introduction

A never-ending race of technical progress, globalization, and modernization has rewarded us with modern civilization, but it has also introduced increased environmental degradation and socioeconomic and political inequality. Ecological degradation and socio-political disparity have resulted in a significant change in the global climate and socio-political power contest, resulting in a rise in natural and man-made disasters (Komolafe et al. 2014; Willow 2014). Aside from environmental damage and economic losses, disasters substantially impact people's physical and mental health, increasing death and morbidity (Satcher et al. 2007). The World Meteorological Organisation (WMO) has reported more than 11,000 disasters caused over 2 million deaths globally, and more than 91% of those deaths occurred in developing countries (World Meteorological Organisation, 2019). There were 3454 disasters in Asia during 1970–2019, with floods and storms being the most prominent causes of these disasters (36%). Storms claimed the lives of the most people, accounting for 72% of all deaths, and floods inflicted the most economic damage (57%) (World Meteorological Organisation 2019).

As a result, disasters disproportionately harm vulnerable populations such as the elderly, women, children, and the disabled (Kar 2010). Older people are more likely to be negatively affected by disasters due to poor physical and mental health conditions such as co-morbidities, difficulties in daily activities of living (ADLs), declining vision and hearing impairments, and cognitive disabilities (Bei et al. 2013; Malik et al. 2018). Aging demographics with physical and cognitive limitations are at a higher risk of dying, suffering injuries, and experiencing post-disaster health problems than younger adults (Gibson & Hayunga 2006). Many studies have indicated that disasters cause severe psychological distress with adverse health outcomes in these susceptible groups (Kar 2010; Delisi 2005). Infectious diseases due to contaminated food and water, vector and insect-borne diseases, and infections due to wounds and injuries are common health issues resulting from disasters (Morgan 2004; Nasci and Moore 1998; Sever et al., 2006). Natural disasters, including floods,

DOI: 10.4324/9781003275916-9

tsunamis, earthquakes, and cyclones, have been secondarily associated with the following infectious diseases: diarrheal diseases, acute respiratory infections, malaria, measles, dengue fever, typhoid fever, and meningitis (Ivers & Ryan 2006; Robinson et al. 2011). Drought-stricken areas confront drinking water shortages, agricultural failure, and excessive heat waves, which significantly influence people's physical and mental health (Gautam & Bana 2014). Harada et al. (2015) explored the link between earthquakes and the risk of mental well-being and found that many individuals experienced considerable psychological distress. Natural disasters and, at the same time, human-made disasters (i.e., environmental degradation, riots) have been associated with increased anxiety, sadness, and post-traumatic stress disorder (Gautam & Bana, 2014; Habiba et al., 2013). Terrorism impacts people's mental health and behavior patterns and has been identified as a risk factor for psycho-physical well-being. Similarly, communal, racial, and social violence have been identified as significant social and public health issues that have long-term health and economic consequences (Minayo 1994; Farooqui & Ahmad 2021). Previous radioactive industrial disasters have shown that all accidents are linked to short-term and long-term health hazards (Keim 2011). For example, in Fukushima, radioactive releases instigated huge fear and uncertainty in many people, spurring massive distress among the affected residents, especially among mothers of young children and nuclear plant workers. Fire accident catastrophes substantially impact physical and psychological casualties and are a vital cause of health disasters in lower- and middle-income countries (LMICs) (Sanghavi et al. 2009). An assessment of the literature indicated that road traffic injuries significantly strain the health-care system (Garg & Hyder 2006).

India is one of the world's ten most disaster-prone countries. Because of its unique geo-climatic and societal characteristics, India has been exposed to a wide range of natural and man-made calamities. Most states are affected by natural disasters such as floods, droughts, landslides, cyclones, heat waves, and earthquakes. Floods, droughts, and earthquakes affect 12%, 68%, and 59% of India's total land area (Chakraborty & Joshi 2016). Cyclones and tsunamis pose a hazard to 80% of India's coastline. In hilly terrain, landslides and avalanches are widespread. Floods and flash floods are frequent in the Middle Ganges plain, while cyclones are common in South Bengal, Odisha, Andhra Pradesh, and Kerala (Pandey et al. 2010). In the hilly terrain of the Himalayas, severe landslides and avalanches have occurred.

Simultaneously, man-made disasters such as road accidents, building collapses, riots, and terrorism are common in various parts of India. In India, the problem of road safety is significant; about half a million people have been injured in traffic accidents. From 1970 to 2008, the number of accidents tripled, with a more than seven-fold increases in injuries. Annually, 2600 individuals die in India due to building and other structural collapses. Since its independence, India has experienced a substantial number of deaths, injuries, and mental anguish due to communal and socio-political riots. As indicated above, natural and man-made disasters enhance health vulnerabilities, particularly among India's elderly. Elderly people are more vulnerable to health and livelihood resulting from natural and man-made disasters because they are often physically weak and sluggish to respond to disasters. They have more co-morbidities than young adults (Ngo 2001; Phifer 1990). As a result, several previous pieces of research focused on the impact of disasters on elderly health in India and elsewhere (Ngo 2001; Phifer 1990; Donner & Rodríguez 2008).

India is undergoing a tremendous demographic transformation, with the senior population growing at an alarming rate. As the aging population grows, studying their health and well-being becomes more important. In India, several previous studies focused on contextualizing the impact of disasters on older adults' (elderly person aged 50–80 years) health and the general population's health. Some important existing literatures are: Patel et al. (2015) found that posttraumatic stress disorder is more severe among older adults immediately after a flood and reduces over time. Similarly, drought-stricken places face a lack of drinking water, agricultural failure, and extreme heat, all of which substantially impact people's physical and mental health (Gautam & Bana 2014). Harada et al. (2015) found a positive correlation between earthquakes and psychological well-being and discovered that hypertension, anxiety, and stress were all linked to earthquakes. Landslide were a risk factor for physical health, such as injuries, disability, and functional limitation. The study conducted by Bhagat (2011) on Hindu–Muslim riots in India suggested that communal riots resulted in displacement, forced migration, loss of livelihood, and public health challenges. Garg and Hyder (2006) reviewed existing literature on road traffic injuries in India and suggested that road traffic injuries are a tremendous burden on the health care system in India. Sanghavi et al. (2009) highlighted fire-related health challenges in India and found a significant number of deaths and injuries occurred due to fire incidents.

As a result, it is essential to contextualize the impacts of natural and man-made calamities on India's senior citizens' physical and mental health. Most of the previous studies were limited to scenarios of disasters' effects on elderly health and failed to elaborate on how it varies with space and socio-demographic settings in India. Therefore, to fill this research gap, the present study aims to examine the spatial distribution and socio-economic variations of the prevalence of self-reported health problems due to disasters in India. Due to climate change, the frequency and spatial distribution of disasters have become more dynamic, which has resulted in severe public health challenges worldwide, especially in lower-and-middle-income countries. From this point of view, this chapter will be helpful in formulating different policies and programs to minimize the risk of disaster-induced health problems and develop coping strategies for climate change and its health effects.

7.2 Data and Methodology

This chapter used data from the first wave of the Longitudinal Ageing Study in India (LASI), conducted during 2017–18. The LASI is a large-scale survey that was implemented and aimed to understand the health status of the aging population and associated socio-economic, demographic, and geographical factors. In LASI, a multistage stratified sampling design was used to select the study population. In the sample frame, a three-stage sample strategy was used for rural areas, and a four-stage sample strategy was used for urban areas. A household with at least one respondent aged 45 or older has been selected as a sample household from all Indian states and union territories, excluding Sikkim (IIPS & USC, 2020). CAPI-based schedules were used to gather data, which were conducted by a trained interviewer. The International Institute for Population Sciences (IIPS) was the nodal agency that maintained and monitored the data collection and quality. The survey's total sample size is 72,250 older adults.

The outcome variable was health problems due to natural disasters and man-made disasters. The prevalence of health problems due to natural disasters was measured based on the LASI question: In the last five years, has your health been severely affected by disasters such as floods, landslides, extreme cold, and hot weather, cyclone/typhoons, droughts, earthquakes, tsunamis, or any other natural calamities? Then, a subsequent question was also asked to those who responded "yes" to the above question: Which of these natural disasters affected your health? Please list all-natural disasters that affected you. The responses were: (a) floods, (b) landslides, (c) cyclone/typhoon, (d) earthquakes, tremors, earthquakes, tsunamis, (f) droughts, (g) other. Similarly, the prevalence of health problems due to man-made disasters is measured based on the LASI question: In the last five years, has your health been severely affected by man-made incidents such as riots, terrorism, building collapses, fires, traffic accidents, or any other man-made incidents? And the response was: yes or no. Then, a subsequent question was also asked to those who responded "yes" to the above question: Which of these man-made disasters affected your health? Please identify all man-made incidents that affected you. The responses were: (a) riots, (b) terrorism, (c) building collapses, (d) fires, € traffic accidents, (f) other. Finally, another question was asked of those who experienced health problems due to natural disasters or man-made disasters: What were the health consequences you suffered from these disasters or incidents? And the responses were: (a) permanent physical disability, (b) psychological trauma and mental health problems, (c) chronic illness, (d) other.

The three main outcome variables were: (1) health problems due to natural disasters (yes, no); (2) health problems due to man-made disasters (yes, no); and (3) total health problems due to both natural and man-made disasters (yes, no). We also categorized major health problems due to any disaster as (a) permanent physical disability, (b) psychological trauma and mental health problems, (c) chronic illness, and (d) other. The outcome variable was categorized based on the LASI report.

Predictor variables were assessed using several existing studies, including geographical, demographic, and socio-economic factors. Geographical factors include the place of residence (rural and urban). Demographic factors include demographic parameters like age (18–44 years, and 45 and above years), sex (male, female), as well as socioeconomic factors like education (illiterate, primary, secondary, higher secondary, and graduate and above years), and wealth quintile (poorest, poorer, middle, richer, and richest), religion (Hindu, Muslim, Christians, and others), and caste (scheduled caste, scheduled tribe other backward class, and no caste).

7.3 Analysis

The unweighted sample and weighted percentage distribution of the outcome variables and respondents background characteristics were presented by descriptive statistics. Bivariate analyses using chi-square significance tests were performed to present patterns of health problems due to disasters by background characteristics. Spatial patterns of health problems due to disasters also presented with the help of Arc GIS technique. For all analyses, STATA SE version 14.1 was used (Stata Corporation, College Station, Texas, USA). The figure was created using the Microsoft Excel tool.

7.4 Results and Discussion

Table 7.1 shows the sample distribution of the study population. More than half of the respondents lived in rural areas. The percentage of female respondents was 10% higher than their male counterparts. The majority of respondents belonged to the Hindu religion and lower castes. Almost half of the respondents were illiterate, and only 5% completed graduation and above. Only 20% of the respondents belonged to the richest quintile.

Table 7.1 Sample distribution of outcome variables and respondent's socio-demographic characteristics

	Total population		*Age 45 & above*	
Age group				
18–44	6,688	9.3	-	-
45 and above	65,562	90.7	65,562	100
Place of residence				
Rural	46,534	64.4	42,424	64.7
Urban	25,716	35.6	23,138	35.3
Sex				
Male	30,569	42.3	30,479	46.5
Female	41,681	57.7	35,083	53.5
Religion				
Hindu	53,118	73.5	48,231	73.6
Muslim	8,667	12.0	7,803	11.9
Christians	7,215	10.0	6,536	10.0
Others	3,250	4.5	2,992	4.6

(*Continued*)

Table 7.1 (Continued)

	Total population		*Age 45 & above*	
Caste				
Scheduled caste	12,046	16.7	10,959	16.7
Scheduled tribe	12,509	17.3	11,365	17.3
Other backward class (OBC)	27,184	37.6	24,629	37.6
None of them	20,511	28.4	18,609	28.4
Level of education				
Illiterates	33,210	46.0	30,820	47.0
Primary	17,736	24.6	16,096	24.6
Secondary	13,949	19.3	12,126	18.5
Higher secondary	3,585	5.0	3,126	4.8
Graduation & above	3,770	5.2	3,394	5.2
Wealth quintile				
Poorest	14,158	19.6	12,941	19.7
Poorer	14,530	20.1	13,190	20.1
Middle	14,537	20.1	13,163	20.1
Richer	14,686	20.3	13,210	20.2
Richest	14,339	19.9	13,058	19.9
Total (n)	**72,250**	**100**	**65,562**	**100**

Source: Authors' own calculation using Longitudinal Ageing Study in India (2017–18)

Figure 7.1 shows the prevalence of health problems caused by disasters by age group in India. The overall prevalence of any health problem due to any disaster was 3.9% among older adults aged 45 and above. At the same time, 2.8% and 1.2% of older adults experienced health problems due to natural and man-made disasters, respectively. The prevalence of any health problems due

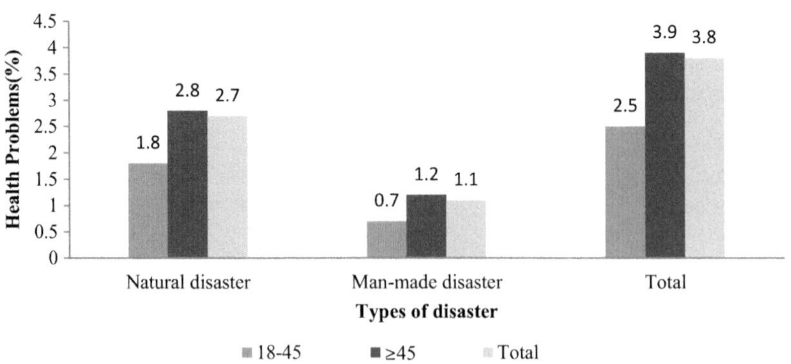

Figure 7.1 Prevalence of self-reported health problems due to disasters by age group in India, 2017–18

Source: Authors' own calculation using Longitudinal Ageing Study in India (2017–18)

to any disasters was found to almost double among older adults compared to adults (aged 18–45).

Figure 7.2 depicts the percentage distribution of natural disasters associated with any health problems among the older adults who experienced health difficulties due to natural disasters. Drought was the most common natural calamity that harmed the health of older adults. The prevalence of any health difficulties resulting from drought was 35% in older adults. Floods were the second most common natural disaster, accounting for almost 30% of any health problems among older adults. Other natural catastrophes were responsible for 24% of the health problems among older people, whereas earthquakes and cyclones were responsible for 18% and 15%, respectively. Tsunamis and landslides have less impact on the health of older adults (1.4 % and 1.3%).

Figure 7.3 shows the percentage distribution of any health problems among older adults who experienced health difficulties due to man-made disasters. The most common man-made disaster among older individuals was traffic accidents; nearly 67% of older adults with man-made disasters caused health problems mentioned traffic accidents as a reason for their health problems. Other man-made disasters accounted for 21% of all cases. Riots were also associated with health issues in older adults, accounting for 13% of any health

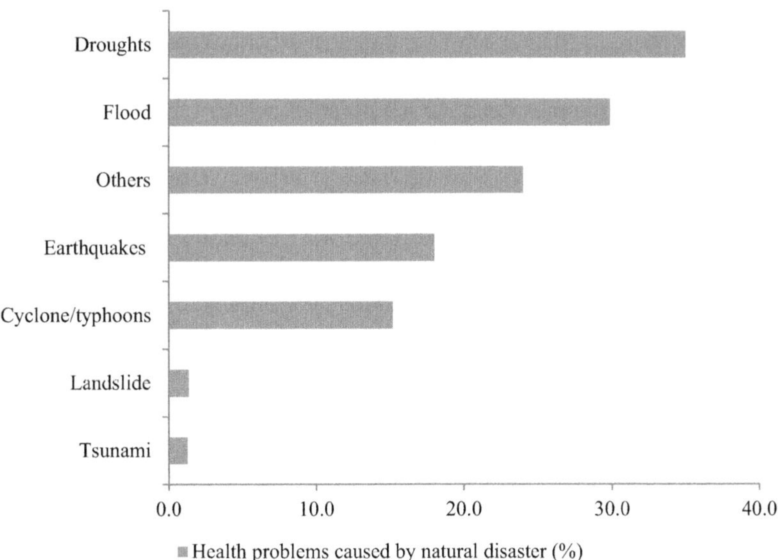

Figure 7.2 Percentage distribution of natural disasters associated with any health problems among the older adults in India, 2017–18 (n = 1,523)

Source: Authors' own calculation using Longitudinal Ageing Study in India (2017–18)

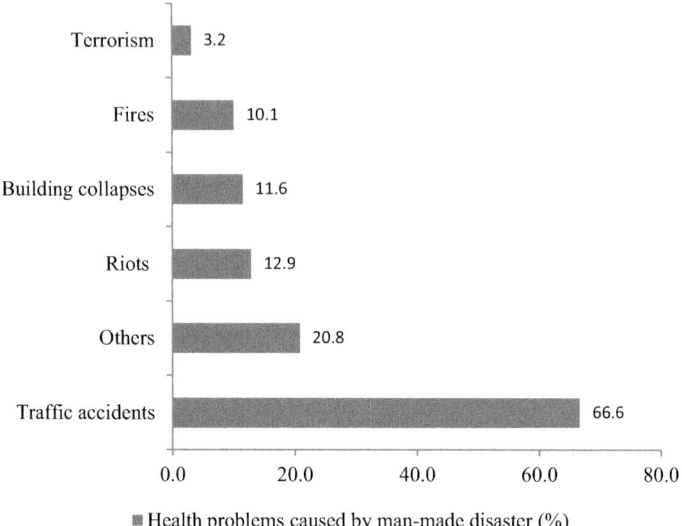

Figure 7.3 Percentage distribution of man-made disasters associated with any health problems among the older adults in India, 2017–18 (n = 1,523) (n = 775)

Source: Authors' own calculation using Longitudinal Ageing Study in India (2017–18)

difficulties. Building collapses (11.6%), fires (10.1%), and terrorism (3.2%) were all factors that contributed to health problems among older adults.

Table 7.2 shows the prevalence of health problems by types of disasters and the respondent's background characteristics. There was a significant gap in the prevalence of health problems due to disasters between rural and urban areas. The prevalence of self-reported health problems due to natural disasters was found highest among Muslims (3.4%) and lowest among others (0.6%). The elderly who belonged to the scheduled caste and other backward classes experienced more natural disaster-induced health problems than their counterparts. Older adults living in rural areas experienced significantly higher health problems due to disasters than their urban counterparts. In particular, the prevalence of health problems due to natural disasters, man-made disasters, and any disasters was found to be three times, two times, and four times, respectively, higher in rural areas compared to urban counterparts. Male older adults reported slightly higher health problems caused by any disaster (4.1%) than females (3.4%). In terms of man-made disasters, older male adults faced health problems almost two times higher than females (1.4% vs. 0.9%). Man-made disasters have a more significant impact on any health problems among Muslims (1.6%) than among Hindus (1.2%) or other communities (1.0%). Health problems are distributed differently depending on caste and other socio-demographic and economic factors. Disasters have a higher impact on scheduled castes (4.2%) than on scheduled tribes (3.4%) and other backward classes (4.1%). People with no education

Table 7.2 Patterns of self-reported health problems (SRHP) due to disasters among older adults by background characteristics in India, 2017–18

	SRHP caused by natural disasters (%)	*SRHP caused by man-made disasters (%)*	*SRHP caused by any disasters (%)*
Place of residence	p = 0.000	p = 0.000	p = 0.000
Rural	3.5	1.3	4.5
Urban	1.1	0.7	1.8
Sex	p = 0.267	p = 0.000	p = 0.000
Male	2.4	1.4	4.1
Female	2.3	0.9	3.4
Religion	p = 0.010	p = 0.000	p = 0.000
Hindu	2.7	1.2	3.7
Muslim	3.4	1.6	4.4
Christians	2.7	0.8	3.6
Others	0.6	1.0	1.3
Social group	p = 0.000	p = 0.000	p = 0.000
Scheduled caste	2.6	1.2	4.2
Scheduled tribe	1.7	0.9	3.4
Other backward class	2.9	1.3	4.1
None of them	1.7	1.0	2.8
Level of education	p = 0.060	p = 0.040	p = 0.010
Illiterates	1.8	1.2	4.3
Primary	2.1	1.1	3.3
Secondary	1.8	1.2	3.2
Higher secondary	1.7	1.1	2.6
Graduation & above	1.2	0.7	2.0
Wealth index	p = 0.371	p = 0.051	p = 0.000
Poorest	2.2	1.0	3.2
Poorer	2.3	1.2	3.7
Middle	2.5	1.0	4.3
Richer	2.3	1.1	3.4
Richest	2.4	1.5	3.9

Source: Authors' own calculation using Longitudinal Ageing Study in India (2017–18)

were more likely to experience health problems as a result of disasters (4.3%) than those who have completed at least their primary education (3.3%), secondary education (3.2%), or higher secondary education (2.6%). Middle-class adults (4.3%) reported having more health problems than the poorest (3.2%), poorer (3.7%), richer (3.4%), and richest (3.9%), correspondingly.

Figure 7.4 illustrates significant cross-state variability in health problems due to natural and man-made calamities. The prevalence of health problems due to any disaster was found to be highest in Jammu and Kashmir (8.6%), Uttar Pradesh (8.6%), and Bihar (8.6%), followed by Madhya Pradesh (7.5%), Jharkhand (5.8%), and Himachal Pradesh (5.8%). At the same time, Jammu and Kashmir were found to be states/Union Territories (UTs) where both natural and man-made disasters were associated with a higher risk of health problems. Health problems due to natural disasters were found to be comparatively

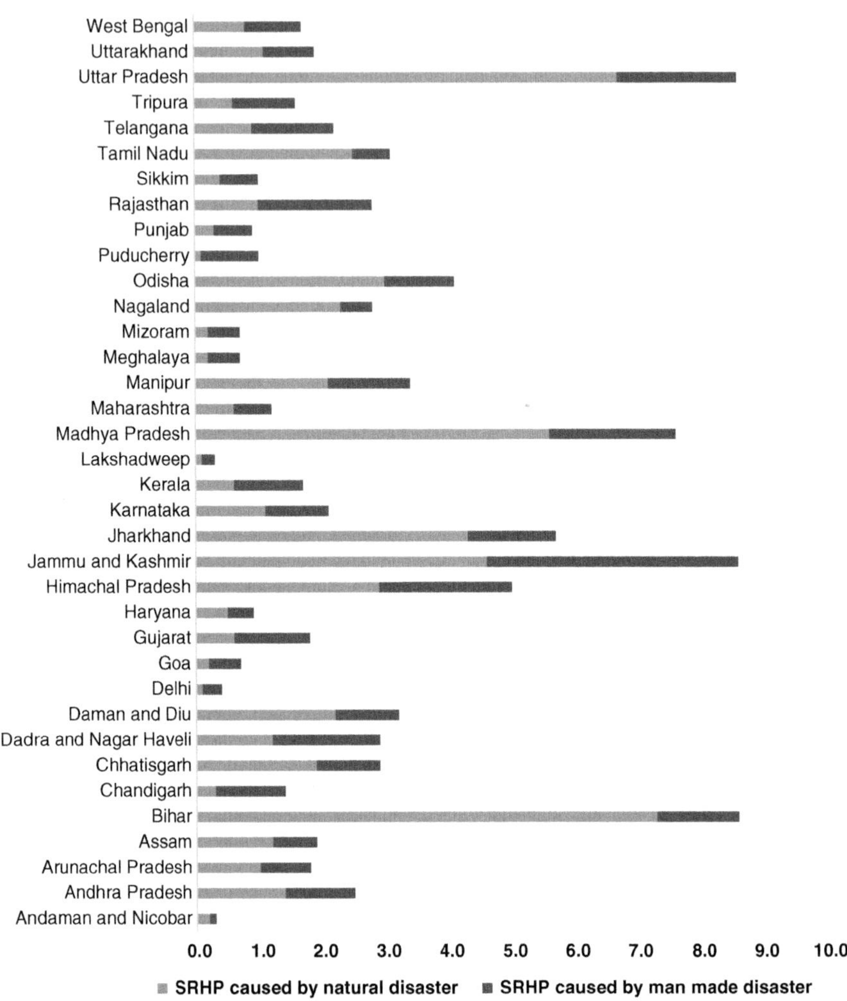

Figure 7.4 Spatial patterns of self-reported health problems (SRHP) due to disasters (%) among total population in India, 2017–18

Source: Authors' own calculation using Longitudinal Ageing Study in India (2017–18)

high in Uttar Pradesh and Bihar than in their counterparts. In particular, natural and man-made disasters appeared to have a greater health impact on adults in Jammu and Kashmir (4.6% & 4.0 %). Natural disasters had a significantly greater impact on health conditions than man-made disasters in states like Bihar (7.3% versus 1.3%), Uttar Pradesh (6.7% versus 1.9%), Madhya Pradesh (5.6% versus 2.0 %), Jharkhand (3% versus 1.4%), Nagaland (2.3% versus 0.5 %), and Mizoram (2.1% versus 0.7%). The states of Chhattisgarh (3%), Tamil Nadu (3.1%), Manipur (3.5%), and Nagaland (3.5%) have reported a moderate prevalence of health concerns as a result of any disasters. People in Andhra Pradesh (2.5%) and Karnataka (2.2%) in the south, Rajasthan (1.8%), Gujarat (1.8%), and Maharashtra (1.2%) in the west, Punjab (1%), and Haryana (0.9%) in the north, Arunachal Pradesh (1.8%) and Assam (1.9%) in the northeast, and West Bengal (1.9%) in the east have reported less prevalence of health problems due to disasters.

Figure 7.5 shows the health effects of natural and man-made disasters among older adults aged 45 and above by state and UT. The prevalence of

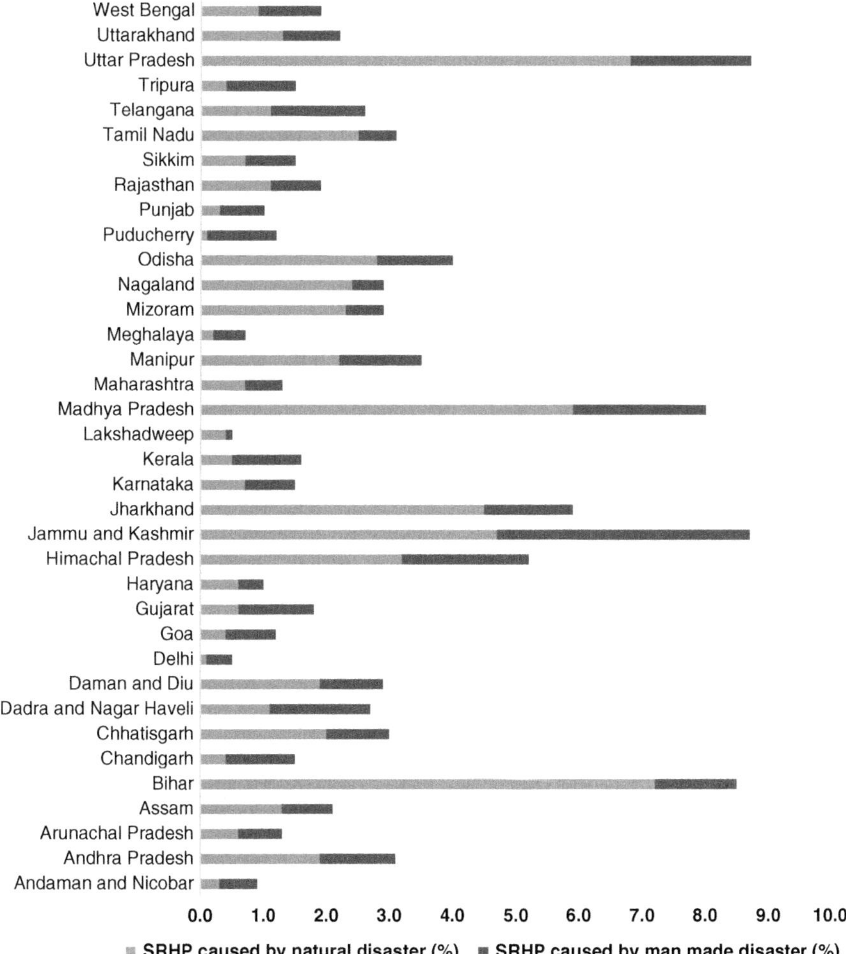

Figure 7.5 State level patterns of health problems (SRHP) due to disasters (%) among older adults in India, 2017–18

Source: Authors' own calculation using Longitudinal Ageing Study in India (2017–18)

health problems among older adults due to any disasters was found highest in Uttar Pradesh (8.7%), followed by Jammu and Kashmir (8.6%), Bihar (8.5%), Madhya Pradesh (7.9%), Jharkhand (5.9%), and Himachal Pradesh (5.9%). At the same time, the prevalence was found to be moderate in Manipur (3.5%), Tamil Nadu (3.1%), Chhattisgarh (2.9%), Nagaland (2.9%), Andhra Pradesh (2.8%), and Mizoram (2.8%). But, Telangana (2.5%) in the south, Rajasthan (1.9%) and Gujarat (1.8%) in the west, West Bengal (1.9%) in the east, and Assam (2.1%) in the northeast had a lower impact on health among older adults. In Jammu and Kashmir, both natural and man-made disasters increased the burden of health problems among older adults (4.7 % and 4.0 %). In states like Uttar Pradesh (6.8% versus 1.9%), Bihar (7.2 % versus 1.3 %), Jharkhand (4.5 % versus 1.4 %), Odisha (2.8 % versus 1.2 %), Nagaland (2.4 % versus 0.5 %), Chhattisgarh (2.0 % versus 1.0 %), Madhya Pradesh (5.9% versus 2.1%), Manipur (2.2% versus 1.3%) and Mizoram (2.3% versus 0.6%) were more vulnerable in terms of natural disaster caused health problems among older adults.

Out of the total respondents, 0.36% of respondents experienced physical disability due to a disaster (Table 7.3). Psychological trauma was found to be a significant health problem due to a disaster among older adults as well as adults. Less than 1% of older adults reported they faced chronic illness due to any disaster in India.

This study presents recent evidence on India's regional and socioeconomic dynamics of disaster-related health concerns. The current study is purely descriptive, contextualizing only spatial patterns and demographic and socioeconomic variability in the levels of disaster-caused health problems in India, focusing on older adults based on recent national level LASI data. The present study found that older adults are more vulnerable compared to adults, similar to many prior studies. Existing studies suggest that older adults are more susceptible to health problems in the aftermath of natural and man-made disasters because of their physical and cognitive limitations (Ngo 2001; Phifer 1990). Droughts and floods in India are the leading causes of health concerns among natural catastrophes. Many previous pieces of research showed similar findings (Pandey et al. 2010). Both disasters negatively affect the health of individuals and the elderly in two ways. First, both caused increased drinking water scarcity, food scarcity, and public health infrastructure challenges, which increased the risk of malnutrition, water-borne diseases, and other diseases. Second, disasters cause sudden socioeconomic loss,

Table 7.3 Prevalence of major health issues due to any disasters in India, 2017–18

Health problems	*Total (%)*	*Older adults (%)*
Physical disability	0.34	0.36
Psychological trauma	2.35	2.40
Chronic illness	0.68	0.72
Others	0.35	0.38

Source: Authors' own calculation using Longitudinal Ageing Study in India (2017–18)

which leads to unemployment and poverty and increases the risk of hypertension, mental trauma, and other mental disturbances. The study discovered that the prevalence of health problems caused by natural disasters is considerable in India and varies widely across states and union territories. In Jammu and Kashmir, Himachal Pradesh, Uttar Pradesh, Bihar, Odisha, Madhya Pradesh, and Tamil Nadu, natural disaster-related health problems were found to be more prevalent. Prior studies suggested that geo-hazards like landslides, flash floods, floods, cyclones, etc., are common in the above states (Batar & Watanabe 2021).

Man-made health problems were found to be less common in India than natural calamities. However, compared to developed countries, India has seen a disproportionately high number of man-made disasters and associated health problems. Health issues are commonly related to man-made disasters, particularly traffic accidents. India has one of the highest rates of traffic accidents, deaths, injuries, and mental illness among developing countries. India's leading causes of high traffic accidents include excessive traffic and poor transportation facilities.

In India, socio-political and communal riots pose a considerable health risk. India has had a lot of social, political, and communal conflict, resulting in many deaths, injuries, and mental illnesses. Following riots, many people cannot forget the brutality, resulting in long-term emotional trauma (Farooqui & Ahmad 2021). Terrorism is another major contributor to man-made catastrophes in India, particularly in Jammu and Kashmir, the central area, and the northeast. Terrorism is also responsible for a considerable number of deaths (Hussain et al. 2012). Jammu and Kashmir is the only state in the country where both natural and man-made catastrophes have hit. It occurs mainly due to harsh physiography, internal civic turmoil, and Indo-Pak tensions. Every day, a large number of people experience socioeconomic hardship and health issues as a result of communal and terrorist unrest.

Aside from geographic disparity, the risk of health problems caused by disasters varied significantly depending on the place of residence (Chan et al. 2019). According to the current study's findings, the prevalence of health problems caused by natural and man-made catastrophes is substantially higher in rural areas than in urban areas. Previous studies have also revealed that rural India has a higher risk of calamities. Many rural inhabitants have formed habitats alongside river banks and in high-risk areas of mountainous areas, which are typically affected by floods, flash floods, and cyclonic hazards. Similarly, rural residents are primarily engaged in agricultural activities. Due to draughts, farmers and farming laborers face severe socioeconomic constraints and food scarcity due to crop failure, which has resulted in death, injury, hunger, and mental illness. Simultaneously, the risk of man-made disasters is high in rural areas due to socioeconomic backwardness, rigid caste, and religious practices, all of which contribute to socio-political and communal tension (Bose, 1981).

Inadequate cooking facilities, poor housing, lack of knowledge about fire safety, and a lack of emergency services facilities are all factors that contribute to fire accidents in rural areas. Males experienced much more health difficulties due to man-made disasters than females. It occurs because men are the family's primary breadwinner, engage in hazardous occupations, and have higher daily activity than women, making them more susceptible to man-made disasters. Muslim minorities reported much more health difficulties as a result of both natural and man-made disasters than other religious groups. Many prior studies have suggested that the Muslim minorities are heavily involved in 3D jobs, which are dangerous occupations with low wages, increasing their risk of health problems due to man-made disasters. At the same time, the Muslim minorities have been subjected to a high level of communal violence, resulting in significant physical and mental health problems. Recent evidence, such as the Ram-Navami episodes, the Delhi riots, and other mob lynchings, reveals that Muslims are particularly vulnerable. Compared to the general caste, the scheduled caste and other disadvantaged classes experienced more health difficulties due to man-made disasters. Both castes are socioeconomically disadvantaged and endure severe societal discrimination. They primarily engage in lower-level activities, exposing their health to danger from man-made calamities. The current study also suggested that with an increasing level of education, the risk of health problems due to man-made disasters decreases. Educated individuals have better jobs, live in a healthier environment, are less likely to get involved in socio-political and community conflicts, and are less likely to be affected by natural disasters. Natural disasters affect all equally without judgment on socioeconomic background. However, significant differences based on religion and castes are found in the present study, perhaps due to variability in place in residence and livelihood and overall health conditions.

7.5 Findings and Policy Implications

1 In total, the prevalence of self-reported health problems for natural and man-made disasters was 2.7% and 1.1%, respectively, and older adults were more vulnerable.
2 The self-reported health problems due to disasters were found significantly high in the northern Himalayan states and the middle Ganges plain (Bihar and Uttar Pradesh Ganga basin), particularly in rural settings.
3 There is a considerable gender divide in health problems caused by man-made disasters, with males having a significantly higher prevalence.
4 Disasters, such as droughts and floods, and man-made disasters, such as traffic accidents and riots, have been identified as significant causes of health problems.
5 Psychological trauma has been linked to severe health problems in India as a result of both natural and man-made disasters.

The present study recommends some strategies to overcome the risk of health problems due to disasters in India; there are following-

1 The current study recommended that effective disaster management programs are needed in India, particularly in disaster-prone areas.
2 Special rationing and health programs, such as mobile health facilities, daily rations during disasters, physiological counselling services, and alternative job schemes, are demanded in high-disaster-prone areas.
3 There is a need to arrange regular seminars, talk shows, street plays, or community-wide discussion programs focusing on traffic safety, fire safety, communal harmony, and unity in diversity. These initiatives will help to prevent man-made disasters such as riots, traffic, work-related, and fire accidents.
4 Psychological trauma is the most common health concern among people who have experienced disasters; hence, mental health care facilities need to be strengthened in disaster-prone areas.

7.6 Conclusion

Disaster-related health concerns were found to be most widespread in North, Middle Ganges Plain, and Eastern India. Floods, droughts, road accidents, and riots, among other disasters, have been linked to severe health consequences. In comparison to young individuals, the elderly had more health difficulties, particularly illiterates, males, Muslims, and those who lived in rural areas. A multi-sectoral action plan is required to reduce the risk of health problems due to disasters.

Acknowledgements

The authors are grateful to ISEC for their continuous support during this venture.

Funding

None.

Availability of Data and Material

The study uses secondary data which is available on reasonable request through www.iipsindia.ac.in/content/lasi-wave-i

Declarations

Ethics Approval and Consent to Participate

The Central Ethics Committee on Human Research (CECHR) under the Indian Council of Medical Research (ICMR) extended the necessary guidance, guidelines and ethics approval for conducting the LASI survey. All methods were carried out in accordance with those relevant guidelines and regulations. The survey agencies that conducted the field survey for the data collection have collected prior informed consent (signed and oral) for both the interviews and biomarker tests from the eligible respondents (aged 45 years and above) in accordance with Human Subjects Protection.

Bibliography

Batar, A. K., & Watanabe, T. (2021). Landslide susceptibility mapping and assessment using geospatial platforms and weights of evidence (WoE) method in the Indian Himalayan Region: Recent developments, gaps, and future directions. *ISPRS International Journal of Geo-Information*, 10(3), 114.

Bei, B., Bryant, C., Gilson, K. M., Koh, J., Gibson, P., Komiti, A., ... Judd, F. (2013). A prospective study of the impact of floods on the mental and physical health of older adults. *Aging & Mental Health*, 17(8), 992–1002.

Bhagat, R. B. (2011). Hindu–Muslim riots in India: A demographic perspective. Lives of Muslims in India: Politics, exclusion and violence, 163–186. www.taylorfrancis.com/chapters/edit/10.4324/9781351227629-9/hindu%E2%80%94muslim-riots-india-demographic-perspective-bhagat

Bose, P. K. (1981). Social mobility and caste violence: A study of the Gujarat riots. *Economic and Political Weekly*, 16, 713–716.

Chakraborty, A., & Joshi, P. K. (2016). Mapping disaster vulnerability in India using analytical hierarchy process. *Geomatics, Natural Hazards and Risk*, 7(1), 308–325.

Chan, E. Y. Y., Man, A. Y. T., & Lam, H. C. Y. (2019). Scientific evidence on natural disasters and health emergency and disaster risk management in Asian rural-based area. *British Medical Bulletin*, 129(1), 91–105. https://doi.org/10.1093/bmb/ldz002

DeLisi, L. E. (2005). The New York experience: Terrorist attacks of September 11, 2001. *Disasters and Mental Health*, 01, 167–178.

Donner, W., & Rodríguez, H. (2008). Population composition, migration and inequality: The influence of demographic changes on disaster risk and vulnerability. *Social Forces*, 87(2), 1089–1114.

Farooqui, N., & Ahmad, A. (2021). Communal violence, mental health and their correlates: A cross-sectional study in two riot affected districts of Uttar Pradesh in India. *Journal of Muslim Minority Affairs*, 41(3), 510–521.

Garg, N., & Hyder, A. A. (2006). Road traffic injuries in India: A review of the literature. *Scandinavian Journal of Public Health*, 34(1), 100–109.

Gautam, R. C., & Bana, R. S. (2014). Drought in India: Its impact and mitigation strategies – A review. *Indian Journal of Agronomy*, 59(2), 179–190.

Gibson, M. J., & Hayunga, M. (2006). We can do better: Lessons learned for protecting older persons in disasters, 1–86. Retrieved from https://assets.aarp.org/rgcenter/il/better.pdf

Habiba, U., Hassan, A. W. R., & Shaw, R. (2013). Livelihood adaptation in the drought prone areas of Bangladesh. In Rajib Shaw, Fuad Mallick & Aminul Islam (eds) *Climate change adaptation actions in Bangladesh*. Springer, Tokyo, 227–252.

Harada, N., Shigemura, J., Tanichi, M., Kawaida, K., Takahashi, S., & Yasukata, F. (2015). Mental health and psychological impacts from the 2011 Great East Japan earthquake disaster: A systematic literature review. *Disaster and Military Medicine*, 1(1), 1–12.

Hussain, J., Iqbal, S., Taj, R., & Khan, A. M. (2012). Impact of terrorism on mental health. *Annals of Pakistan Institute of Medical Sciences*, 8(1), 46–49.

IIPS, NPHCE, HSPH and USC. (2020). *Longitudinal Ageing Study in India (LASI) India Report*. International Institute for Population Sciences, Mumbai; National Programme for Health Care for Elderly, Ministry of Health & Family Welfare;

Harvard T. H. Chan School of Public Health; and the University of Southern California, Wave 1, 2017–18, https://iipsindia.ac.in/sites/default/files/LASI_India_Report_2020_compressed.pdf

Ivers, L. C., & Ryan, E. T. (2006). Infectious diseases of severe weather-related and flood-related natural disasters. *Current Opinion in Infectious Diseases*, *19*(5), 408–414.

Kar, N. (2010). Indian research on disaster and mental health. *Indian Journal of Psychiatry*, 52(Suppl1), S286.

Keim, M. E. (2011). The public health impact of industrial disasters. *American Journal of Disaster Medicine*, 6(5), 265–274.

Komolafe, A. A., Adegboyega, S. A. A., Anifowose, A. Y., Akinluyi, F. O., & Awoniran, D. R. (2014). Air pollution and climate change in Lagos, Nigeria: Needs for proactive approaches to risk management and adaptation. *American Journal of Environmental Sciences*, 10(4), 412.

Malik, S., Lee, D. C., Doran, K. M., Grudzen, C. R., Worthing, J., Portelli, I., ... Smith, S. W. (2018). Vulnerability of older adults in disasters: Emergency department utilization by geriatric patients after Hurricane Sandy. *Disaster Medicine and Public Health Preparedness*, 12(2), 184–193.

Minayo, M.C. (1994). Inequality, violence, and ecology in Brazil. *Cadernos De Saude Publica*, 10(2), 241–250.

Morgan, O. (2004). Infectious disease risks from dead bodies following natural disasters. *Revista Panamericana de Salud Pública*, 15, 307–312.

Nasci, R. S., & Moore, C. G. (1998). Vector-borne disease surveillance and natural disasters. *Emerging Infectious Diseases*, 4(2), 333.

Ngo, E. B. (2001). When disasters and age collide: Reviewing vulnerability of the elderly. *Natural Hazards Review*, 2(2), 80–89.

Pandey, A. C., Singh, S. K., & Nathawat, M. S. (2010). Waterlogging and flood hazards vulnerability and risk assessment in Indo Gangetic plain. *Natural Hazards*, 55(2), 273–289.

Patel, F., Oswal, R., & Mehta, R. (2015). Posttraumatic stress disorders in adult victims of 2006 Flood in Surat, Gujarat. *Journal of Research in Medical and Dental Science*, 3(4), 303. https://doi.org/10.5455/jrmds.20153413

Phifer, J. F. (1990). Psychological distress and somatic symptoms after natural disaster: Differential vulnerability among older adults. *Psychology and Aging*, 5(3), 412.

Robinson, B., Alatas, M. F., Robertson, A., & Steer, H. (2011). Natural disasters and the lung. *Respirology*, 16(3), 386–395.

Sanghavi, P., Bhalla, K., & Das, V. (2009). Fire-related deaths in India in 2001: A retrospective analysis of data. *The Lancet*, 373(9671), 1282–1288.

Satcher, D., Friel, S., & Bell, R. (2007). Natural and manmade disasters and mental health. *JAMA*, 298(21), 2540–2542.

Sever, M. S., Vanholder, R., & Lameire, N. (2006). Management of crush-related injuries after disasters. *New England Journal of Medicine*, 354(10), 1052–1063.

Willow, A. J. (2014). The new politics of environmental degradation: Un/expected landscapes of disempowerment and vulnerability. *Journal of Political Ecology*, 21(1), 237–257.

World Meteorological Organisation. (2019). Atlas of mortality and economic losses from weather, climate and water extremes (1970–2019). Retrieved from https://library.wmo.int/doc_num.php?explnum_id=10989

8 Indigenous and Scientific Based-Resilient Strategies to Pre and Post Disasters Management in Cameroon

Insights from the Lake Nyos 1986 Gas Disaster in the Boyo-Menchum Basin

Augustine Toh Gam, Mbanga Lawrence Akei, and Mofor Gilbert Zechia

8.1 Introduction

Cameroon's geographical and geologic configuration exposes her to several types of natural and human-induced hazards occurring in all its ten regions, following the triangular structure, climatic zonation, multiculturalism, spatial development infrastructures, sites of settlements, and the morphological shape. Among these natural hazards observed, Tchindjang (2013); Ndille and Belle (2014) identified earthquakes, volcanoes, floods, landslides, and drought, with volcanism and landslides appearing the most preponderant. The spatiotemporal distributions are seen as follows: the most recent 2019 Gouache landslide in the West Region, 2016 Eseka train accident in the Centre Region, the volcanic eruptions of Mount Cameroon in 1982, March/April 1999 and May 2000, about seven major floods: Kribi (1998), Lagdo, Maga (1998), Far North (Diamare, 1996, 1998, 1999) and Limbe (2001), over nine major landslides: South West (Bafaka Balue, 1997), Centre (Yaounde, 1998), 20 fire disasters: Nsam (1998), Bafoussam market (1999), Mokolo (1998), Limbe and three plane crashes (Bang, 2009).

In the Western Highlands region of Cameroon lies a volcanic zone characterized by many faulted structures and, above all, it constitutes the most active volcanic region in Cameroon (Tchindjang, 2018). In this region, volcanic, floods, and landslide activities constitute the most geomorphic hazards in recent decades with the preeminent one being the 2003 Wabane landslide that took the lives of 22 persons and the 2012 Babessi floods (Tchindjang, 2013; Balgah et al., 2015). In the North-Western part of the region lies the Bamenda escarpment, which is the most hazardous zone along the North-West stretch of the Cameroon volcanic line (Afungang et al., 2017).

DOI: 10.4324/9781003275916-10

The study area, Boyo-Menchum Basin (BOMBA), has an outstanding geographical landscape with crater lakes among which are Lake Nyos, Lake Wum, Lake Ilum and Lake Kuk. Lake Nyos, located 1200 m high in the northern flanks of Mount Oku, is normally deep blue, and peaceful. On one Thursday evening, 21 August 1986, a gas outburst occurred at this idyllic lake that made it known throughout the world as a "killer Lake" (Michele et al., 1987). Members of families and animals were killed and others were left helpless and confused. Over 1,800 persons and 3,000 cattle died near the Lake and along drainage up to 10 km semi-circumference northward away from the Lake. The birds and insects populations were significantly reduced for at least 48 hours, but plant life remained essentially unaffected (UNDP, 1987; Michele et al., 1987).

For 35 years, the local population, the Cameroon government alongside foreign bodies have continued to strengthen their efforts to address the Lake Nyos post-disaster issues in a proactive way. However, these have yield minimal successes as communities especially those that were devastated in the Lake Nyos vicinity and those around the resettlement camps have resorted to indigenous technological knowledge in order to prepare for, cope with, or survive the disaster alongside the scientific technics. This is because the coping and managing strategies of the Lake Nyos gas disaster by the state failed to incorporate the long-term sustainable unique characteristics that are peculiar to the local population and to their environment. It is from this context that this chapter seeks to bring to focus to the indigenous technological resilience and scientific strategies integrated to control and cope with in the post-disaster landscape and the return of the survivors.

8.2 Study Area and Methods

8.2.1 Location of the Study Area

Boyo-Menchum basin consists of the section of both the Boyo and Menchum Divisions that witnessed the Lake Nyos gas disaster. These are two Divisions out of the seven administrative Divisions that make up the North-West region of Cameroon. At the sub-divisional level, the study involves three subdivisions, viz. Wum Central, Fungom, and Fonfuka subdivision. The Basin is found within the Bamenda Highlands and is home to several volcanic lakes, viz. Lake Nyos, Lake Wum, Lake Ilum, Lake Kuk and Lake Benakuma. It is situated between latitude 6°20′0″N–6°50′8″N of the Equator and longitude 10°00′E–10°30′E of the Greenwich Meridian.

This basin consists of seven different resettlement camps, viz. Kimbi, Buabua, Ukpwa, Esu, Komfutu, Ipalim, and Yemnge, and three disaster affected villages (Subum, Nyos and Chah) all in the band of the Boyo and Menchum Divisions (Figure 8.1).

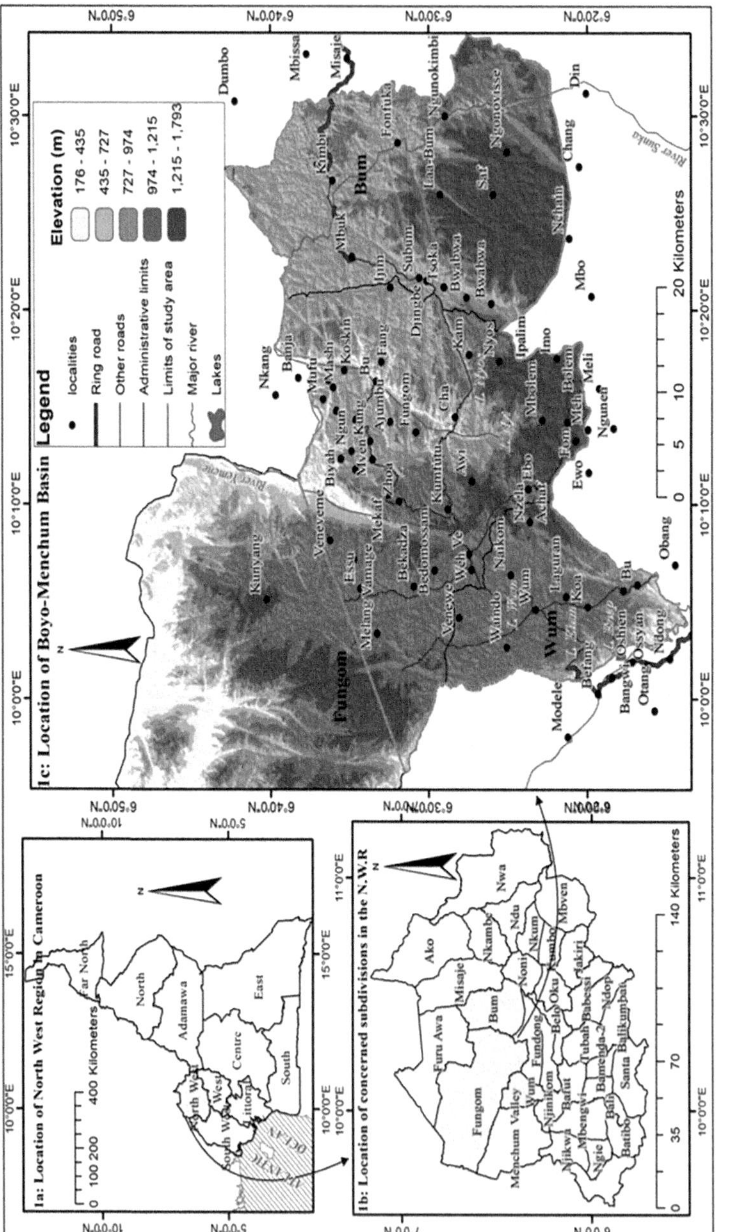

Figure 8.1 Location of the Boyo-Menchum Basin

8.3 Methods

Initially, contacts were established with the representatives of the resettlement and self-relocation communities, leaders of cultural associations, disaster victim representatives as well as some renowned politicians, chiefs who helped to identify priority issues in relation to indigenous technics. Secondary data were collected from published and unpublished documents related to the study, administrative records, workshop reports, institutional reports, speeches, formal complaints, projects reports, and conference papers. Some were accessed in websites through requests from the administrators concerned. Minutes concerning the planning of the emergency stage of the disaster in 1986 were taken from the Menchum Presbytery archives. Information from BUCREP and PNDP and Wum Central, Fonfuka, and Zhoa Councils was also vital for the acquisition of population data as far as the human landscape was concerned.

A field survey was carried out and 397 questionnaires were conducted at ten villages researching the relocation, un-resettled, and the resettlement villages. Transect works, focus group discussions, and participatory rural appraisals were applied to better understand the disaster landscape and the Global Position System (GPS) was used to collect waypoints and altitudes of study area were not left out. The analysis permitted the identification of varied local and scientific technics specific to the Boyo-Menchum Basin. The obtained data were analysed using statistical tools such as Microsoft Excel and open source Quantum Geographical Information System (QGIS 2.18) use to establish the digital elevation points of the study area.

8.4 Improvise and Scientific Resilient Strategies to Post-Disaster Management in Cameroon

8.4.1 Manifestation of the Lake Nyos Gas Disaster

According to the field evidence, on Thursday 21 August 1986, the population of Nyos, Chah, Subum, Mbonge, and Fang woke up to harrowing images. A huge explosion of water mixed with carbon dioxide gas occurred killing over 1800 people, over 3000 cattle, flooded the lake buffer zone, submerged farm parcels around the lake and destroyed other sources of livelihood. Topographically, Lake Nyos, the epicentre of the disaster, is located about 1114 m above the sea level. Nyos Valley (940 m), Chah (600–750 m), and Subum (750–900 m), the devastated villages, were all areas located in the downstream of the Lake and in relatively lower altitudes. During the explosion, the CO_2 gas relatively heavier than the surrounding air accompanied with water droplets flew down to these valleys with the aid of gravity. This justifies the reason why some hill tops and relatively high location villages were able to survive the disaster. Nyako, a farmer located just about 100 m up the Lake with his herds, was not affected as well as some quarters such as Pe-tsi, Pale-Nyos, Kwe-Nyos, Abar and the greater section of Fang located on top of hills though closer to

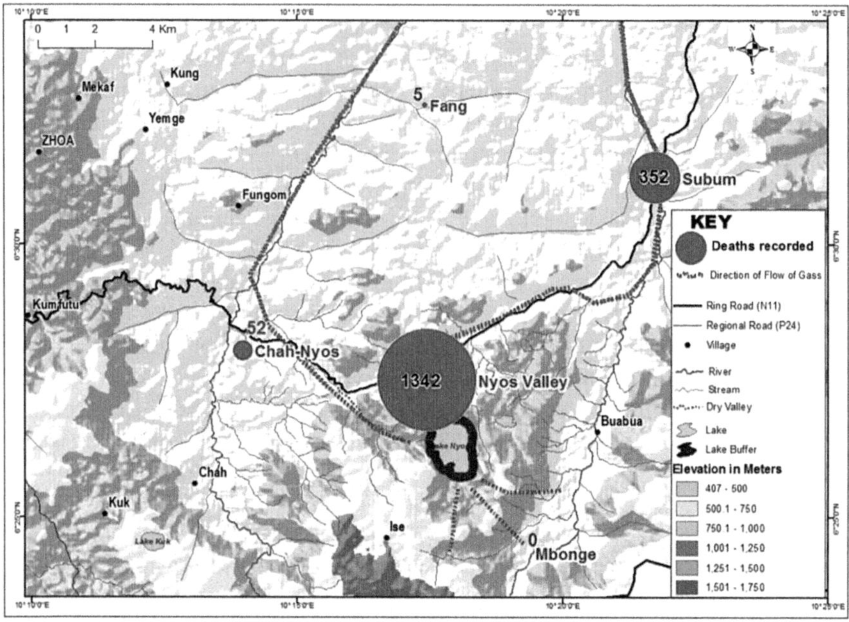

Figure 8.2 Traces of gas flow and death toll per village across the study area in 1986

Source: Field work 2021

the Lake were not also affected. Baxter (1986) further posits that the best estimate to better understand the flow of the gas as conditioned by the relief was based on the location of human corpses, cattle, and wild animal carcasses whose distribution covered an area about 15 km wide and 25 km long following the valley.

In terms of impacts, the 1986 Lake Nyos gas disaster was highly population related because of the huge victims of the population in Nyos, Chah, and Subum who were highly affected and the death tolls were disproportionately enormous per village: about 1342 (Nyos), 352 (Subum), and 52 deaths in Chah.

These nearly 2000 deaths in aggregate were attributed to the effects of the gas disaster in the study area. Insight from the population revealed that, these villages were significantly reduced to zero population as from 1986 until the year 2000 when the first farmer arrived at Nyos (Figure 8.2).

8.4.2 Indigenous Resilient Strategies to the Lake Nyos Gas Disaster

Local knowledge after the Lake Nyos gas disaster has quickly gained ground. This knowledge is varied and judged based on its origin, its relation to their natural environment and the history passing on from generation to generation. Table 8.1 presents varied indigenous strategies used by the local population to adapt to post-disaster situations in the Boyo-Menchum Basin.

Table 8.1 Indigenous improvised safety strategies to the Lake Nyos gas disaster

Variables	*Scaling (%)*				
	SA	*A*	*D*	*SD*	*U*
Physical observation of the state of livestock and aquatic species	57.7	17.1	5.3	10.1	9.8
Relocations of settlements on hilltops	40.8	26.4	17.6	9.1	6.0
Sensitivity of the population to strange sounds, and unusual smells	36.5	19.6	20.9	18.4	4.5
Pouring of libations and incantation by the traditional rulers	26.7	26.7	30.2	9.3	7.1
Uncoordinated yelling of the population found around the lake	32.0	31.2	23.4	10.1	3.3
Leaking of palm oil believed by the local population to boost the immune system	27.7	25.2	27.5	14.6	5.0
The burying of faces on heaps of sand to reduce the concentration of the poisonous gas inhale in case of another gas explosion disaster	33.0	32.5	15.6	14.4	4.5

Source: Field survey, 2020

Key: SA = Strongly agree, A = Agree, D = disagree, SD = Strongly disagree, U = undecided

One of the main adaptation strategies put forth by the local population after the disaster is physical observation of livestock and aquatic species within Lake Nyos and its vicinity strongly agreed by 57.7% of the population (Table 8.1).Fauna species were subdued by the disaster such as reptiles (snakes and African rock pythons), mammals (African palm Civet, rat moles squirrel, cane rats, short horn buffalos, chimpanzee, bushbuck and monkeys), and bird species such as pied crow, grey headed sparrow, swallow, collared sunbird coming from the Kimbi-Fungom National Game Reserve. Fisheries species locally known as *njaa si nguisi* in Mmen language (translated in English as meat from the Lake) and tadpoles were found floating on top of the lake waters even before the outburst of the gas disaster the next day. The local population believed that the temperature of the lake prior to the disaster was hot and accompanied by the traces of the poisonous carbon-dioxide gas. These bi-conditions suffocated the fishes and crabs found on the lake. This became enough of a signal for the local population to conclude that the presence of any other dead animals and fishes in the Boyo-Menchum basin notifies the occurrence of another disaster. This has compelled the returning population to the disaster affected areas to frequently observe the behaviour of the animals around their environs and the fisheries (tilapia) reintroduced into the lake by the state.

During the gas explosion, only villages downstream of the lake (Nyos, Chah and Subum) were devastated while those in relatively higher locations and distant to the lake vicinity (Pale-Nyos, Pere tsi-Nyos, Ise, Bafmen, Konene …) were spared. Today, the majority of the population has resorted to hilltop locations. This is peculiar with the herdsmen and farmers as most affected after the disaster have relocated to hilltops for fear of another disaster. The colonization of hilltops today will aid in saving people in the event of another tragedy, as they were not injured during the gas disaster. Results from field observation have proved that the survivors live in higher altitudes compared to the lake and carry out farming and grazing on the hilltop, slopes, and the valleys.

The population also ascertained that strange sounds, noises, and smells were unusual environmental conditions felt during the occurrence of the disaster (Table 8.2).

Slightly above 34% of the population in the study area believed that the disaster was associated with a noise that could only be compared to that of a distant thunder, while the smallest percentage (1.1%) indicated it was marked by the presence of white smoke covering the lake surface and the immediate river channels. The greater proportions of the population indicate that the disaster was associated with distant thunder attributed to the loud noise produced by the gas as well as the flooding that accompanied the gas during the explosion. This noise was heard far beyond the nearby villages. Field interviews also indicated that this dreaded noise was accompanied by the trembling of the earth. Besides thunder, the localities found at the lower part of the lake were also not exempt from the smell of rotten eggs as revealed by the inhabitants of Subum, Mbonge, Lower Nyos, and Fang located closer to the banks of river Jongha. Written evidence from Freeth *et al.* (1987) also indicated that the majority of the population relate the circumstance to a strong malodour, mostly as gun powder accompanied by loud rumbling noise. To this, the Lake Nyos survivors and the nearby population have become knowledgeable about these unusual elements and hold that on the observation of any of these signs, the population living on relatively lower areas automatically takes refuge on hilltops in order to save their lives.

Other indigenous resilient strategies improvised to manage the Lake Nyos post-disaster in the Boyo-Menchum basin by the population are the pouring of libations and incantation by the traditional rulers, believed to appease the gods, the burying of faces in sand heaps to reduce the concentration of the

Table 8.2 Environmental conditions during the disaster

Conditions	*Frequency*	*Percentage*
Strong winds	18	4.5
Distant thunder	137	34.5
Smell of gun powder	96	24.2
Smell of bad egg	121	30.5
White smoke	4	1.1
No prior signal	21	5.2
Total	397	100.0

Source: Field survey, July 2020

poisonous gas inhalation, use of palm oil believed by the local population to boost the immune system and that palm oil has anti-poison properties. The population believes that after processing, palm oil remains pure and does do not mix with any other substance. Thus taking in palm oil will help occupy the space formerly filled by the poisonous gas and other substances and force the poisonous gas to come out through vomiting. Also, the uncoordinated yelling of the population found around the lake by the population also denotes something unusual on the Lake and is thus a signal for the entire community to climb mountains or relocate further away from the Lake and its outlets before proper investigations.

8.4.3 Scientific Resilient Strategies to Post-Disaster Management in the Study Area

Alongside the practising of indigenous adapting strategies to post-disaster, scientific adaptations strategies also co-exist in the Boyo-Menchum basin.

8.4.3.1 Degasification of Lake Nyos

After the occurrence of the Lake Nyos gas disaster, scientists for a while become speechless. Evidence from the International Association of Volcanology and Chemistry of the Earth Interior (2016) revealed that an international conference was held by Cameroon government and UNESCO for the world scientists to determine the cause of the disaster. After the international conference, the scientists came to a conclusion that the disaster resulted in the explosion of carbon dioxide gas (280×10^6 m^3 of dissolved CO_2) accumulated beneath the Lake waters, also that the concentration of CO_2 in Lake Nyos has increased from January to March 2001. To dispose of this gas, the first degassing programme was initiated and implemented in Nyos in 2001. A 203 m long pipe which runs to the bottom of the lake and allows the CO_2 concentrated Lake waters to escape at a regular rate and the other two degassing programme were completed in 2011 supported by the Cameroon government, the United State of America, Japan, Belgium and France.

Between 2004 and 2021, it was discovered that the degassing project's height consistently decreased, as seen by the water jets shooting out of the lake's surface. Further information from the stakeholders revealed that at the beginning of the degasification process, the gas venting cowls propelled high jets of water above 50 m, which proves the huge quantities of gas trapped at the bottom of the lake by the degassing pipes and the low heights of the water jets today show the positive advances of the degasification today. These jets are spontaneously fractioned (Halbwachs et al., 2004) to help dissolved the salts, mainly siderite (iron carbonate-$Fe^{++}CO_3^-$) precipitate forming solid particles of iron hydroxide which sink down to the bottom of the Lake as well as the CO_2 still dissolving from the water, which extracted through the pipes is liberated to the air. Still according to Halbwachs et al., 2004), the water that falls on the surface of the lake having lost its dissolving gasses and irons, has a low density and mixes easily with the surface water. Stratification occurs when water drawn from deep seas is released to the surface, thus reducing the concentration of CO_2 in the lake.

8.4.3.2 *Installation of an Automatic Alarm Beside the Lake*

A solar powered monitoring alarm system was also installed to monitor carbon dioxide levels in the lake waters ensuring that if the Lake is to explode again, this alarm will alert the population an hour before about either strange happenings or a further disaster occurrence. Information from focus group discussions further revealed that a signal from this alarm is responded to by further yelling of the population as they take to the direction of the hill for collective safety and for others who are farther away from the lake to also be aware. Some inhabitants of this locality have a negative perception about this monitoring alarm as it is sensitive to all human and natural activities. This is because during the dry season, even fires set by either farmers because of transhumance or hunters cause this automatic alarm to produce a signal thereby causing the population to flee their homes for not reasons linked to the risk of disaster.

8.4.3.3 *Reinforcement of the Natural Dam to Retain the Lake Waters*

The United Nations Development Programme (UNDP), Japan Science and Technology Agency (JST) and the Japanese International Corporation Agency (JICA) found that if the dam retaining waters from the lake collapsed, over 50,000,000 m^3 of water will be released which may further lead to another catastrophe. Thus, the reinforcement of the natural dam with the construction of a spillway has been carry out by the European Union. This helped to reduce surface erosion of the dam and desolation of soft rocks by the water that constantly flows on the surface of these rocks.

8.4.3.4 *Organisation of Civil Protection Safety Campaigns with the Affected Population*

Social outreach activities to over 275 people around the Lake were organised on the manifestation of the gas disaster and participants trained on what measures to take in the event of such a disaster. These include groups such as the women group leaders, religious groups, common initiatives groups, farmers groups, social welfare groups, and students based on what happened at Nyos, what measures were taken to manage the disaster, what kind of signs indicate danger from carbon dioxide gas, what to do if you observe signs of danger. Also, automatic lake observation equipment was installed to observe the behaviour of the lake waters after the disaster. These plethoric measures have gone a long way to restore hopes to the local population and thus the resurgence of the survivors back to the disaster-affected areas of Nyos, Chah, and Subum villages.

To reduce the risk associated with traveling from the study area to the capital of Cameroon (Yaounde), additional measures were taken. These included the installation of two degassing cones, a side valve, a shut-off valve, a 10-metre-deep Diabolo, an aluminium raft, a depth meter, ten pressure sensors, two temperature sensors, a speed sensor and data storage. There was also the ability to view images captured by two webcams in Chambery or Yaounde, and the ability to transmit data gathered by the Inmarsat system regarding the situation of the lake to Yaounde.

8.4.4 Unmet Resilient Strategies to the Post-Disaster Management in the Study Area

Nigerian experts arrived with a machine to totally demolish the little barrier holding the lake in place, allowing the lake's water to entirely drain out and reducing the chance of flooding in the future. This was because the Nigerian scientists believed that the outflow of the Lake waters will help to keep under control the further accumulation of the carbon dioxide gas. Also, another proposal was that a drilling machine should be stationed at the Lake site to constantly drain the Lake's waters to reduce it to it a minimum. This was to control flood, reduced the CO_2 gas accumulation as well as to completely limit the volume of water that spilled over the tiny natural dam returning water from the lake.

8.5 Discussion

During a disaster, people are disproportionately killed with those at the lower altitudes (valleys and river banks) experience severe damages and tolls. These findings are in line with that of Michele et al. (1987) and Freed et al. (1987) who found that during the Lake Nyos gas disaster, Nyos, Chah, and Subum located at lower altitudes compared to the lake experienced great damages and tolls. Upper Nyos, located uphill above the Lake, was safe from the disaster devastation. Also this study is in line with that of Mbanga et al. (2021) on post-disaster settlements dynamics but differs in that in a bid to prevent disaster devastations, the population choose to relocate in the uplands to avoid the flow of gas and water should in case another disaster occur. The relocation of the population to a relatively geophysical position is also in line with the findings of Mavhura et al. (2013) who state that the extent to which indigenous knowledge enhanced resilience to floods was influenced by geophysical locations, exposure to flooding, and socio-economic abilities. Enormous scientific mitigations strategies are put in place in Lake Nyos to prevent the further occurrence of another gas disaster. These findings neatly fit into the reports of the UNDP (2010), which highlighted the installations of degassing pipes at Lake Nyos and the reinforcement of the natural dam (Figure 8.3).

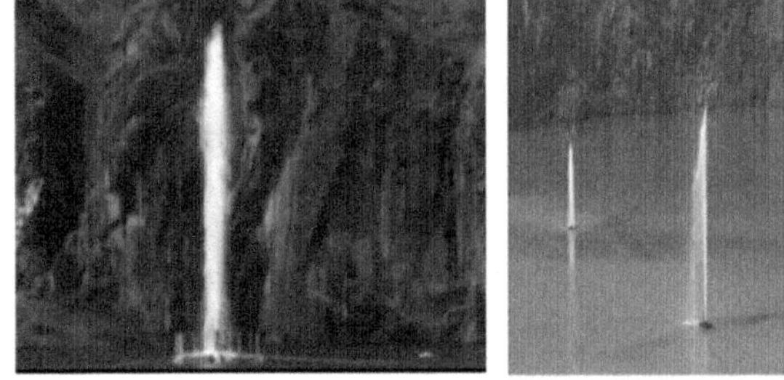

Figure 8.3 Trends of water jets from the Lake Nyos degassing programme

8.6 Conclusion and Recommendations

The focus of this chapter was to examine the manifestations of natural disasters, and bring to focus both the local and scientific resilient strategies used to adapt to disaster devastations and anticipated disaster occurrence. The geography, geology, and human imprints have exposed Cameroon to varied natural and human-induced disaster. For three and a half decades, disaster occurrences have maintained an increasing trend with devastation to infrastructure and animals. Local and scientific resilient measures are blended to better manage post-disaster situations and mitigate further disaster occurrences. The survivors of the Lake Nyos gas disaster have resorted to being sensitive to strange sounds, and unusual smells, relocations of settlements on hilltop, physical observation of livestock and aquatic species, pouring of libations and incantation by the traditional rulers, and the uncoordinated yelling of the population found around the Lake while on-going scientific improvement consist of the degasification of the Lake, installation of a solar alarm to detect strange happenings in the Lake, and the construction of a spillway over the natural dam to reduced surface erosion. These scientific moves towards disaster management in the Boyo-Menchum basin are influenced by the West and not familiar to the local population. Thus, promoting the local adaptation strategies alongside the use of local resources by the local population to mitigate natural disasters will help to palliatively ameliorate and reduce the dynamic risk of another disaster.

Bibliography

Afungang, N.R., Nkwemoh, A.C. & Ngoufo, R. (2017). Spatial Modelling of Landslide Susceptibility Using Logistic Regression Model in the Bamenda Escarpment Zone, North West Cameroon. *National Journal of Innovative Research and Development* *6*(12), 187–199.

Balgah, A.R., Buchenrieder, G. & Mbue, N.I. (2015). When Nature Frowns: A Comprehensive Impact Assessment of the 2012 Babessi Floods on People's Livelihoods in Rural Cameroon. *Jàmbá: Journal of Disaster Risk Studies*, *7*(1), 8.

Bang, H. (2009). *Natural Disaster Risk, Vulnerability and Resettlement: Relocation Decisions Following Lake Nyos and Monoun Disasters in Cameroon*. Published PhD Thesis, School of International Development, University of East Anglia UK, 314.

Baxter, J.P. (1986). Medical Report of British Technical Team on Visit to Lake Nyos, 19.

Freeth, J.S., Kay, F.L.R., & Baxter, J.P. (1987). Reports by the British Scientific Mission Sent to Investigate the Lake Nyos Gas Disaster, 86.

Halbwachs, M., Sabroux, J.-C., Grangeon, J., Kayser, J. G., Tochon-Danguy, J.-C., Felix, A., Beard, J.-C., Villevieille, A., Vitter, G., Richon, B., Wuest, A. & Hell, J. (2004). Degassing the "Killer Lakes" Nyos and Monoun, Cameroon. *EOS*, 85(30), 281–288.

International Association of Volcanology and Chemistry of the Earth Interior. (2016). Commission on Volcanic Lakes CVL9-Cameroon, 14–24 March 2016 "30 Years After the Lake Nyos Disaster" Scientific Program, 14p.

Mavhura, E., Manyena, B., Collins A., & Manatsa, D. (2013). Community Resilient to Disaster in Muzarabani, Zimbabwe. *International Journal of Disaster Risk Reduction*, 5, 38–48.

Mbanga, J., Amoako, D.G., Abia, A.L.K., Allam, M., Ismail, A. and Essack, S.Y. (2021) Genomic Analysis of Enterococcus spp. Isolated From a Wastewater Treatment Plant and Its Associated Waters in Umgungundlovu District, South Africa. *Front. Microbiol.*, 12, 648454. doi: 10.3389/fmicb.2021.648454

Michele, L., Michael, A., Compton, R.H., Devine, D.J., William, C., Alan, M., Kling, W.G., Koenigsberg, E.J., Lockwood, J.P., & Wagner, N.G. (1987). Final Report of the United States Scientific Team to the Office of U.S. Foreign Disaster Assistance of the Agency for International Development, 39.

Ndille, R., & Belle, J. A. (2014). Managing the Limbe Floods: Consideration for Disaster Risk Reduction in Cameroon. *International of Risk Science*, 5, 147–156.

Tchindjang, M. (2013). Mapping of Natural Hazards in Cameroon. *Actes de la 26ème International Cartographic Conference (Dresde, Allemagne), Buchroithner MF.* Édit. ICC, 13 p

Tchindjang, M. (2018). Lake Nyos, a Multirisk and Vulnerability Appraisal. *Geosciences*, *8*(312), 29.

UNDP (United Nations Development Program). (1987). *Assistance in the Organization of International Conference on the Lake Nyos Disaster*, Final Reports No. 86, Cameroon, 59.

UNDP (United Nations Development Programme). (2010). *Human Development Report 2010: The Real Wealth of Nations: Pathways to Human Development*. New York.

9 Depopulated Region in Mirik Municipality of Darjeeling District, West Bengal

A Demographical Analysis

Ershad Ali and Bipul Chandra Sarkar

9.1 Introduction

Depopulation means a shrinking population that is strongly linked to an aging population. It is determined by a birth deficit with deaths and/or a negative net migration (Reynaud & Miccoli, 2018). Some cities in the world have experienced negative growth rates. Detroit in Michigan, New Orleans in Louisiana, and Leipzig in Germany are some prominent examples of such shrinking cities. Similar to these cities, the city of Mirik in West Bengal, India, is now experiencing a negative growth rate, but in a particular ward (Ward No. 2). Though the population and density of the population in Mirik Municipality have deteriorated in recent years, there is one ward (Ward No. 2) that is experiencing a negative growth rate. The name Mirik is derived from the Lepcha words Mir-Yok, which means something like a place burned by fire. The Mirik Municipality has an area of 6.25 km^2 and is located at an altitude of 1700 m above sea level; it is a small hamlet. This Mirik community was founded in 1984 as the Mirik Notified Area. From the land of the Thurboo Tea Garden and the land of Khasmahal by agreement between the State Govt. and Private Tea Company, this community was agglomerated with nine wards. Mirik Bazar started as the region's trading center, where people from the surrounding villages and tea gardens used to come to trade and buy their necessities. In today's garden area there was a playground where the British officers played polo. In 1969, the West Bengal Tourism Department began acquiring 335 acres of land from the neighboring Thurbo Tea Estate. The transformation of this land into a tourist destination began in 1974 when Mr. Siddhartha Shankar Ray was the Chief Minister of West Bengal. The tourist site, which included the newly built lake (called Sumendu Lake) and the day center, was inaugurated in April 1979 by the next Chief Minister of West Bengal, Mr. Jyoti Basu (Draft Development Plan of Mirik Municipality, 2008–2013). With the blossoming of tourism, the village of Krishnanagar at the other end of the lake with hotels and restaurants for tourists is being developed.

DOI: 10.4324/9781003275916-11

9.2 Aim and Objective

The area chosen for the study has an important cultural region in which each of the cultural elements is closely linked. The urbanization in the Mirik Municipality has a unique meaning in the sense that it is a touristic place where people come from different places. Historical development and records clearly show that this city has significant potential to become a wealthy hill station. The aim of the field study is therefore to find out the reason for the negative growth rate and population density of a ward of the Mirik Municipality.

9.3 Study Area

The Mirik Municipality is located in the Mirik sub-division of the Darjeeling district (West Bengal, India) in the hilly area of the Sikkim-Darjeeling-Himalayas. Geographically, it is located in between 26°54′N to 26°57′N and 88°10′E to 88°13′E.

9.4 Database and Method

The urbanization information regarding population growth, population density, etc. was collected from primary and secondary sources. Primary data come from sample surveys (scheduled questionnaire) and secondary data come from census data of the Mirik Municipality , 2001 and 2011, and the draft of the Mirik Municipality development plan, 2008–2013. The data were analyzed using various statistical and mathematical formulas like-

$$\text{Population density} = \frac{\text{Total population in the ward}}{\text{Total Area}} \times 100,$$

$$\text{Decadal Growth Rate} = \frac{\text{Population in earlier Census Year}}{\text{Population in later Census Year}} \times 100,$$

and also interpreted in terms of their relevance. Charts were prepared on Microsoft Office Excel 2007 and maps on GIS software. The statistical technique is also used to determine the change in concentration of the urban population, i.e., the negative population growth of a ward in the Mirik Municipality.

9.5 Results and Discussion

Depopulation is a process in which the population density and size of an area steadily decrease over time.

9.5.1 Growth of Population of Mirik Municipality in 2001 to 2011

According to the 2011 census, the population of Mirik Municipality was 11,513 (of which 5,688 were men and 5,825 were women) compared to 9,112 in

Table 9.1 Ward-wise decadal growth rate of population in Mirik Municipality, 2001–2011

Ward No.	*Population in 2001*	*Population in 2011*	*Number of increasingpopulation*	*Decadal growth rate in %*
1	831	1113	282	25.34
2	1414	1247	−167	−13.39
3	1421	1516	95	6.27
4	772	1096	324	29.56
5	591	857	266	31.04
6	720	982	262	26.68
7	1032	1319	287	21.76
8	1270	1856	586	31.57
9	1090	1176	86	7.31
Total	9112	11,513	2401	20.85

Source: Census of Mirik Municipality, 2001–2011

the 2001 census (District Census Handbook of Darjeeling, 2011). Table 9.1 clearly shows that the population growth rate of the Mirik Municipality (+) is 20.85%.

Figure 9.1 shows the district-related growth of the municipality of Mirik, with every ward (1, 3, 4, 5, 6, 7, 8, and 9) showing a positive trend between 2001 and 2011, except for Ward No. 2. But at Ward No. 2 there is a negative growth rate. The highest rate of population growth is found in Ward No. 8 (31.57%), followed by Ward No. 5 (31.04%), 4 (29.56%), 6 (26.68%), 1 (25.34 %), 7 (21.76%)), 9 (7.31%), and (6.27%), respectively. In addition, there is a negative population growth rate in ward 2 (−13.39%).

9.5.2 Population Density of the Mirik Municipality in 2001 & 2011

Population density measures the variation in population distribution (Khullar, 2014). It is expressed as the number of people per unit area. According to the 2001 census, the population density of the municipality of Mirik is 1862 people/km^2 and 2154 people/km^2 in the 2011 census, i.e., 292 people/km^2 will be added in one decade for all wards (Table 9.2). Compared to some hilly cities like Darjeeling (15,554 inhabitants/km^2), Kalimpong (5393 inhabitants/km^2), which are in the same physiographic region of the lower Himalayas, the municipality of Mirik has a relatively low density of further development due to its historical background.

9.5.2.1 Density Distribution in 2001

All wards of the Municipality are moderately dense. Based on the 2001 census, the Mirik Municipality can be divided into four density zones. These are High-Density Zone, Medium-Density Zone, Low-Density Zone, and Very Low-Density Zone (Figure 9.2a).

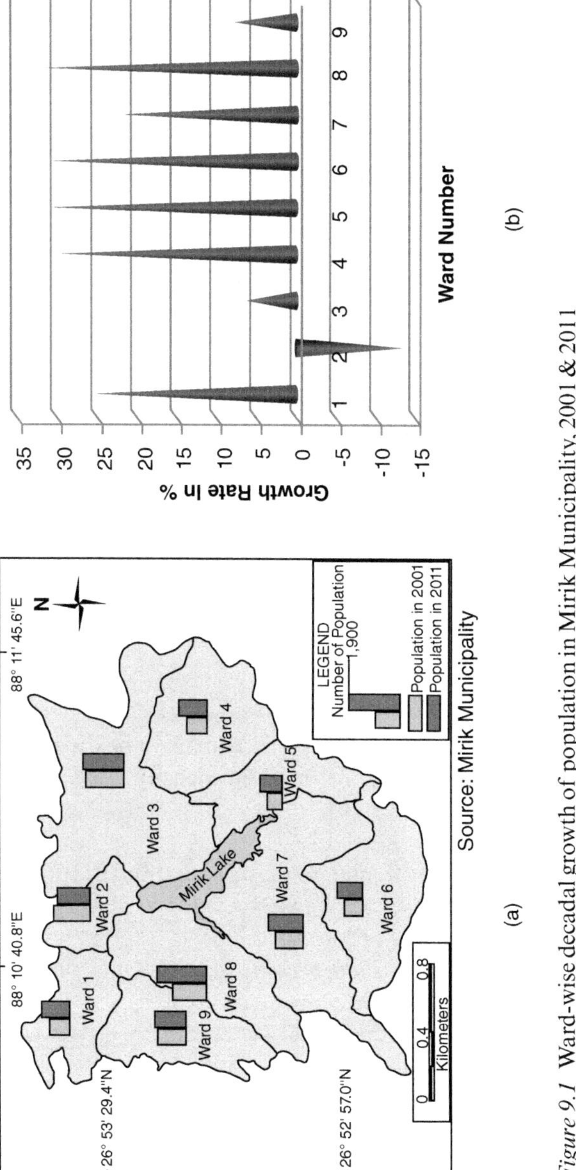

Figure 9.1 Ward-wise decadal growth of population in Mirik Municipality, 2001 & 2011

Table 9.2 Ward-wise population density of Mirik Municipality in 2001 and 2011

Ward No.	*Area (km²)*	*Population in 2001*	*Density in 2001 (per km²)*	*Population in 2011*	*Density in 2011 (per km²)*
1	0.43	831	1955	1113	2619
2	0.24	1414	5860	1247	5166
3	1.33	1421	1072	1516	1144
4	0.72	772	1075	1096	1527
5	0.45	591	1326	857	1922
6	0.74	720	977	982	1333
7	1.18	1032	876	1319	1120
8	0.74	1270	1726	1856	2522
9	0.58	1090	1887	1176	2036

Source: Census report of Mirik Municipality, 2001 and 2011

- *Very High-Density Zone (>2000 persons/sq.km)*: In 2001, only 1 ward comes under this category i.e., Ward No. 2 which was 5860 persons/sq.km.
- High Density *Zone (1500–2000 persons/sq.km)*: 3 wards are coming under this category. These are Ward No. 1(1955 persons/sq.km), 9 (1887 persons/sq.km) and 8 (1726 persons/sq.km).
- *Moderate Density Zone (1000–1500 persons/sq.km)*: 3 wards are under this group. These are Ward No. 5 (1326 persons/sq.km), 4 (1075 persons/sq.km) and 3 (1072 persons/sq.km).
- *Low Density Zone (<1000 persons/sq.km)*: 2 wards are under this category. These are 6 (977 persons/sq.km) and 7 (876 persons/sq.km).

9.5.2.2 Density Distribution in 2011

Based on the 2011 census, the municipality can again be divided into 4 density zones. These are high-density zone, medium-density zone, low-density zone, very low-density zone (Figure 9.2b).

- *Very High-Density Zone (>2500 persons/sq.km)*: In 2011, three wards come under this category. These are Ward No. 2 (5166 persons/sq.km), 1 (2619 persons/sq.km), and 8 (2522 persons/sq.km).
- *High Density Zone (2000–2500 persons/sq.km)*: Only one ward comes in this category, i.e., Ward No. 9 (2036persons/sq.km).
- *Moderate Density Zone (1500–2000 persons/sq.km)*: two wards are come under this category. This are Ward No. 5 (1922 persons/sq.km) and 4 (1527 persons/sq.km).
- *Low Density Zone (<1500 persons/sq.km)*: three wards are come in this category. These are Ward No. 6 (1333 persons/sq.km), 3 (1144 persons/sq.km) and 7 (1120 persons/sq.km).

9.5.3 Depopulation in a Ward of Mirik Municipality and Its Reason

The concept of population change is often used to refer to the change in the population of territory over a period of time, regardless of whether the fact is

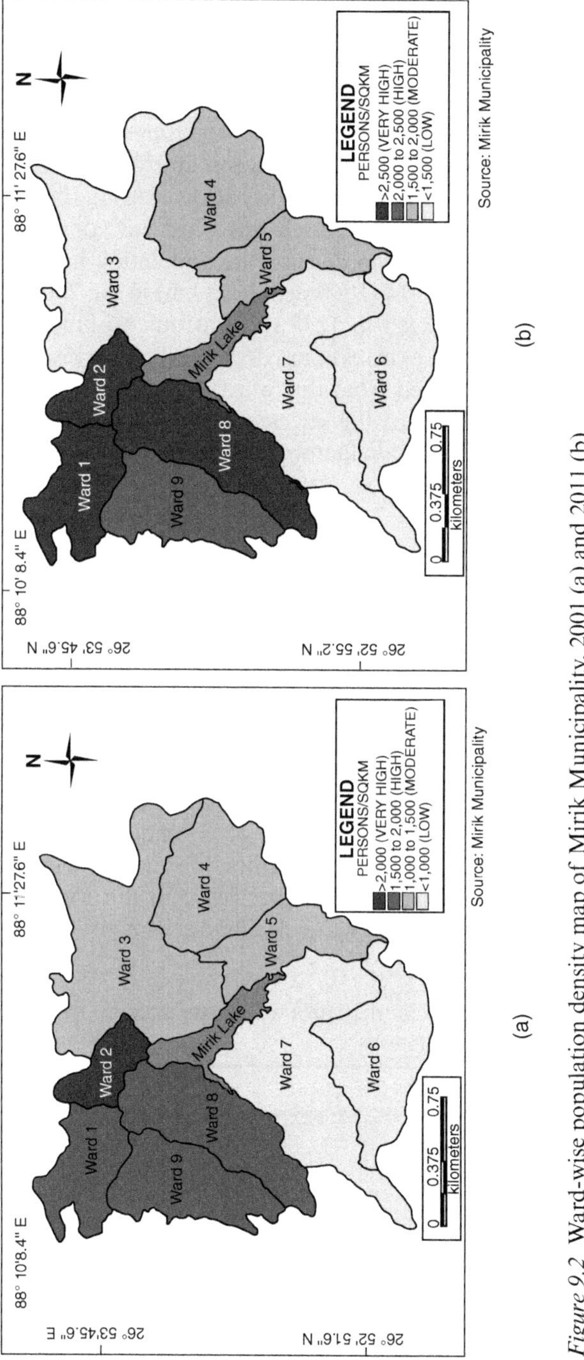

Figure 9.2 Ward-wise population density map of Mirik Municipality, 2001 (a) and 2011 (b)

positive or negative (Chandna, 2015). Resistance to urbanization in general and resistance to urban growth, in general, indicate rising land values and reluctance to provide land for the poor in urban areas (Ramachandran, 2015). One of the keys to analyzing depopulation is understanding that the nature of the problem has fundamentally changed. In the later years of the 20th century, emigration due to negative natural growth was no longer the main cause of the demographic decline in many cities (Collantes et al., 2013). This affects the size of the community. The Mirik Municipality in Darjeeling district has seen a negative growth, Ward No. 2. Although this ward has consistently taken its first place in terms of population density, the population has declined. Ward No. 2 has a population of 1414 (5860 people/sq km) in the 2001 census, but in 2011 the population decreased to 1247 (5166 people/sq km), i.e. 147 people were reduced (694 people/sq km). km are decreases). A survey was carried out among those who had moved from this ward to another to find out the real reason for the depopulation of the selected ward. According to the response perceived on this survey, the main factors behind depopulation are listed below.

9.5.3.1 Migration

From the primary survey, it was observed that the population of Ward No. 2 has decreased mainly due to emigration (Ali, 2018a) Human migration is the movement of people from one place to another with the intention of settling permanently or temporarily in a new location. People can migrate as individuals, in family units, or large groups from one city to another city or from one community to another within the city (Bhende & Kanitkar, 2010). A primary survey was conducted in this relevance and the results of this survey are dramatic. Table 9.3 shows that many families were transferred from Ward No. 2 to different wards. We can therefore speak of a tendency towards inner-city migration. Due to the lack of private sectors, jobs in the Mirik Municipality are very limited. Migration is also taking place in the Mirik Municipality. Public health and higher education are also inadequate. In order to have better

Table 9.3 Factors for depopulation of studied ward of Mirik Municipality

Sl No.	*Factors*	*% Of Vote (%)*
1	Migration	77
2	Congestion	38
3	Lack of drinking water	22
4	Poor drainage system	14
5	Expensive life style	09
6	Lack of proper urban planning	03
7	Landslide	02

Source: Primary survey

(Vote = % of respondents support the reason)

Table 9.4 Intra-urban migration from Ward No. 2

Number of family	*Year of shifting*	*Shifted Ward No.*
02	2003	1
03	2006	4
02	2009	5
03	2010	8

Source: Primary survey

health and medical facility and better educational facility, some of the residents are emigrating from both Ward No. 2 and Mirik Municipality Table 9.4.

9.5.3.2 Congestion

In recent years the problem of increasing road traffic has attracted more public attention than any other problem in urban areas (Hudson, 1981). It leads to the destruction of a city's buildings and populations. In the municipality of Mirik, the settlement pattern of the municipality has grown arbitrarily and unscientifically, which can lead to an overload of the municipality. While there is a sufficient path to get to Ward No. 2, most of the roads are very narrow, which makes the area congested and unsanitary. The Mirik bazaar or supermarket is another major reason for the congestion of such a ward.

9.5.3.3 Lack of Drinking Water

Mirik town is famous for its lake, namely Mirik Lake or Sumendu Lake, which is located in the central part of the municipality. Besides this lake there are several streams (jhoras) like Kali Jhora (between the boundaries of Ward Nos. 7 and 8), another one is between the boundary of Ward Nos 8 and 9 and another is from Devisthan to Kowlay Road. Roughly speaking, Mirik Lake is the most important body of water in this city (Table 9.5), but the water quality of Lake Sumendu is very poor. Table 9.5 shows that the seawater contains 1100 mg/l potash and <5 mg/l oil and fat.

Tube wells and streams (jhoras) are the main sources of drinking water in the municipality of Mirik. There is a great shortage of drinking water at Ward No. 2, but there is a lack of pure drinking water in the entire community. Many families have to get their daily drinking water from either Ward No. 3 or Ward

Table 9.5 Water quality of Mirik Lake

Name of the water body	*Date of collection*	*Date of testing*	*BOD (mg/I)*	*TSS (mg/I)*	*Caliform (mg/I)*	*Oil & grease (mg/I)*
Mirik Lake	07.02.2008	08.02.2008	10	30	1100	<5

Source: Draft Development Plan of Mirik Municipality, 2008–013

No. 8. Therefore, many families are dissatisfied with their current living situation on this ward.

9.5.3.4 Poor Drainage System

The municipality of Mirik has a 6.78 km long drainage network in the municipality. The municipality of Mirik suffers from the lack of a suitable drainage network (approx. 75%). Much of the surface runoff from the various districts flows directly into Mirik Lake as it is in the lowest part of the city (Figure 9.3). The waters of Lake Mirik are diverted through Kali Jhora and the rest of the water is diverted through local streams. Most of the drains are Kuchcha (made of a weak material) in nature and a few of the new ones are Pucca (made of a strong material) drains. There is only one ward, namely Ward No. 7 has an extensive drainage network of 3.37 km. The drainage system is severely affected due to its small proportion (0.99 km) of built drainage channels in ward no. 2 (Table 9.6). As a result, in the rainy season, the wastewater from the drainage channel runs off everywhere in the municipality.

9.5.3.5 Expensive Lifestyle

India has made significant strides in reducing the proportion of its population living in poverty, as India officially defined it, over the past quarter-century. In 1983, nearly 46% of the rural population and 41% of the urban

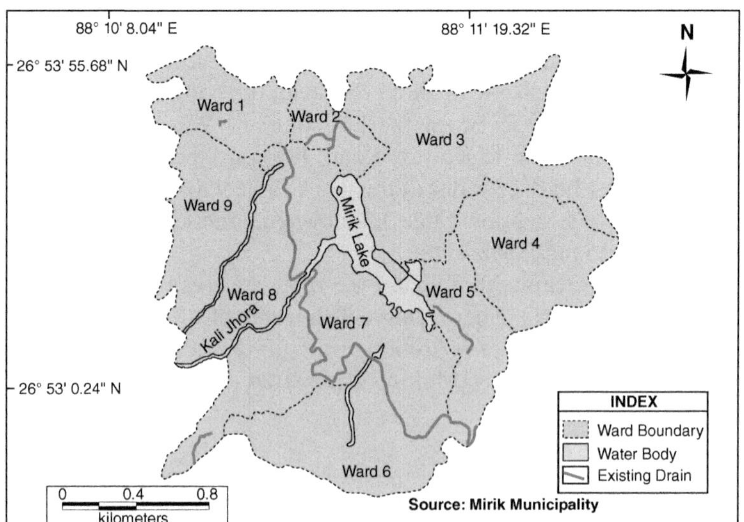

Figure 9.3 Drainage network map of Mirik Municipality

Table 9.6 Ward-wise length of the drains of the Mirik Municipality, 2011

Ward No.	*1*	*2*	*3*	*4*	*5*	*6*	*7*	*8*	*9*	*Total*
Length of drains (in km)	0.08	0.99	0.00	0.00	1.04	0.00	3.37	1.31	0.00	6.78

Source: Draft Development Plan of Mirik Municipality, 2008–2013

population were below the poverty line, and by 2004–2005 the incidence of rural and urban poverty had decreased to 28.3% and 25.7% respectively (Chandrasekhar & Montgomery, 2010). The poor people cannot live in the city center. They can't compare their lifestyle with their high-profile neighbors. In Mirik town, some of the residents who migrated from the ward no. 2 said that they could not sustain their livelihoods because of the expensive lifestyle. Since it is a hill station, the price for each product is far high. In addition, high consumption of daily necessities, the effects of high taxes, etc. make their lives difficult.

9.5.3.6 *Lack of Proper Urban Planning*

Urban planning deals with such programs of different types, whereby a distinction can be made between operational planning (planning for the better operation of the part of the existing city), development planning (with regard to the development of an urban area), and restorative planning (similar to the development, but through the restoration of existing facilities) (Northam, 1979). In the Mirik Municipality, proper urban planning concerning all types of planning is lacking throughout the Municipality, especially in Ward No. 2. Therefore, most of the residents of this ward live in unsanitary conditions. According to Smt. Jayanti Mukherjee, the chairman of the Mirik Municipality (2016), the growth of the Mirik town is not coming as expected due to its poor urban planning.

9.5.3.7 *Landslide*

Ward No. 2 is located in the northern part of the municipality, where the surface drops steeply about 2 km (Figure 9.3). There is also a newly formed sandy bottom that was likely formed in the Carboniferous. Because of this new sandy soil formation, there are sometimes few landslides. The residents of this ward have said that in the northern part of this ward there have been landslides every 4–5 years, although the number of damages is very small. In 2003 there was a landslide on Ward no. 2 and the number of households affected was 04. In 2010 another landslide occurred and 05 households were affected by this hazard (Figure 9.4).

FIGURE 7: POINT OF LANDSLIDE IN MIRIK MUNICIPALITY

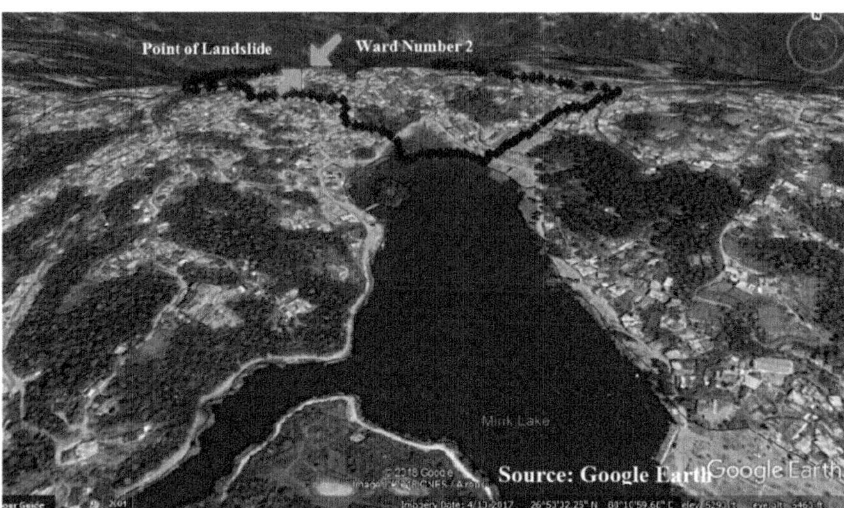

Figure 9.4 Point of Landslide in Mirik Municipality showing in Google Earth

9.5.4 Impacts of Depopulation of Ward No. 2 in Mirik Municipality

1 **Decreased tax base**: As fewer people pay property and other taxes, they have less money to spend on new roads, bridges, power lines, and other public services or to pay off their debts. It is also less economical for the provincial government to maintain schools, health facilities, and other resources in places with shrinking and dispersed populations in the municipality of Mirik. Private companies, including grocery stores, restaurants, and other commercial establishments, are also less interested in operating in areas with declining populations.

2 **Shifting demographics**: The potential impacts of intra-urban migration on long-term population growth are conditioned first and thereby the demographic profile is started to shifting (Pinilla et al., 2008). A major problem is now experienced by Mirik Municipality as it seemingly has suffered loss of its young population of the studied ward. Most of the migration was by young adults, between the age of 15 and 24, or families with young children. Although they leave for a variety of reasons, the most common are: scarcity of drinking water, expensive lifestyle, lack of urban planning.

3 **Social consequences**: The negative growth rate of Ward No. 2 of the Mirik Municipality has an impact on the emotional and social well-being of those who stay alive as well as those who leave them. Many people have lost their friends or family members by leaving the ward. This disrupts social bonds and reduces support from emotional, financial, and others.

4 **Infrastructural consequences**: Shrinking wards, i.e., Ward No. 2, have seen dramatic social changes due to declining fertility, changes in life expectancy, population aging, and household structure. This leads to different requirements of households and represents a challenge for the urban housing market and the development of new land or urban planning. Decreasing the population of a Municipality does not inspire confidence in a city, especially if the city is in a hilly area, and often degrades the morale of the Municipality.

9.5.5 Policy Implications to Depopulated Region of Mirik Municipality

a **Plan-wise settlements**: The housing problem is due to population growth, unplanned industrialization, profit-making, and the widening gap between population and house construction (Mandal, 2000). Infrastructure and housing planning are a must for any city's development. The Mirik Municipality is suffering a lot from the unplanned settlements, especially in Ward No. 2. More improvements and planned living space is needed.

b **Improve proper urban planning**: The urban system needs to be improved in terms of its topography. Each planning region is an organism and had a dynamic unit (Chandna, 2000). Urban health often depends on strong urban planning that can maintain economic balance and control increasing land-use pressures and urban populations (Misra, 2002). So, without proper urban planning, urban health can conquer. Ward No. 2 needs much better planning to reduce depopulation. Since inner-city migration takes place in the municipality, the other district also needs much better urban planning as a cure for further depopulation in the future.

c **Developing poor drainage system**: There is a lack of drainage systems in Ward No. 2 in particular. Drainages need to be much wider and deeper so that the water can drain easily in the rainy season.

d **PHE water**: Some of the hill wards in India are facing the major problem of insufficient water supply. The quality of the water supply is also very poor and unhygienic (Agarwal & Taneja, 2005). The Mirik Municipality is also struggling with the water problem. Therefore, the community has to provide running water two to three times a day. In this regard, the municipality can take advantage of the Jhoras lake and water as it is a potential resource. Since it is a hilly city, the Mirik Municipality can set up a tube in each district and especially focus on congested areas.

e **Economy**: Population and economic development are correlated. Population size, growth rate, and population composition are important factors in determining infrastructures such as food supply, education, housing, and the economy (Sharma, 2008). Such basic amenities are withheld from the Mirik Municipality. The mining towns require large working and storage areas. The inhabitants of the hilly areas suffer from their poor economic situation and yet their income is often limited (Gallion & Eisner, 2005). Most of the people in Mirik Municipality live in the low-income group. Most of the offices and banks are located near Ward No. 5. Therefore, the municipality needs to introduce new rules for the distribution of government offices and banks to stop the urban migration trend.

f **Developed transportation facility**: The traffic jams on the streets of Indian cities must be seen to be believed. The mix of vehicles causes uncontrollable chaos on the streets (Sidhartha & Mukherjee, 2000). Mirik is a popular hill station in West Bengal and the traffic congestion is increasing day by day. So, in terms of traffic, the roads need to be wider so that there is better accessibility. Part of the municipality suffers from poor accessibility due to the lack of road networks. A rail service is not available here (Ali, 2018b). So, if the railroad or toy train can go over Mirik city, it will be a more attractive tourist spot compared to other tourist spots and it will make a name for itself on the map of India.

9.6 Conclusion

From the various results of the study, it can be said that the Mirik Municipality has been of great importance since ancient times. This community has a historical background and has different demographic characteristics. Although the town has high and increasing trade and economic activity based mainly on tourism, the problem of the depopulated region in the Mirik Municipality will be exacerbated in the future when the municipality will act as stagnated work. The increase in human population is certainly a threat to environmental sustainability, but local phenomena of depopulation can also be viewed as threats to local environmental sustainability. Demographic change, effects of emotional and social well-being, the decline in fertility, changes in life expectancy, population aging, household structure, etc. are the likely consequences of depopulation. Therefore, the main problem is depopulation and the municipality must take appropriate steps for future urban policy. However, the challenges facing the region have already been fairly well recognized. It is necessary to conduct a reasonable policy towards this issue and good and efficient organization of activities. An important argument in the process of transformation of this region should be needed the financial support for any remedial project.

Acknowledgement

The authors thank Dr. Indira Lepcha (nee) Lama (Associate Professor), Department of Geography & Applied Geography, University of North Bengal, and Mr. Manas Paul (Assistant Professor), Department of Geography, Rammohan College, Kolkata for their support and guidance.

Bibliography

Agarwal, Siddharth & Taneja, Shivani (2005): All Slums Are Not Equal: Child Health Conditions among the Urban Poor. *Indian Pediatrics*, 42 (Online), pp: 233–244.

Ali, Ershad (2018a): Demographic Characteristics of Mirik Municipality with References to 2001 and 2011 Census Data, *International Journal of Research in Geography*, Vol. 4, No. 1, pp: 50–66.

Ali, Ershad (2018b): Spatio-Temporal Changes of Land Use and Land Values of Mirik Municipality of Darjeeling District, West Bengal, *International Journal of Scientific Research in Science and Technology*, Vol. 4, No. 5, pp: 1228–1243.

Bhende, A.A. & Kanitkar, T. (2010): *Principles of Population Studies*, Himalaya Publishing House, Mumbai, pp: 622–626.

Census of India. (2011): *District Census Handbook Darjeeling*, Directorate of Census Operations, West Bengal, Series- 20, Part XII-B, pp: 9–16 & pp: 36–41.

Chandna, R.C. (2000): *Regional Planning and Development*, Kalyani Publishers, New Delhi, pp: 294–297.

Chandna, R.C. (2015): *Geography of Populations: Concepts Determinants and Patterns*, Kalyani Publishers, Kolkata, pp: 146–158.

Chandrasekhar, S. & Montgomery Mark, R. (2010): *Broadening Poverty Definitions in India: Basic needs and urban housing*. (Online), Human Settlements Programme International Institute for Environment and Development (IIED), pp: 1–47. http://books.google.com/books?id=Q95DvczX-4AC&pgis=1

Collantes, F., Pinilla, V., Seaz, L. & Silvestre, J. (2013): Reducing Depopulation in Rural Spain: The Impact of Immigration, *Population, Space & Place*, Vol. 20, pp: 605–621.

Gallion, Aurther B. & Simon Eisner (2005): *The Urban Pattern City Planning and Design*, J.S. Offset Printers, Delhi, pp: 263–286.

Hudson, F.S. (1981): *A Geography of Settlements*, Macdonald & Evans Ltd., UK, pp: 267–286.

Khullar, D.R. (2014): *India A Comprehensive Geography*, Kalyani Publishers, New Delhi, pp: 322–368.

Mandal, R.B. (2000): *Urban Geography A textbook*, Concept Publishing Company, New Delhi, pp: 479–489.

Misra, R.P. (2002): *Regional Planning – Concepts, Techniques, Policies and Case Studies*, Concept Publishing Company, New Delhi, pp: 795–799.

Mirik Municipality. (2013): *Draft Development Plan of Mirik Municipality (2008–2013)*, Mirik Municipality, West Bengal, pp: 127–186.

Northam, R.M. (1979): *Urban Geography*, John Wiley & Sons, New York, pp: 472–485.

Pinilla, V., Ayuda, M. & Saez, L. (2008): Rural Depopulation and the Migration Turnaround In Mediterranean Western Europe: A Case Study of Aragon, *Journal of Rural and Community Development*, Vol. 3, pp: 1–22.

Ramachandran, R (2015): *Urbanization and Urban Systems in India*, Rakmo Press, New Delhi, pp: 334–341.

Reynaud, Cecilia & Miccoli, Sara (2018): Depopulation and the Aging Population: The Relationship in Italian Municipalities, *Sustainability*, Vol. 10, p: 1004. https://doi.org/10.3390/su10041004

Sharma, K.L. (2008): *Indian Social Structure and Change*, Rawat Publications, New Delhi, pp: 275–280.

Sidhartha, K. & Mukherjee, S. (2000): *Cities, Urbanisation & Urban Systems (Settlement Geography)*, Kitab Mahal Printing Division, Allahabad, pp: 357–382.

10 Assessment of a Sea Turtle Nesting Beach, a Tropical Habitat

Rushikulya, East Coast of India

Umakanta Pradhan, Pravakar Mishra, and Pratap Kumar Mohanty

10.1 Introduction

The sea turtle plays an essential role in marine ecosystems. The olive ridley sea turtle is one of the most abundant sea turtle species compared to the other six. Mostly, they are found in the tropical water of the eastern Pacific Ocean, northern Indian Ocean, and along the African coast of the Atlantic Ocean. These are schedule-I species under the wildlife (protection) act and listed as "endangered" in the ICUN Red Data Book, in the Appendix-I of CITES (Convention on International Trade in Endangered species of Wild Flora and Fauna), also listed in the CMS (Convention on Migratory species). In recent years, the effective management of marine species has been challenging because of the sea turtle's mysterious nature of diversity and complex life stages. The sea turtle migrates thousands of kilometers from the feeding ground to the nesting beaches to lay its eggs. This is the beauty of these animals that they choose the nesting site, usually where they were born. Research has been focused for decades on evaluating the sea turtles' migration and migration route; forage behaviors; nesting site selection with their characteristics; hatchling variability in point of mortality, gender and climate change, with natural and natural and anthropogenic pressure. The Odisha coast along the western boundary of the Bay of Bengal is known worldwide as one of the famous nesting sites of the endangered olive ridley turtle. The nests on the mainland shore are mostly near the mouths of rivers or estuaries and, therefore, are closely associated with low salinity, high turbidity, and productive water.

Three major mass-nesting sites or rookeries of the olive ridley along Odisha coast, east coast of India are identified, namely Gahirmatha, Devi and Rushikulya (Shanker et al., 2003). These rookeries along the Odisha are reported as the world's famous nesting site. Sea turtles arrived in the coastal waters of Odisha in November and started the mass nesting "arribada" during February and March. The mass hatching occurs after 45–50 days from the mass nesting. The Rushikulya rookery is the latest and was noticed in 1994 (Pandav et al., 1994). The numbers of olive ridley sea turtles nesting fluctuate year to year, which is significant, but on average, the nesting is increasing,

DOI: 10.4324/9781003275916-12

which shows the rookery is healthy and protected. The present chapter attempts to explain the vulnerability of the sea turtle rookery through field photography, which includes the influence of both natural and human impacts.

The Rushikulya nesting site is identified as a healthy and safe rookery for sea turtles. On and average, the nesting population increased from 1994 to 2020. The turtle nesting, health of habitat and related issues are discussed with positive, negative feedback, and both.

10.2 Positive Feedback

10.2.1 Forage

The sea turtles migrate thousands of miles in a year from the different oceanic regions and mostly prefer the estuarine or mangrove area for nesting because of the availability of good foraging. They are omnivorous, primarily carnivores: their food includes mollusks, fish, fish eggs, crabs, shrimps, rock lobsters, jellyfish, and occasionally, algae. In general, the southwest monsoon rainfall brings a considerable amount of land runoff to the coastal water along India's east coast, which contains the high nutrient and causes the bloom conditions and higher chlorophyll-a concentration. The Rushikulya coastal water has a higher biomass (Chl-a concentration) during the winter season (Baliarsingh et al., 2015), which indicates a suitable forage system for migratory sea turtles. The higher concentration of Chl-a is an indicator of zooplankton abundance.

10.2.2 Nesting Beach

Changes in coastal geomorphology such as beach changes, sediment erosion/accretion, shoreline change, sediment grain size of nesting beaches and associated oceanographic and meteorological conditions are considered important factors that significantly influence the nesting and hatchling behavior of the olive ridley turtle (Barik et al., 2014). The Rushikulya nesting beach is sited on the north of the inlet while sand spits on the south. The north of Rushikulya beach is identified as a recycling beach with a medium size of sand, which is favorable for the nesting processes. The southern part of the nesting beach is highly eroded with the northerly elongation of sand spit (Pradhan et al., 2020). The backshore beach elevation is more than 2 m while the tide is micro (<2 m) in nature (Pradhan, 2019) favors nesting.

10.2.3 Governance/NGOs

Proper management and supportive governance is a significant and influential component to protect sea turtles and their habitat. In India, the Ministry of Environment Forest and Climate Change (MoEF&CC) is the prime authority to conserve marine turtles and their nesting ground. Recently, the MoEF&CC

has released "Marine Mega Fauna Stranding Guidelines" and "National Marine Turtle Action Plan." India has identified its essential sea turtle nesting habitats as "Important Coastal and Marine Biodiversity Areas" and included them in the Coastal Regulation Zone (CRZ)-I. Apart from the government, non-governmental organization (NGOs) are also actively working for sea turtle protection. For proper management of sea turtles at Rushikulya, both government (central and state) and Rushikulya sea turtle protection committee, NGOs have worked together for a long time. They are participating in awareness, providing education through different community levels to protest sea turtles and habitats.

10.3 Negative Feedback

10.3.1 Beach Erosion

From the shore, where sea turtles begin their lives, gender selection and nesting are greatly influenced. Coastal erosion has been a major environmental concern throughout the world, which causes the loss of coastal habitats and biodiversity (Malini and Rao, 2004). Particularly, because to the reversing monsoon wind, the coastline geomorphology of India's east coast is extremely dynamic (Mohanty et al., 2008). The erosion of the nesting beach is prominent during high wave conditions such as monsoon and cyclonic periods (Mishra et al., 2011: Mohanty et al., 2012) and depends upon the northward extension of sand spit at the south of the inlet (Pradhan et al., 2015). Not at the study area, northward growth of the sandspit is a conspicuous feature along the east coast of India and other parts of the Orissa coast (Mohanty et al. 2008). As the sand spit traps the sediment flowing from south to north, erosion mainly occurs to its immediate north.

10.3.2 High Wave

The coastal erosion has resulted from high waves. Generally, the high waves occur during the southwest monsoon and cyclonic storm periods along the coast (Patra et al., 2016). Based on observation, the average wave height observed ranges from 0.25 m in December to 0.97 m in July. During our field survey, high wave activity (2.3 m) was observed on March 13, 2009, which caused severe erosion and resulted in the destruction of the turtle nesting beach. The high waves cause coastal erosion, which creates less space in the beach for the nesting and steeper beach profile.

10.3.3 Cyclone/Flood

In two decades, an increase in the intensity of tropical storms and cyclones over the Bay of Bengal wreaked havoc along India's east coast and in Bangladesh, killing dozens of people, flooding low-lying areas, and affecting the power supply. The beach experienced erosion due to complex characteristics in the nearshore region occurred by storm-generated waves, locally generated wind, and increase in sea level (storm surge), which are threats to the life and wealth of the coastal community (Rao et al., 2007; Amrutha et al., 2014). The cyclone-driven rainfall (stormwater) brings more debris and plastic to the adjacent coast and sea through the river. Debris along the coastlines causes harm or even death to nesting and hatchling sea turtles through ingestion, entrapment, or entanglement (Martin et al., 2019). The excess debris at the coast was the biggest problem for turtles digging their nesting pits. It was reported that the coastal flood caused by cyclone "Titli" brought a massive amount of debris to the coast. With better management and public support, the debris is removed and a better and clean beach is prepared for the upcoming nest activities.

10.3.4 Fishing

Fisheries have an important contribution to socioeconomic condition which provides food and livelihood around the world. But the recent fishing practice such as overfishing and the use of non-degraded fishing materials are threats to marine sources (Gallo et al., 2006). Many reported that the recent fishing activities are increasing plastic pollution in the marine environment. The major threat to most sea turtles is fishing gear. Thousands of turtles are accidentally caught and injured by nets, ropes, and fishing hooks. Physical observation from the ground also found that the old net and fishing materials are dumped near the fishing hamlet adjacent to the Rushikulya nesting ground. Still, these are cleaned from time to time by the authorities and the public. The concerned authorities banned fishing activities during the nesting and hatchling period and provided the minimum support price to affected fishermen for their livelihood.

10.3.5 Animal

The sea turtles face many death obstacles to survive in the water as well as inland. Predators such as raccoons, foxes, coyotes, feral dogs, ants, crabs, armadillos and mongooses can unearth and eat sea turtles during nesting and the hatchlingperiod; crabs and birds can eat when they run from the nest to the ocean. Fish and dolphins can eat juveniles when they move from coastal waters to the open ocean. Also, threats are at the land when the eggs are in incubated phase. It was observed, once the nesting pit is disturbed, the eggs inside the pit will be damaged. Thus the government, sea turtle protecting groups and local people are very much concerned and provide the necessary arrangement to save turtles during the nesting period to hatchling and eggs in the pit at the nesting ground at Rushikulya rookery. They adapt the net barricade along the backshore region to restrict the entry of animals to the nesting ground and outward movement of juvenile turtles from the beach.

10.3.6 Artificial Light

Light pollution along the nesting beach can impact fundamental biological processes for marine turtles. The apparent brightness and glare of artificial lighting often lead to both nesting and hatchlings' activity. Generally, the sea turtles like nesting on a dark beach while the juveniles after the hatchling are attracted towards the brighter region. The presence of artificial light at beaches and nearshore waters may also alter the predator-prey dynamics, influencing the behavior of species that predate on sea turtle eggs and hatchlings (Silva et al., 2017; Colman et al., 2020), as well as attracting hatchlings dispersing from natal beaches (Thums et al., 2016). Though the national highway (NH 5) is about 3 km away from the Rushikulya nesting ground and the fisherman hamlet is near the nesting beach, still there is no light pollution that has an influence on turtle activity. The coastal vegetation (Casurina forest) triggers the darker beach while governance and NGOs are encouraging to coastal villagers to switch off lights when not required during nighttime.

10.3.7 Marine Litter/Plastic

Not only regionally, global plastic pollution is a complex and multi-dimensional problem. As a result, it has a significant impact on flora and fauna in the terrestrial as well as the marine environment. Marine litter and plastic pose a significant threat to marine mammals, birds, turtles, and fish due to entanglement and ingestion. Plastic items have become the major component of marine debris, which are originated mainly from landfill sites through flood/stormwater, human activities such as tourism, fishing, etc. Numerous studies reported that the marine debris and plastic in seawater and beaches threaten sea turtle forage and nesting and are identified as negative feedback to coastal biodiversity (Fujisaki and Lamont, 2016). The flood due to heavy rain or storm brought a huge amount of debris to the adjacent nesting beach at Rushikulya. The debris is both degradable and non-degradable (plastic and thermocool), which is a major problem for nesting. It was reported that a huge amount of debris came to the nesting beach after the cyclonic storm "Titli," but the administration and public cleaned the nesting beach.

10.3.8 Backshore Vegetation

Sand dune/backshore vegetation comprises vital components of coastal habitats owing to their role in sediment binding, accumulation, land building, increases water holding capacity, development in the food chain and act as a buffering zone during cyclones, coastal flood and Tsunamis (Desai, 2000; Ramarajan and Murugesan, 2014). The coastal vegetation along the Odisha coast is well studied and reported, mainly trees, shrubs and herbs. The Rushikulya nesting beach is available primarily with the Casuarina and Cashew (Anacardium occidentale L.) trees behind the backshore, while some herbs and shrubs are present at the beach. Mainly, Kansai lata (Ipomoea pes-caprae L.), Gudukanka Lata (Spinifex Litorex), Whiteweed/Goat weed (Ageratum conyzoides L.), Crown flower (Calotropis gigantea L.) dry land, Bara Gokhru (Pedaliummurex L) shrubs and herbs are found throughout the year. The existence of trees (Casuarina and Cashew) at the rear of the backshore limit is good for the sea turtle activities at the beach by restricting the artificial light pollution. While the shrubs and herbs are sand binders in nature on the beach but they are found as a negative impact during nesting and hatchling because of problems with crawling and digging.

10.3.9 Human and Settlement

Sea turtles are considered a food source in many countries around the world. During the nesting period, people hunt turtles for meat and eggs also. Additionally, people may use other turtle parts for products, including oil, cartilage, skin, and shell. Overexploitation is a major threat to sea turtle conservation, while better management can be made by regional collaboration regarding sea turtle traffic and law enforcement (Joseph et al., 2019). But in India, sea turtles have spiritual or mythological importance, which plays a major role in protection and rescue. The sea turtle is invited as a guest by people along the Odisha coast. People are incredibly kind and contribute to the preservation of the environment in the following ways.

- Sea turtles are not disturbed by local people during the nesting and hatchling period.
- People rescue the disoriented juvenile sea turtle.
- The fisherman community forbids fishing during the turtle nesting and hatchling period.
- Local peoples are participating in awareness, coastal cleaning, and sea turtle protecting campaigns.

The present work highlights the important issues related to sea turtle vulnerability, which may help in conservation and monitoring on a regional to a global scale. Results of our analysis show that the protection of sea turtles during nesting to the hatchling period needs to increase the coastal vegetation, awareness in the local community, protection from the predators, frequently cleaning the nesting ground and decreasing of fishing activity, light pollution, marine litter/plastic, coastal settlement, tourism. The protection and conservation of any habitats is a major part of the sustainable development goal (SDG); hence this work may bring an overall idea of sea turtle conservation and their habitat.

Acknowledgment

This study was undertaken as part of the National Centre for Coastal Research (NCCR), Ministry of Earth Sciences, Government of India. The authors are thankful to Mr. Rabindranath Sahu, Rushikulya Sea Turtle Protection Committee, Ganjam for providing some field photos.

Bibliography

Amrutha, M.M., Sanil Kumar, V., Anoop, T.R., Balakrishnan Nair, T.M., Nherakkol, A., and Jeyakumar, C., Waves off Gopalpur, northern Bay of Bengal during Cyclone Phailin, *Annales Geophysicae*, 32, (2014), 1073–1083. DOI 10.5194/angeo-32-1073-2014

Baliarsingh, B.K., Lotliker, A.A., Sahu, K.C., and Kumar, T.S., Spatio-temporal distribution of chlorophyll-a in relation to physcio-chemical parameters in coastal waters of the northwestern Bay of Bengal. *Environmental Monitoring and Assessment*, 187, (2015), 481. DOI 10.1007/s10661-015-4660-x

Barik, S., Mohanty, P.K., Kar, P., Behera, B., and Patra, S., Environmental cues for mass nesting of sea turtles. *Ocean and Coastal Management*, 95, (2014), 233–240.

Colman, L.P., Lara, P.H., Bennie, J., Broderick, A.C., de Freitas, J.R., Marcondes, A., Witt, M. J., and Godley, B. J. Assessing coastal artificial light and potential exposure of wildlife at a national scale: The case of marine turtles in Brazil. *Biodiversity and Conservation*, (2020). DOI: 10.1007/s10531-019-01928-z

Desai K.N., Dune vegetation: Need for a reappraisal, Coast. *A Coastal Policy Rese Newslett*, 3, (2000), 6–8.

Fujisaki, I., and Lamont, M.M., The effects of large beach debris on nesting sea turtles. *Journal of Experimental Marine Biology and Ecology*, 482, (2016), 33–37, ISSN 0022-0981. https://doi.org/10.1016/j.jembe.2016.04.005

Gallo, Berenice M.G., Macedo, S., Giffoni, Bruno de B., Becker, J.H., Barata, Paulo C.R., Sea turtle conservation in Ubatuba, Southeastern Brazil, a feeding area with incidental capture in coastal fisheries. *Chelonian Conservation and Biology*, 5(1), (2006). https://doi.org/10.2744/1071-8443(2006)5[93:STCIUS]2.0.CO;2

Malini, B.H. and Rao, K.N., Coastal erosion and habitat loss along the Godavari delta front—A fallout of dam construction. *Current Science*, 87(9), (2004), 1232–1236.

Martin, J.M., Jambeck, J.R., Ondich, B.L. and Norton, T.M., Comparing quantity of marine debris to loggerhead sea turtle (*Caretta caretta*) nesting and non-nesting emergence activity on Jekyll Island, Georgia, USA. *Marine Pollution Bulletin*, 139, (2019), 1–5. https://doi.org/10.1016/j.marpolbul.2018.11.066

Mishra, P., Patra, S.K., Ramana Murthy, M.V., Mohanty, P.K., and Panda, U.S., Interaction of monsoonal wave, current and tide near Gopalpur, east coast of India and their impact on beach profile; a case study. *Natural Hazards*, 59(2), (2011), 1145–1159.

Joseph, J., Jolis, G., Nishizawa, H., Alin J.M., Othman, R., Isnain, I., Nais, J., Mass sea turtle slaughter at Pulau Tiga, Malaysia: Genetic studies indicate poaching locations and its potential effects. *Global Ecology and Conservation*, 17, (2019), e00586. https://doi.org/10.1016/j.gecco.2019.e00586

Mohanty, P.K., Panda, U.S., Pal, S.R., and Mishra, P., Monitoring and management of environmental changes along Orissa Coast, *Journal of Coastal Research*, *24*(2A), (2008), 13e27.

Mohanty, P.K., Patra, S.K., Bramha, S., Seth, B., Pradhan, U.K., Behera, B., Mishra, P., and Panda, U.S., Impact of groins on beach morphology: A case study near Gopalpur Port, east coast of India. *Journal of Coastal Research*, 28(1), (2012), 132e142.

Pandav, B., Choudhury, B.C. and Kar, C.S., A status survey of olive ridley sea turtle (*Lepidochelys olivacea*) and their nesting beaches along the Orissa coast, India. Unpublished Report. Wildlife Institute of India, Dehradun, (1994), 48 pp.

Patra, S.K., Mishra, P., Mohanty, P.K., Pradhan, U.K., Panda, U.S., Murthy, M.V.R., Sanil Kumar, V. and Nair, T.M.B., Cyclone and monsoonal wave characteristics of northwestern Bay of Bengal: Long-term observation and modeling. *Natural Hazards*, (2016). https://doi.org/10.1007/S11069-016-2233-0

Pradhan U.K., Mishra, P., Mohanty, P.K., and Behera, B., Formation, growth and variability of sand spit at Rushikulya river mouth, south Odisha coast, India. *Procedia Engineering*, 116, (2015), 963–970.

Pradhan, U. K., Coastal geomorphology and oceanographic conditions near Rushikulya estuary. Ph.D. Thesis, Berhampur University, (2019), pp. 184.

Pradhan, U.K., Mishra, P., Mohanty, P.K., Panda, U.S. and Ramanamurthy, M.V., Modeling of tidal circulation and sediment transport near tropical estuary, east coast of India. *Regional Studies in Marine Science*, (2020), 101351. https://doi.org/10.1016/j.rsma.2020.101351

Ramarajan, S. and Murugesan, A.G., Plant diversity on coastal sand dune flora, Tirunelveli district, Tamil Nadu. *Indian Journal of Plant Sciences*, 3(2) (2014, April–June), 42–48.

Rao, A.D., Dash, S., Babu, S.V. and Jain, I., Numerical modeling of cyclone's impact on the ocean—A case study of the Orissa super cyclone. *Journal of Coastal Research*, 23(5), (2007), 1245–1250.

Shanker, K., Pandav, B., and Chodhury, B.C., An assessment of the olive ridley turtle (*Lepidochelys olivacea*) nesting population in Orissa, India. *Bilogical Conservation*, 115, (2003), 149–160.

Silva, E., Marco, A., da Graça, J., Pérez, H., Abella, E., Patino-Martinez, J. and Almeida, C., Light pollution affects nesting behavior of loggerhead turtles and predation risk of nests and hatchlings. *Journal of Photochemistry and Photobiology B: Biology*, 173, (2017), 240–249. https://doi.org/10.1016/j.jphotobiol.2017.06.006

Thums, M., Whiting, S.D., Reisser J., Pendoley K.L., Pattiaratchi, C.B., Proietti, M., Hetzel, Y., Fisher, R. and Meekan, M.G., Artificial light on water attracts turtle hatchlings during their near shore transit, *Royal Society Open Science*, 3, (2016), 160142. https://doi.org/10.1098/rsos.160142

11 Risk and Livelihood Pattern of Informal Workers

Household Evidence from Kolkata City

Alisha Safder and Md. Julfikar Ali

11.1 Introduction

In recent decades all around the world a massive amount of workers are forced to live in the informal work sector without any social protection. The informal sector, in recent decades, has taken a leading part in developing countries especially in India and simultaneously attracted the attention of academicians, researchers, social development activists, and policy planners (Papola, 1981). India is one of the leading developing countries in the world with an enormous population and the majority of them contribute to the informal sector while this sector not only embraces the informal workers of this sector but also constitutes the casual labourers of the formal sector (Shonchoy and Junankar, 2014). The crucial reasons behind this informality and its rapid growth are, growing population, urban poverty, illiteracy, growing unemployment, etc. The India Labour and Employment Report (2014) estimates about 92% of national workers are involved in informal occupation (I.L.E. Report, Alakh, 2014) while NSSO Employment and Unemployment Survey (2011–12) highlighted that 87.82% of workers in West Bengal are employed in informal work (NSSO, 2013). This sector constitutes about 62% of GDP, 50% of national savings, and 40% of national exports (ILO 2002, report VI). As the fast pace of urbanization persists in most developing countries, the urban dwellers earn their living with informal working, which is consistently become significant for the expansion of most developing countries in South Asia, like India (Lim, 2015; Mazumdar and Sarkar, 2015). Those urban dwellers are forced to work and live in a vulnerable urban environment, which leads to a worse livelihood pattern deprived of basic amenities (Bose and Roy, 2019). Simultaneously, the majority of urban informal workers in emerging countries are excluded from regional economic development plans whereas they are more vulnerable to exploitation regarding their rights and security (Lim, 2015) as they are employed according to their education and skill (Timalsina, 2011).

Though many authors focused on the very crucial determinants of the informal economy and its workers, there existed a gap in the issues discussed and the issues not discussed. However, they did not focus on the magnitude of

DOI: 10.4324/9781003275916-13

the dominancy of informal employment at the micro level especially in slum areas, as informality is a specific characteristic of slums. They didn't focus on the livelihood pattern of informal workers among slums and non-slums. Informal labours works in a susceptible condition, which was not deliberately focused on in the past studies at grassroot level.

Kolkata is one of the biggest metropolitan cities of India. Additionally, it contains enormous concentrations of urban poverty, with slums serving as symbols of the region's character (Bose and Ghosh, 2015) by the scarcity of basic needs of appropriate housing, drinking water and sanitation (Lim, 2015). The aim of this work is to portray the appearance of the socio-economic condition of Kolkata's informal workers who are deprived and marginalized throughout their livelihood. They face lots of discrimination relating to their working environments with very little or no access to social protections. Such information at micro level is not recorded in official documents nor any research earlier work, leading to deficiency in knowledge. Here out of the total sampled households (490 households), 336 were found to be informal work dominated households which are the major concern in this study and the rest of the sampled households are formal work dominated. Among the total sampled population, 565 are informal workers. On the basis of the empirical data of the sampled informal work dominated households and individual workers, this study evaluates the pattern of livelihood and compares it with formal work dominated households.

11.2 Database and Methodology

The analysis of the research work is mainly based on the primary sources of data. As there is a scarcity of secondary data on the information of informal workers at a micro-level, the analysis is based on other sources. The informal economy is a complex area of study where it is difficult to find the actual figure of informal units and its workers. In spite of these complexities some qualitative inferences and quantitative analysis has been done on the basis of empirical observation and the primary data analysis. Primary data have been collected through a field survey using a survey schedule and direct interviews with the respondents at sample households. Furthermore, stratified random sampling has been used for a further sample selection.

11.2.1 Sample Selection for Field Study

The municipal wards of KMC have been arranged in descending order on the basis of share of slum households of each ward. Initially, 42 wards, or 30% of the 141 wards, were selected, equally from top to bottom and bottom to up of the wards that were organized descendingly according to slum share. Finally, 10% of wards from initially selected wards (i.e. 14 wards of KMC) have been taken into consideration for the final sample survey on the basis of major land use character. The final selected wards are 58, 137, 80, 22, 23, 54, 66, 29, 108,

65, 59, 123, 96, and 138. Since the ward-wise data on slum population and slum households are not provided by the Census 2011, the selection procedure is pursued on the basis of Census 2001 data.

11.2.2 Selection of Households for Primary Survey

After the selection of 14 wards, a procedure of sampling of households for survey has been done. In this study, the 35 households have been randomly selected in each of the 14 wards. The number of slum and non-slum households in each selected wards follows a corresponding percentage of slum and non-slum households based on slum data of KMC wards-wise (Census of India, 2001). For example in ward number 137, the percentage of slum and non-slum households is 98.49% and 1.51%, respectively. Following this, the total sampled 35 households comprise 34 slum and 01 non-slum households for the survey. This same approach has been applied in the selection of sample households of all 14 sampled wards. As per the selection procedure, 216 slum households and 274 non-slum households have been identified for further field survey.

11.3 Livelihood Framework

Livelihood has become a common topic in the perspective of development literature. The term 'livelihood' depicts the intricacies of survival in low-income countries in a more comprehensive way. The concept of livelihood incorporates both non-economic and economic aspects of survival and also comprises the social interactions and associations that mediate people's access to different assets and income level (Ellis, 2000). Livelihood approaches are ways where households make their living by uniting and transmuting various abilities and assets into activities which bring valued outcomes. As there is a lack of a well-developed social security system, the concept of household strategy becomes imperative as it defines the range of actions that households or families undertake to deal with or overcome challenges that they face in their lives (Narayan, Patel, Schafft, Rademacher, & Koch-Schulte, 2000; Uphoff, 1986). In general, the livelihood of dwellers is determined by their state of living and what resources they have for use in terms of social, economic, environmental, and political condition (Meikle, 2002). This livelihood crisis is indeed noticed among poverty-stricken developing countries while the major reason for this meagre economic condition and poverty in those countries is insufficient access to financial facilities while it is essential for low-earning group of people (Mazumdar and Sarkar, 2015). Rural migrants as well as poor urban inhabitants are also involved in informal work where they transact their livelihood with informal activities which has consistently become significant in most developing countries in South Asia, like India (Lim, 2015; Mazumdar and Sarkar, 2015). In urban areas, slum dwellers especially are mostly engaged in informal work as they have a lack of skills and improper educational qualifications (Mitra, 1990) and are forced to work and live in an vulnerable urban environment which leads to a worse livelihood pattern deprived of basic amenities (Bose and Roy, 2019). Moreover, the majority of urban informal workers in emerging countries are excluded from regional economic development plans whereas they are more vulnerable to exploitation regarding their rights and security (Lim, 2015) as they get employed according to their education and skill (Timalsina, 2011).

11.3.1 Profile of Kolkata

Kolkata city has been selected as a study area for pursuing the present research work. It is the capital of West Bengal and one of the larger as well as older metropolitan cities in India, and the home of millions. Kolkata has experienced a huge occupational shift in different economic sectors. Slums are one of the important demographic entities in Kolkata (Chaudhuri, 1990), occupying a 31.35% share of the total population (Census of India, 2011). The continuance of the slum population depends on the nature and pattern of economic activity in which a hefty amount of city population find employment (Mitra, 1990). Slum regions are heavily inhabited places where the majority of unofficial economic activity takes place (Romatet, 1983). As per empirical observation and literature reviews, slum population are mainly engaged in low-paid informal work like contractual labour, daily wage labour, domestic worker, and self-employed (Mazumdar and Sarkar, 2015). Along with the slum population many non-slum populations are also involved in informal work but the share of slum dwellers is higher in informal work. This study tries to identify the magnitude of informal work in slum and non-slum households who lag behind and depend on the informal economic system.

11.3.2 Household as a Unit of Research and Analysis

The household is frequently used as the basic unit analysis. However, many authors would agree that the concept of household is rather intricate and hard to combine into one clear definition as it comprises many aspects (Van Vuuren, 2003). The classic definition of household applied in research for survey purpose contains co-residence and consumption 'from a common pot' (Chaudhuri, 2013). This definition applies to most of Kolkata's households, except migrants and those whose families are settled in villages, homeless, and have food from roadside eateries. For data collection, a questionnaire was directed to each household. Certain selected variables of total sampled population for analysis and their frequency distribution are depicted in Table 11.1.

Urban households highly depend on cash as it remains the major constituent of urban livelihood, and hence access to it becomes a fundamental need for their continued existence. In the study carried out, it is found that the distribution of the male and female sampled population in the study area is almost equal in both slum and non-slum areas (Table 11.1).

Table 11.1 Selected variables and their frequency distribution

Variable	*Categories*	*Slum (N = 996)*	*Non-slum (N = 1156)*
Gender of sampled population	Male	502 (50.4)	579 (50.09)
	Female	494 (49.6)	577 (49.91)
Household size	Up to 02	24 (11.11)	29 (10.58)
	3 to 4	100 (46.30)	151 (55.11)
	5 to 6	64 (29.63)	61 (22.26)
	7 to 8	23 (10.65)	21 (7.66)
	above 8	5 (2.31)	12 (4.38)
Educational status	Pre-primary	13 (1.3)	10 (0.87)
	Primary	104 (10.45)	55 (4.75)
	Upper primary	227 (22.79)	141 (12.2)
	Secondary	206 (20.68)	250 (21.62)
	HS	98 (9.84)	230 (19.9)
	Graduation	53 (5.32)	233 (20.16)
	Above	2 (0.20)	42 (3.63)
	Not enrolled yet	59 (5.92)	46 (3.98)
	LWFS	46 (4.62)	59 (5.1)
	Illiterate	188 (18.88)	90 (7.79)
Marital status	Married	461 (46.29)	608 (52.6)
	Unmarried	482 (48.39)	500 (43.25)
	Divorced	4 (0.4)	3 (0.26)
	Separated	5 (0.5)	11 (0.95)
	Widow/Widower	44 (4.42)	34 (2.94)

Note: Figures in parenthesis are percentages

Source: Field survey (done by author), 2019, LWFS: Literate without formal schooling, HS: Higher Secondary

The number of family members in a household more or less defines the household size (Table 11.1). A small household was considered to have one or two family members, a medium-sized household between three and four members, an above medium-sized between five and six members and a large household was seen to have seven or more members (Chaudhuri, 2013). In this study, the majority of sampled households (both slum and non-slum) were found with medium-sized households (46.30% slum and 55.11% non-slum respectively) while the percentage of large sized households is found higher in slums (10.65%) than non-slums (7.66%), which depicts the vulnerability of their survival.

Among the sampled slum population, the majority of people were found to have completed their upper primary (22.79%) education followed by secondary (20.68%) education while the percentage with a higher education is minor. However, the share of illiterate persons (18.88%) is noticeable among the slum population. On the other hand, a large number of the sampled non-slum population were found to have attained up to secondary education (21.62%) followed by graduation (20.16%) and HS (19.9%) level of education. Moreover, the percentage of illiterate (7.79%) among the non-slum population is much smaller than the slum population.

In the study, a hefty amount (48.39%) of the slum population was found to be unmarried while 46.29% of the sampled slum population was married which depicts the higher number of younger people who are dependent on family heads. Moreover, among sampled non-slum population the share of married (52.6%) people is higher than unmarried (43.25%). These variables represent a glance of the sampled population distribution among slums and non-slums to identify their pattern of livelihood.

11.4 Occupational Profile

The Indian economy constitutes a divergent size and magnitude in terms of formal and informal workforce which contributes to its development and growth (Bhosle, 2014) in which the prevalence of the informal sector is found at a larger scale. Employers who manage their means of subsistence through this are vulnerable to working conditions due to the incapacity of the formal or organized sectors to absorb the whole workforce and unorganized sector (Mohapatra, 2012). Furthermore, the majority of slum households work in the informal sector (Mazumdar and Sarkar, 2015) with irregular earning and protection as either casual labour or self-employed (Bhowmik and More, 2001). In this study a variety of informal work has been found where a large number of populations are involved. Furthermore, this study has also found some formal workers while survey and it is used in analysis for the comparison with informal workers at some points. The analysis is based on the consideration of dominant households' occupation of informal and formal work (i.e. occupation with higher income share in household). As per the study the dominant household occupation is regarded as informal and formal work dominated households. On the other hand, individual workers in formal and informal work have also been evaluated. In this study, informal work dominated households and its workers are the major concern.

This study reveals that, out of total sampled population only 34.85% are employed while a large share (65.15) are not employed (Table 11.2). The share

Table 11.2 Share of workforce by slum and non-slum households

Workforce	*Sampled workforce*				*Total population (%) (N = 2152)*
	Slum pop. (%) (N = 996)	*Non-slum pop. (%) (N = 1156)*	*Male (%) (N = 1081)*	*Female (%) (N = 1071)*	
Working population	37.65	32.44	55.13	14.38	34.85
Non-working population	62.35	67.56	44.87	85.62	65.15
Total (%)	**100**	**100**	**100**	**100**	**100**

Note: pop. – Population

Source: Field survey (done by author), 2019

of total workforce out of total sampled population is 37.65% in slums and 32.44% in non-slums. Out of the total sampled working population the share of males is much higher (55.13%) than females (14.38%).

As the slums have large number of family members, and the majority of them work to earn their daily bread and butter so the share of working population must be higher in slums than non-slums.

In this study, the dominant household occupation is assumed in terms of largest income of a household from a particular occupation (i.e. informal or formal occupation). For example, if a household earn a higher income from an informal occupation then it is considered as an informal work dominated household and so on. Table 11.3 illuminates the dominant household occupations of the total sample where out of the total sampled households (i.e., 490 households) 68.57% of sampled households are considered as informal work dominated households and the rest (31.43%) are formal work dominated households. Among slums only 13.02% of sampled slum households are recorded as formal work dominated while the share is quite higher in non-slum households (46.15%). Besides, the majority (86.98%) of sampled slum households have chosen informal employment as the dominant household occupation though the share of informal work dominated households among non-slums is 53.85%. Predominance over informal work is higher among slum households than non-slums as they are already deprived of basic amenities and this informal dominant occupation makes them more vulnerable.

Table 11.4 shows the distribution of the sampled population in dominant household occupations by slum and non-slum households. Here only 24.67% out of the total sampled working population has been found as formal workers and 75.33% of the population work as informal workers. Furthermore, only 8.82% of the sampled working populations of slums work as formal workers while 40.64% of the sampled non-slum population has recorded working as formal labour. A large number (91.18%) of the sampled working population of slums work as informal labour and the share of informal labour is much lower (59.36%) in case of the sampled non-slum population. As stated earlier the number of informal workers is also higher among the slum population than non-slums. As the informal sector easily gain employment and also do not need any specific professional qualification it is quite easy to engage in this

Table 11.3 Dominant occupations of sampled households by slums and non-slums

Dominant occupation	*Sampled households*		*Total HH (%) (N = 490)*
	Slum HH (%) (N = 216)	*Non-slum HH (%) (N = 274)*	
Formal	13.02	46.15	31.43
Informal	86.98	53.85	68.57
Total (%)	**100**	**100**	**100**

Source: Field survey (done by author), 2019

Table 11.4 Surveyed population employed in dominant household occupation by slum and non-slum

Dominant occupation of household	*Surveyed working population*		*Total working population (%) (N = 750)*
	Slum pop. (%) (N = 375)	*Non-slum pop. (%) (N = 375)*	
Formal	8.82	40.64	24.67
Informal	91.18	59.36	75.33
Total (%)	**100**	**100**	**100**

Source: Field survey (done by author), 2019

sector for the slum population which further makes their living condition vulnerable.

Table 11.5 signifies the working categories of sampled informal workers by slum and non-slum households. According to this survey, a sizable portion of informal workers (39.47%) are self-employed, and the percentage of people living in non-slum areas is larger (48.43%) than that of those living in slums (33.63%). On the other hand, regular/permanent wage earners among sampled informal workers are found to be almost insignificant (i.e. 0.29% in slums and 0.45% in non-slums). Furthermore, out of the total sampled informal workforce 31.15% of workers work as temporary monthly wage earners. However, in this category the sampled informal workers of slums and non-slums accounted for 27.78% and 36.32% respectively. Out of the total sampled working population the percentage of weekly wage earners is found to be only 15.22%. Therefore, in this study 18.42% informal workers of slums and 10.31% informal workers of non-slums have been observed as weekly wage earners. Additionally,

Table 11.5 Working category of surveyed informal workers

Working category		*Slum population*	*Non-slum population*	*Total workers (%) (N = 565)*
		Informal workers (%) (N = 342)	*Informal workers (%) (N = 223)*	
Regular/permanent wage earner		0.29	0.45	0.35
Self-employed		33.63	48.43	39.47
Temporary wage earner	Monthly wage earner	27.78	36.32	31.15
	Weekly wage earner	18.42	10.31	15.22
	Daily wage earner	19.88	4.49	13.81
Total (%)		**100**	**100**	**100**

Source: Field survey (done by author), 2019

out of the total working population only 10.26% is found to be daily wage earners where the share of the sampled slum population is 19.88% and non-slum is 4.49%. This study observed a large amount of self-employed out of the total sampled informal workforce. The share of weekly and daily wage earners is found to be higher among informal workers of slums than non-slums. This condition of informal workers depicts their poor working status with major insecurities which in turn affects their economic as well as social well-being.

11.5 Education and Employment

Education is a way of getting better jobs and also a source of gaining knowledge and skill. According to UNESCO Education for All Report 2008, India has only a 66% literate population (76% men and 54% women) which exposes a huge extent of illiteracy. It increased to 74% in 2011. There is a strong relationship between education and skill level of worker as the formal sector places a great significance on formal education and knowledge skills in order to be recruited in the sector. Usually recruitments are mainly based on educational credentials of workers while promotions are based on experience and efficiency. Workers of organized/formal sectors work with high expertise, better earnings and constancy whereas the workers in unorganized/informal sector get low wages with unstable employment as they do not have proper education and skills (Varghese, 1989). On the other hand, the literacy rate of Kolkata city is 86.31% (Census of India, 2011) which is much higher than the national average (74.04%) but still the surveyed informal workers and their households are less literate.

In this study (Table 11.6), the majority (24.42%) of the sampled workforce is found to attain up to upper primary level of education followed by secondary (21.95%) and HS (10.97%) level. Though a large number (30.41%) of

Table 11.6 Educational attainments of sampled informal and formal workers

Educational level	*Informal workers*		*Total workers (%) (N = 565)*
	Slum pop. (%) (N = 342)	*Non-slum pop. (%) (N = 223)*	
Primary	3.80	2.24	3.19
Upper primary	27.48	20.18	24.42
Secondary	18.42	27.35	21.95
HS	8.20	15.25	10.97
Graduate	3.80	8.52	5.66
Post graduate	0.00	0.90	0.54
LWFS	7.89	11.21	9.20
Illiterate	30.41	14.35	24.07
Total	**100**	**100**	**100**

Note: LWFS- Literate without formal schooling

Source: Field survey (done by author), 2019, H.S. - Higher Secondary, pop. - population informal workers of slums were found to be illiterate, among them 27.19% were found to have completed their upper primary level of education followed by secondary (18.42%) and HS (8.20%) and graduation (3.80%). Among the informal workers of non-slums a large share (27.35%) is found to have attained up to secondary education followed by upper primary, HS, and graduation, 20.18%, 15.25%, and 8.52% respectively.

The share of illiterate people (14.35%) among non-slum informal workers is much less than in the slums. Despite this, few sampled informal workers were also to be found literate without formal schooling (LWFS) system, i.e., 7.89% in slums and 11.21% in non-slums. In the study area highly educated people are almost negligible among the sampled informal workers. The majority of informal workers try to engage in informal work at an early age for means of their livelihood. Among the informal workers a large share of the slum population have less education which affects their capacity to earn money.

11.5.1 Earning According to Skills

Working skill is gained through specific professional qualifications and training after completing a certain level of education otherwise workers without any professional certificates are known as semi-skilled (i.e. professional skill acquired by another person and done with proficiency) or unskilled workers. Informal workers work according to their acquired skills, i.e. higher skills with higher earning and vice-versa. A huge number of informal workers stopped their education partway so they are disadvantaged in terms of proper working skills and advance as semi-skilled or unskilled labour. The employers hire informal workers from a poor section of society according to their skills and also mistreat them in terms of working conditions along with security issues (Biswas, 2020; Bhosle, 2014). In this study, workers with some professional qualifications and training have been chosen as skilled labour otherwise semi-skilled or unskilled labour. Table 11.7 shows that, more than a half (57.35%) of sampled informal workers earn Rs. <10000 per month. At this point, among the unskilled workers an enormous number (59.12%) of informal workers have been earning Rs. < 10000 (very low) salary per month while 33.71% of workers have Rs. 10000–20000 monthly salary.

Besides, 57.02% of semi-skilled workers were found to have Rs. <10000 monthly salary whereas 37.13% of semi-skilled workers earn between Rs. 10000 and 20000 per month. The share of higher monthly incomes among semi-skilled and unskilled informal workers is found to be much less. In contrary, 51.90% of sampled skilled workers were found to earn Rs. >40000 monthly salary while only 2.16% found to earn Rs. <10000 monthly salary. This result reveals that the majority of informal workers have been earning very low monthly income depending on their skills which is not enough to sustain a good livelihood status. Moreover, skilled workers have higher income which helps them to lead a better living standard. Informal workers have a low amount of income and living condition which illustrates the vulnerable

Table 11.7 Monthly incomes of surveyed informal workers

Monthly income of informal workers	*Informal workers*			*Total workers (%) (N = 565)*
	Skilled workers (%) (N = 42)	*Semi-skilled workers (%) (N = 342)*	*Unskilled workers (%) (N = 181)*	
> 40000 (Very high)	51.90	0.30	1.66	0.88
30001–40000 (High)	28.57	1.17	2.20	2.30
20001–30000 (Moderate)	11.98	4.38	3.31	4.25
10000–20000 (Low)	5.39	37.13	33.71	35.22
<10000 (Very low)	2.16	57.02	59.12	57.35
Total (%)	**100**	**100**	**100**	**100**

Source: Field survey (done by author), 2019

livelihood status of informal workers. People with higher education work as skilled labour with higher income while low educational qualifications make them semi-skilled or unskilled labour who get insufficient income for survival.

11.5.2 Significance of Informal Work in Households' Livelihood

Informal work is quite significant for the poor section of society because it provides an income source without any specific educational qualification and skills. The majority of poor urban dwellers mainly depend on informal work for their survival. Table 11.8 reveals the significance of informal work among sampled households by slum and non-slum categories.

In this study, a hefty amount (91.67%) of sampled households was found to entirely depend on informal work for their livelihood. Majority (98.94%) of informal work dominated slum households found to entirely depend on informal work for their survival while a negligible (1.06%) amount of households partially depend on informal work. Moreover, 82.43% of informal work dominated non-slum households' livelihood is found to entirely depend on informal work and the rest (17.57%) are partially dependent. The majority of informal work dominated households entirely depend on informal work.

Table 11.8 Significance of informal work in surveyed households' livelihood

Level of significance	*Informal work dominated HH*		*Total HH (%) (N = 336)*
	Slum HH (%) (N = 188)	*Non-slum HH (%) (N = 148)*	
HH livelihood entirely depends	98.94	82.43	91.67
HH livelihood partially depends	1.06	17.57	8.33
HH livelihood not at all depends	0	0	0
Total (%)	**100**	**100**	**100**

Source: Field survey (done by author), 2019

Contrarily, the sampled households who partially depend on informal work may have other sources of income for sustenance.

11.6 Income-Expenditure Pattern

Income, expenditure, and savings are correlated and also reliant on each other. Income-expenditure is a specification to reveal the economic status of a household (Ismail and Bakar, 2012). The expenditure pattern of a household depends on its family size and family expenses (Forsyth, 1960) while it varies in the formal and informal sector. Most of the formal work dominated households with higher income pay out less on daily needs and more on lavish goods while people with low earning expend a higher share of their income on daily needs and some of them save a bit (Ismail and Bakar, 2012).

Table 11.9 shows the average monthly income, monthly expenditure and family size of sampled households depending on their dominant household type. The study included 490 sampled houses in total, of which 154 sampled households (645 persons) had a predominance of formal labor, and 336 sampled households (1507 populations) had a predominance of informal work. The average family size found in formal work and informal work dominated households is 4.20 and 5.51 respectively. On the other hand, the average income of sampled formal work dominated households is found abundantly higher (Rs. 42026/month) than the average income of sampled informal work (Rs. 17430/month) dominated households. According to their household incomes the expenditure patterns also varies, sampled formal work dominated households expends almost 77.59% of their income while the sampled informal workers expends 91.30% of their total household income. This average income-expenditure pattern reveals the income inequality between the formal and informal occupation dominated households on which basis they sustained their livelihood.

Figures 11.1 and 11.2 elucidates the expenditure patterns, in percentage, of formal and informal work dominated households. In comparison, informal work dominated households expend 34% on food while formal work dominated households expends 24% of their income on food on average. On the other hand, the average expenditure on medicine (6%), education (5%), and electricity (5%) was found to be identical for both work dominated households.

Table 11.9 Family size and average income-expenditure of formal and informal work dominated households

Dominant household occupation	*Total no. of households*	*Total sampled population*	*Average family size*	*Average monthly household income (Rs.)*	*Average monthly household expenditure (Rs.)*
Formal	154	645	4.20	42026	32606 (77.59%)
Informal	336	1507	5.51	17430	15913 (91.30%)

Source: Field survey (done by author), 2019

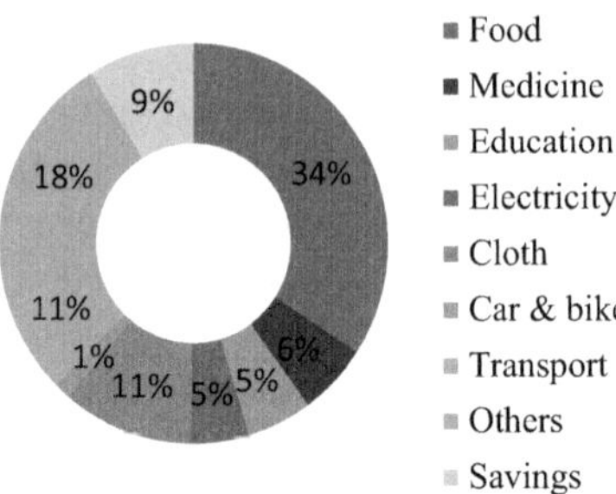

Figure 11.1 Average overall monthly expenditure of informal work dominated households

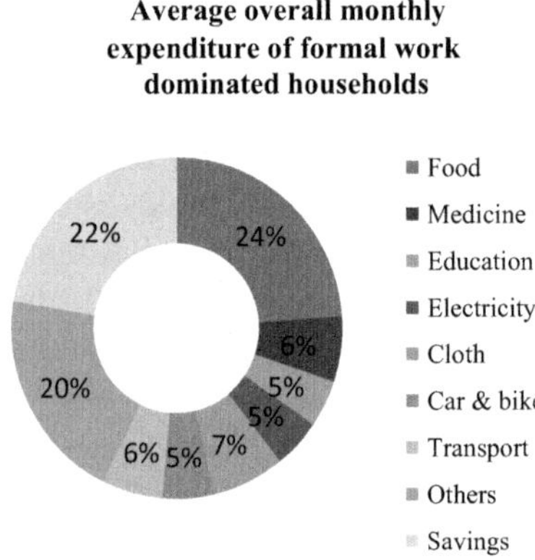

Figure 11.2 Average overall monthly expenditure of formal work dominated households

In spite of this, the foremost contrast has been found in expenditure on transport and car and bike expenses, i.e. 6% and 5% for formal work dominated households though 11% and 1% respectively for informal work dominated households. Furthermore, as the average income is high for the formal work dominated households hence their savings percentage (22%) is also higher than the savings (9%) of informal work dominated households.

Figures 11.3 and 11.4 describes the major average monthly expenditure of sampled households on food, medicine and education. Here, a comparison of expenditure is explained among the sampled formal and informal work dominated households along with the intra-informal sector between sampled slum and non-slum households. Sampled informal work dominated households disbursed monthly an average of Rs. 5954 (24%), 1026 (6%) and 950 (5%) on food, medicine, and education correspondingly while these amounts are almost

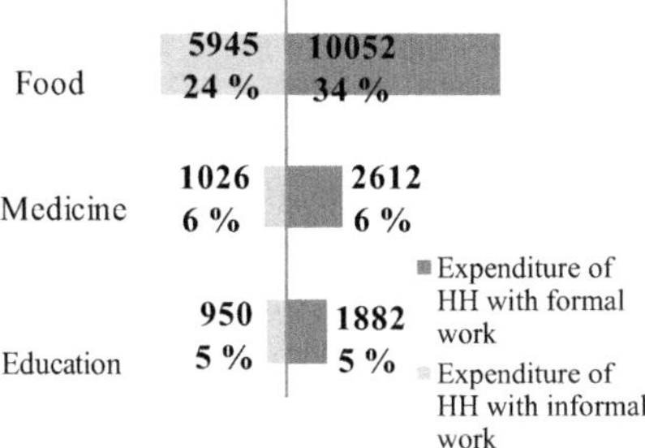

Figure 11.3 Comparison of average monthly expenditure (in Rs. and %) between formal and informal work dominated households

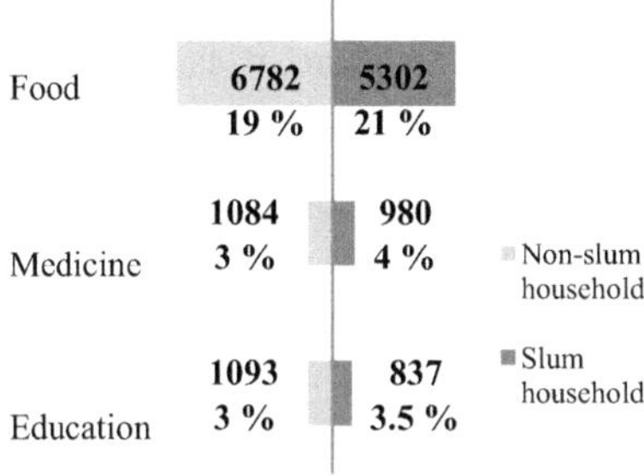

Figure 11.4 Comparison of average monthly expenditure (in Rs. and %) between informal work dominated non-slum and slum households

near to double, i.e, Rs. 10052 (34%) on food, Rs. 2612 (6%) on medicine and Rs. 1882 (5%) on education for the sampled formal work dominated households. In contrast, the average monthly expenditure, for sampled non-slum and slum households within the informal sector, is nearer to each other which reveals that there are not so drastic changes found in slum and non-slum households who work informally.

11.7 Housing Status

Housing conditions and structure varies depending on a household's income level. Slum housing conditions are vulnerable regarding privacy, environment, comfort etc. while the non-slums are quite better than the slums. Kolkata's slums are an essential component of the city, even though certain areas are more crowded and packed than other parts. Here this study investigated some major aspects of housing conditions of sampled slum and non-slum households who engaged in informal work. Housing tenure reveals whether the house is rented or owned, which is also an important factor of assessing the living standard of households. Table 11.10 gives a picture of the housing tenure of sampled slum and non-slum households.

In Table 11.10, near half (47.32%) of the sampled informal work dominated households were found to have their own houses. Out of the total sampled informal work dominated households, non-slum households accounted for the higher (58.78%) share to have their own houses than slum households (38.40%). Moreover, 52.68% of sampled informal work dominated households were found to live on rent basis. In this study, 61.60% of slum households and 41.22% of non-slum households were found to live in rented houses. This study shows that a large number of sampled households were found to live in rented houses, among which the number of informal work dominated slum households is prevailing. As the majority of the slum population are involved in informal work hence with their poor earning they cannot afford their own houses as well.

It is surprisingly observed that, out of total sampled informal work dominated households a large number (60.71%) of households were observed with only one bedroom (Table 11.11). Here it is perceived that, the share of

Table 11.10 Housing tenure of surveyed households by dominant occupation and slum and non-slum households

Housing tenure	*Informal work dominated*		*Total HH (%) (N = 336)*
	Slum HH (%) (N = 188)	*Non-slum HH (%) (N = 148)*	
Own	38.40	58.78	47.32
Rented	61.60	41.22	52.68
Total (%)	**100**	**100**	**100**

Source: Field survey (done by author), 2019

Table 11.11 Number of bedrooms of surveyed households by slum and non-slum

No. of bedroom	*Informal work dominated HH*		*Total HH (%) (N = 336)*
	Slum HH (%) (N = 188)	*Non-slum HH (%) (N = 148)*	
1 bedroom	72.87	45.27	60.71
2 bedroom	20.74	44.59	31.25
3 bedroom	5.85	7.43	6.55
More than 3 bedroom	0.54	2.70	1.49
Total (%)	**100**	**100**	**100**

Source: Field survey (done by author), 2019

sampled slum (72.87%) households is much higher than non-slum households (45.27%). Besides, 44.59% of informal work dominated non-slum households have been found whereas the share of slum households is quite low (20.74%). Moreover, the percentage share of three bedrooms out of the total sampled found only 6.55% among which the share of sampled informal work dominated slum and non-slum households was found to be very low, i.e. 5.85% and 7.4% respectively. Despite this, the study also found some informal work dominated households who have more than three bedrooms whereas the amount is almost negligible (1.49%). In the sampled informal work dominated non-slum households the share is 2.70% while in slum households the amount is insignificant (0.54%).

11.7.1 Household Livelihood Index by Working Category of Informal Workers

In this study an effort has been made to build the Household Livelihood Index (HLI) of sampled informal work dominated households based on Principle Component Analysis (PCA) factor extraction method using significant variables of livelihood indicators. Ultimately, HLI is the combination of the indices of Housing Quality Index (HQI), Basic Amenities Index (BAI), Economic Index (EI), Educational Quality Index (EQI) and Health Index (HI). However, these indices of livelihood have been derived from the selected 24 variables related to overall living standard (Table 11.12).

The index of livelihood is a composite score of indices of each component of livelihood while the index of each component is a composite score of all variables under same component, which can be written as;

$$\mathrm{HLI} = \mathrm{HQI} + \mathrm{BAI} + \mathrm{EI} + \mathrm{EQI} + \mathrm{HI} \tag{11.1}$$

Each index has been constructed on the basis of factor-based weights assigned to each variable with the help of the Principle Component Analysis (PCA) factor extraction method. The model of computation of livelihood index is as follows;

$$\mathrm{HLI} = w_1(x_1) + w_2(x_2) + w_3(x_3) + w_4(x_4) + \ldots\ldots\ldots\ldots + w_{24}(x_{24}) \tag{11.2}$$

where,

HLI denotes household livelihood index in composite form
w denotes component score or coefficient of variability
x denotes Standardized value of variable

In the present analysis coefficient of variability of first component has been considered for measuring the livelihood index, because it alone accounts for 91.48% variance (Table 11.12).

Table 11.12 Variables used for evaluating livelihood indices

Variables		*Variable abbreviation*	*Component matrix*	
			Component 1	*Component 2*
Housing Quality Index (HQI)	Households with own houses	OH	**0.994**	−0.107
	Households with permanent housing structure	PHS	**1.000**	−0.028
	Households with Pucca houses	PH	**1.000**	−0.028
	Households with cement roof	CR	**0.999**	−0.040
	Households with concrete floor	CF	**1.000**	0.022
Basic Amenities Index (BAI)	Households with separate bathroom facility	SB	**0.969**	−0.246
	Households with at least two bed rooms	ALTBD	**0.951**	−0.309
	Households with separate room for each married couple	SRMC	**0.958**	−0.287
	Households with sufficient food throughout the year	SFTY	**0.971**	−0.237
Economic Index (EI)	Working population	WP	**0.968**	0.252
	Female work participation in total workforce	FWP	**0.751**	0.661
	Permanent working status	PWS	**0.757**	−0.653
	Share of skilled and semi-skilled workers	SKSM	**0.902**	0.433
	Households entirely dependent on informal working	EDIW	**1.000**	−0.015
	Households with savings of monthly earning	SAV	**0.923**	−0.384
Educational Quality Index (EQI)	Dropouts in primary and upper primary level (being earliest education level)	DRPU	**0.945**	0.327
	Male literate	ML	**0.995**	−0.104
	Female literate	FL	**0.998**	0.057
	Literacy without formal schooling	LWFS	**0.994**	0.105
	Population have completed 10+2	CMHS	**1.000**	−0.012
Health Index (HI)	Workers health not suffers due to working condition	WHNS	**0.964**	0.267
	Informal working with hygienic working environment	HWE	**0.998**	0.058
	Households first visit medical store for illness	VMSI	**0.869**	0.495
	Households with taken regular ante-natal care during pregnancy	ANC	**0.989**	−0.148
Variance (in %)			**91.48**	**8.52**

Extraction Method: Principle Component Analysis (PCA)

Source: Variables selected and computed by author from forgoing study (2019) using SPSS version 23

The analysis describes (Table 11.13) the self-employed working category recorded in non-slum households and ranked first with highest HLI of 13.10. In addition, the same category of slum households has been shown to have the highest HLI 5.87 and the highest index of all the livelihood components, with the exception of the Educational Quality Index (EQI). The next category of monthly wage earner has been found to represent the different livelihood pattern in slum and non-slum households with HLI 0.84 and 4.12 individually and positioned second in both indexes. Furthermore, the livelihood pattern of weekly wage earners has been found quite similar to monthly wage earners, but it varies in slum and non-slum households with HLI value in slum households is 0.63 while the value of HLI in non-slum household is 1.79 and ranked third in the index. On the contrary, the last category of daily wage earner has recorded the deplorable livelihood index. This category has recorded the least index of livelihood in both slum and non-slum households, i.e. −14.58 and −10.33 illuminating the poorest livelihood standard and ranked third. From the result it seems that the livelihood category differently impacts on different working categories in slum and non-slum households. Self-employed workers don't have any fixed income; it fluctuates with situation rather. The extent of this category varies from street hawkers to small informal businesses. Besides, monthly and weekly wage earners have a fixed income with no workday limits, and no further earnings. This working situation makes their livelihood inferior. Furthermore, the situation of daily wage earners is too vulnerable as they need to earn on a daily basis to sustain their family which exposes them to high risk in livelihood standard. The quality of living of daily wage earners is more or less the same in both the slum and non-slum households. Generally, the different working categories make differences in their living standard because it is somehow related to their income-expenditure pattern which is unanimously acknowledged. The above analysed Livelihood Indices of sampled informal work dominated households on the basis of income range, slum share, and working category of informal workers proved the hypothesis that "informal economy determines the livelihood of informal workers". These indices evaluate the inferior livelihood pattern of informal workers of various category, such as the lowest income worker group and the largest proportion of slum households that primarily engage in informal enterprises and have low living standards.

11.8 Discussion and Conclusion

The informal sector is a heterogeneous sector of various enterprises with both growth and survival assimilation. The significance of informal sector has been progressively comprehended in recent years and various features of the activities studied around the world; perhaps because it became clear that the formal sector service was fundamentally incompetent of engrossing all job seekers (Chukuezi, 2010). The urban economy is quite diverse with a wide range of growth prospects such as enormous number of self-employment and small-scale economic

Table 11.13 Household Livelihood Index by working category of informal workers

Working status	*Housing Quality Index (HQI)*		*Basic Amenities Index (BAI)*		*Economic Index (EI)*		*Educational Quality Index (EQI)*		*Health Index (HI)*		*Household Livelihood Index (HLI)*		*Rank in HLI*	
	Slum	*Non-slum*	*Slum*	*Non-slum*	*Slum*	*Non-slum*	*Slum*	*Non-slum*	*Slum*	*Non-slum*	*Slum*	*Non-slum*	*Slum*	*Non-slum*
Self-employed	1.49	3.58	3.49	4.15	1.12	1.36	−2.05	1.30	1.82	2.71	5.87	13.10	1st	1st
Monthly wage earner	1.14	1.32	−0.12	0.62	0.83	1.33	0.60	1.04	−1.98	−0.19	0.84	4.12	2nd	2nd
Weekly wage earner	0.95	1.13	−1.06	0.48	0.73	−0.38	0.50	1.75	−0.49	−1.19	0.63	1.79	3rd	3rd
Daily wage earner	−3.83	−3.03	−4.11	−3.33	−2.45	−0.85	−2.56	−2.16	−1.64	−0.96	−14.58	−10.33	4th	4th

Source: Computed by author from field survey, 2019

activity with flexible economic returns (Chen and Raveendran, 2011). Informal workers are neglected in every aspect where they are trying to deal with countless challenges every day in their lifestyle and working condition (Gupta, 2009, Pradeep et al., 2017). Besides this, substantial earning prospect in informal sector directed to burgeoning of slums in many Indian cities (Bhat and Yadav, 2017) wherein slum people are engaged in informal working in huge number (Chen and Raveendran, 2011). It has been revealed from this study that the informal workers have deplorable livelihood conditions where they struggle constantly for survival and especially the slum households are at much risk. The findings of this study also reveal that, a large number of sampled households depend on informal work for sustenance. The majority of informal workers and informal work dominated households have a vulnerable socio-economic condition. They are deprived of basic amenities for survival and have very low income wherein they can't afford a good quality of life. Most of the slum people are involved in the informal sector with a wide range of occupations suffering more from vulnerable working circumstances and they lead a vulnerable quality of life. They work in this extreme condition without any social protection. The Household Livelihood Index reveals that the sampled informal workers of slums have the recorded lowest value as the majority of them are engaged in informal activities which further lead to an inferior quality of life. However, it is a challenging task to analyse the livelihood pattern of urban informal workers especially in slums where the unpredictability of informal activities is higher. Despite this, the government and policy-makers should give special attention to improving the overall development of informal workers and their protection to make them stand strong against the livelihood crisis.

Bibliography

Bhat, J. A., & Yadav, P. (2017). Economic Informal Sector and the Perspective of Informal Workers in India. *Art and Social Sciences Journal*, *8*(1), 1–9.

Bhosle, B. V. (2014, October). Informal Sector: Issues of Work and Livelihood. *Yojana*, pp. 36–39.

Bhowmik, S. K., & More, N. (2001). Copying with Urban Poverty- Ex-Textile Mill Workers in Central Mumbai. *Economic and Political Weekly*, *36*(52), 4822.

Biswas, D. (2020). Problems of Unorganized Workers and their Social Security Measures in India: An Analysis. *International Journal of Research and Analytical Reviews*, *7*(1), 931–940.

Bose, M., & Roy, S. D. (2019). Conditions of Employment and Livelihood Security of Informal Workers: A Study of Four Villages in Jalpaiguri District, West Bengal. *Agrarian South: Journal of Political Economy*, *08*, 1–26.

Bose, R., & Ghosh, S. (2015, December). Slums in Kolkata: A Socio-Economic Analysis. *The Empirical Economietrics and Quantitative Economics Letters*, *4*(4), 134–148.

Census of India. (2001). *Town Directory, Kolkata*. West Bengal: Directorate of Census Operations.

Census of India. (2011). *District Census Handbook, Kolkata, Village and Town Wise Primary Census Abstract (PCA)*. New Delhi: Directorate of Census Operations, West Bengal.

Chaudhuri, S. (1990). *Calcutta- The Living City* (Vol. 1). New Delhi: Oxford University Press.

Chaudhuri, S. (2013). Living at the Edge: A Socio-economic Profile of Kolkata's Poor. *South Asian Survey*, *20*(1), 44–58.

Chen, M. A., & Raveendran, G. (2011 (updated 2014)). *Urban Employment in India: Recent Trends and Patterns*. Women in Informal Employment: Globalizing and Organization (WIEGO), Working paper no. 7.

Chukuezi, C. O. (2010, August). Urban Informal Sector and Unemployment in Third World Cities: The Situation in Nigeria. *Asian Social Science*, *6*(8), 1–7.

Ellis, F. (2000). The Determinants of Rural Livelihood Diversification in Developing Countries. *Journal of Agricultural Economics*, *51*(2), 289–302.

Forsyth, F. G. (1960). The Relationship Between Family Size and Family Expenditure. *Journal of the Royal Statistical Society*, *123*, 367–397.

Gupta, K. R. (2009). *Economics of Development and Planning* (Vol. 2). Daryaganj, New Delhi: Atlantic Publishers and Distributors.

I. L. E. Report, Alakh, N. S. (2014). *India Labour and Employment Report: Workers in the Era of Globalisation*. New Delhi: Academic Foundation & Institute for Human Development.

International Labour Organisation. (2002). *Decent Work and the Informal Economy, Report IV, 90th Session*. Geneva: International Labour Office.

Ismail, R., & Bakar, N. T. (2012, January). The Relationship Between Income, Expenditure and Household Savings in Peninsular Malaysia. *Malaysian Journal of Consumer and Family Economics*, *15*, 168–189.

Lim, L. L. (2015). *Extending Livelihood Opportunities and Social Protection to Empower Poor Urban Informal Workers in Asia*. Jakarta: Oxfam for Asia Development Dialogue.

Mazumdar, R., & Sarkar, A. (2015, March). Determinants of Credit Accessibility in Unorganized Sector: Empirical Evidences from Kolkata. *International Journal of Core Engineering & Management*, *1*(12), 110–128.

Meikle, S. (2002). The Urban Context and Poor People. In C. Rakodi, & T. Lloyd-Jones, *Urban Livelihood: A People Center Approach to Reducing Poverty* (pp. 37–51). London: Earthscan Publications Limited.

Mitra, A. (1990). Duality, Employment Structure and Poverty Incidence: The Slum Perspective. *Indian Economic Review, New Series*, *25*(1), 57–73.

Mohapatra, K. K. (2012, November). Women Workers in Infomal Sector in India: Understanding the Occupational Vulnerability. *International Journal of Humanities and Social Science*, *2*(21), 197–207.

Narayan, D., Patel, R., Schafft, K., Rademacher, A., & Koch-Schulte, S. (2000). *Voices of the Poor: Can Anyone Hear Us?* New York: Oxford University Press.

NSSO. (2013). *Key Indicators of Employment and Unemployment in India (2011–12)*. New Delhi: Government of India.

Papola, T. S. (1981). *Urban Informal Sector in a Developing Economy*. New Delhi: Vikash Publishing House.

Pradeep, M. D., Ravindra, B. K., & Ramjani Sab, T. (2017). A Study on the Prospects and Problems of Unorganised Labours in India. *International Journal of Applied and Advanced Scientific Research*, *2*(1), 94–100.

Romatet, E. (1983). Calcutta's Informal Sector: Theory and Reality. *Economic and Political Weekly*, *18*(50), 2115–2128.

Shonchoy, A. S., & Junankar, P. R. (2014). The Informal Labour Market in India: Transitory or Permanent Employment for Migrants? *IZA Journal of Labour and Development*, *3*(9), 173–202.

Timalsina, K. P. (2011, March). An Urban Informal Economy: Livelihood Opportunity to Poor or Challenges for Urban Governance. *Global Journal of Human Social Science, 11*(2), 24–32.

UNESCO. (2008). *Education for All by 2015: Will We Make It?- EFA Global Monitoring Report 2008*. Paris: UNESCO: Oxford University Press.

Uphoff, N. (1986). *Local Institutional Development: An Analytical Sourcebook with Cases*. West Hartford: Kumarian Press.

Van Vuuren, A. (2003). *Women Striving for the Self-relience: The Diversity of Female Headed Households in Tanzania and the Livelihood Strategies they Employ*. Leiden: African Studies Centre.

Varghese, N. V. (1989). *Education and the Labour Market in India* (Vols. IIEP Research Report, 67). France: International Institute for Educational Planning.

12 Fertility Status and Deprivation of Muslim Women Population

A Micro Level Study in Selected CD Blocks of West Bengal

Soleman Khan and A. K. M. Anwaruzzaman

12.1 Introduction

Fertility is the only biological medium which replenishes human society (Kanitkar and Bhende 248). The role of gender equality is very important for various aspects of development including lowering the fertility rate across countries (Mikkola 02; Ghosh 68; Arnaout et al. 02; Anderson and Kohler 382). Low gender equality is witnessed for Muslims in various socio-economic indicators like educational attainment and work participation status (Islam and Siddiqui 1325; Sengupta and Rooj 87; Shehu and Zejno 20). Muslim women are considered as the much deprived section of the society in India (Hussain et al. 311; Jahan 01; Hossain 14; Siddiqui et al. 254). Under the 'parda' system often Muslim men hide the deprivation of Muslim women in terms of gender equality (Mollah 1011). The high fertility rate of Muslims compared to Hindus is a debatable topic in the population study of India (Pasupuleti et al. 147). Although many previous studies have tried to explore the facts behind this high fertility, nevertheless, confusion still remains; still the issue is largely responsible for the disturbance of communal harmony. Therefore, in this chapter an attempt has been made to understand the level of deprivation of Muslim women in terms of gender and how this influences their fertility behaviour. The purpose of the chapter is to assess the level of deprivation among Muslim women in terms of gender equality, and to explore the influence of low gender equality on the high fertility rate of Muslim women.

12.2 Study Area

In the present study, six selected CD blocks of West Bengal have been considered for analysis. The CD blocks include Barasat-II, Englishbazar, Mograhat – II, Nalhati-II, Kaliachak-III, and Suti-II. The latitudinal and longitudinal details of the sample CD blocks have been represented in Table 12.1. According to the Census of India (2011), all the selected six CD blocks have been found to be numerically dominated by a Muslim population (≥ 50 per cent Muslims) when the population of the statutory town is not considered within the

DOI: 10.4324/9781003275916-14

Table 12.1 Geographical coordinates of sample CD blocks

Sl. No.	*Districts*	*Sample CD blocks*	*Latitudinal and longitudinal extent of the sample CD blocks*
1	North 24 Parganas	Barasat - II	22°38′02″N–22°44′39″N 88°26′30″E–88°36′52″E
2	Maldah	Englishbazar	24°49′45″N–25°05′03″N 87°58′42″E–88°10′58″E
3	South 24 Parganas	Magrahat - II	22°12′45″N–22°20′43″N 88°19′32″E–88°26′37″E
4	Birbhum	Nalhati - II	24°10′41″N–24°21′15″N 87°55′38″E–88°01′16″E
5	Maldah	Kaliachak - III	24°40′18″N–24°51′50″N 87°53′58″E–88°08′55″E
6	Murshidabad	Suti - II	24°31′02″N–24°38′11″N 87°54′17″E–88°06′15″E

Source: Based on Census of India, 2011

respective CD block population. These sample six CD blocks comprising 1.16 per cent of the total area of the state account for 2.13 per cent of the total population. Besides, these sample CD blocks account for 3.86 per cent of the total Muslim population of the state. The status of the accessibility and availability of various amenities and services is different among these CD blocks; this helps with understanding the status of gender equality among Muslims in detail in both the backward and the advanced areas (based on accessibility and availability of amenities and services) and its influence on their high fertility (Figure 12.1).

12.3 Methodology

The present study is mainly based on primary data. The primary data were collected through the field survey method (using survey schedule) during 01/01/2018 to 31/12/2018. In the present study 600 (100 respondents from each sample CD blocks) married Muslim women of the reproductive age group, i.e. 15–49 are considered as respondents following the multi-stage random sampling technique. At the first stage Muslim majority CD blocks (≥ 50 per cent Muslim population) are identified based on Census 2011 data (excluding statutory town population). Considering the data on infrastructure, amenities and services in the Census 2011 database, a Z score of all Muslim majority CD blocks is calculated. Then the CD blocks are organised in descending order based on their Z score value. Two CD blocks are selected from the top third Z scoring CD blocks (Barasat – II from North 24 Parganas and English Bazar from Maldah district), another two from the middle third group

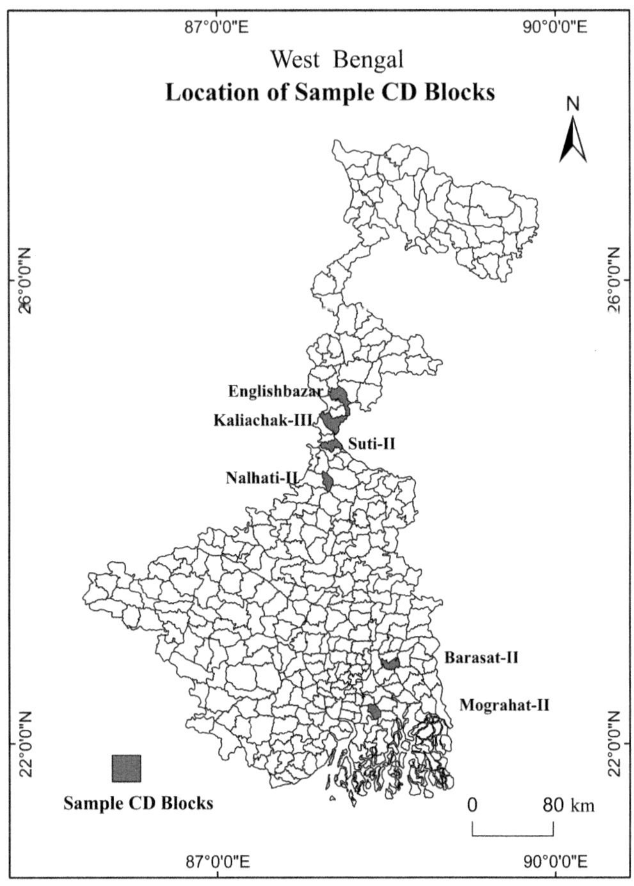

Figure 12.1 Location map of sample CD blocks

(Mograhat – II from South 24 Parganasana, Nalhati– II from Birbhum district), and the remaining two from the bottom third Z score ranking CD blocks (Kaliachak – III from Maldah district and Suti – II from Murshidabad district). After the demarcation of sample CD blocks, in the second stage, based on the same principle (Muslim majority and status of amenities and services) two villages are selected from each sample CD block for final study. While selecting the sample villages from the Z score list of villages of respective sample CD blocks, the first village is selected from the top of the Z score list and the second one is selected from the bottom of the list. At the third stage, the list of Muslim households with married women between the age group 15 and 49 years is collected from the Integrated Child Development Scheme (ICDS) Centres, i.e. Anganwadi Centres and also from the ASHA workers. The respondents are selected randomly (simple random sampling) by using random table method. The population of statutory town is not considered here for study.

In order to achieve the listed objectives, a number of statistical tools and techniques have been applied. The statistical tests correlation and regression have been conducted in order to understand the association between gender deprivation and high fertility rate for Muslims. MS Excel is also used for analysing the data. Necessary mapping has been done using ArcGIS 10.3 software. For representation of data, cartographic techniques like bar, pie, line graph, etc. are used.

12.4 Result and Discussion

To analyse the level of deprivation among Muslim women in terms of gender equality and its association with fertility, the study is divided into a number of sub-points like decision and age at marriage, educational attainment, financial autonomy, contraceptive use and non-consensual sex, gender preference during birth and the prevalence of patriarchy, and domestic violence. The result of the detailed analysis is discussed below under the heading of the respective sub-points.

12.5 Decision and Age at Marriage

The result of the field survey data shows that Muslim women have very low opportunity to express their decision for marriage (Figure 12.2) which is largely associated with the low age at marriage of Muslim women in the state. The average age at marriage of Muslim women is found to be only 16.92 years. As high as 63.72 per cent respondents who were married Muslim women have mentioned that their family members (mainly male members) do not care

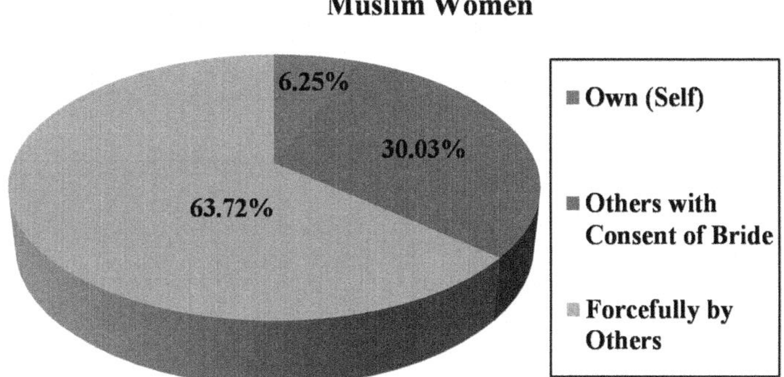

Figure 12.2 Decision of marriage of respondent married Muslim women

about their consent for marriage and they got married at an early age (Khan and Anwaruzzaman 93). This highlights the high prevalence of the practice of child marriage among Muslim women. The survey data reveals that only 30.03 per cent of Muslim brides gave consent for marriage (Figure 12.2) while for the Muslim men the same is around 94.05. This indicates the high gender inequality status for Muslim women regarding consent for marriage. The study reveals that around 61.81 per cent of marriage of the respondents is at an early age (below 18 years), which may largely promote the high fertility rate (Anwaruzzaman and Khan 27). The state's total child marriage rate, as reported by the Census of India (2011), is around 40.27 per cent. As per our prevailing societal norms, marriage is the only process through which couples can participate in child birth. The result derived from the analysis of field data shows that a negative relationship prevails between age at marriage and fertility level (Figure 12.3). The analysis of the sample CD blocks data points out that the women who get married below age 18 have the highest total fertility level (2.96). The increase of the age at marriage gradually decreases the fertility level. The age at marriage ≥25 shows a total fertility rate of 1.81, which is below replacement level (2.1).

The trend line analysis also shows a downward slope between age of marriage and total number of live births/Muslim women. Here the correlation coefficient $r = -0.36$ and the value of p is <0.001. Therefore, it can be concluded that there is a significant negative relationship existing between them. The CD block-wise analysis of the age at marriage and the consent for marriage of Muslim married women (Figure 12.4) shows that it has an association with their total fertility rate (TFR). CD block Barasat – II, Englishbazar, and Nalhati – II represent comparatively higher age at marriage and consent for marriage for Muslim married women and represent a low level of fertility. On the other hand, CD blocks Suti – II, Kaliachak – III and Mograhat – II witnessed comparatively low age at marriage and consent for marriage with a high level of fertility.

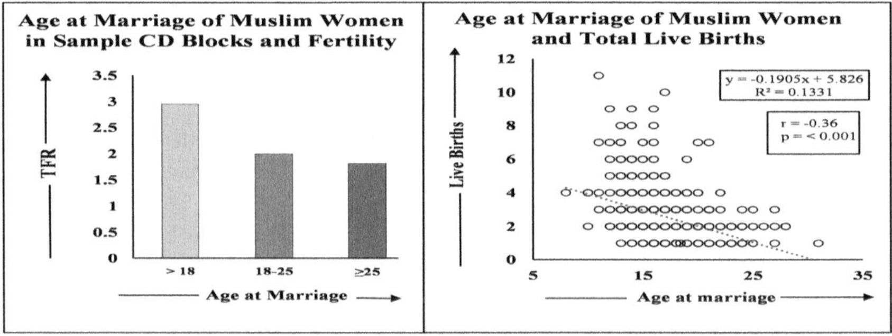

Figure 12.3 Age at marriage data

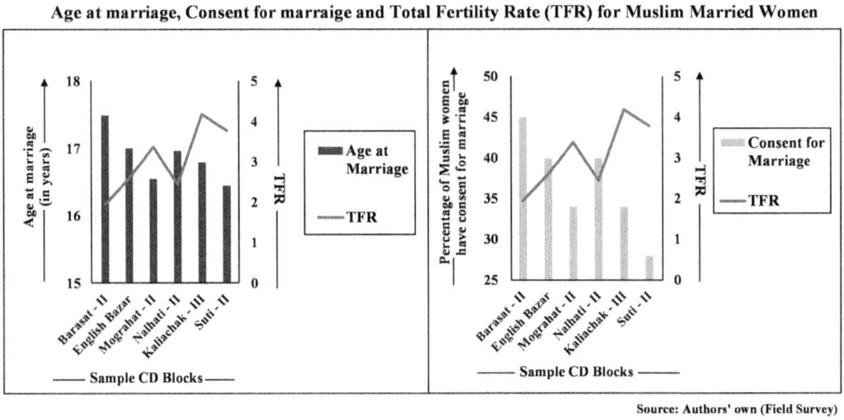

Age at marriage, Consent for marraige and Total Fertility Rate (TFR) for Muslim Married Women

Source: Authors' own (Field Survey)

Figure 12.4 Sample CD blocks

12.6 Educational Attainment

Educational attainment analysis is an important tool for the analysis of gender equality (Sahin 60; Chisamya, DeJaeghere and Kendall 744). According to the Census of India (2011), the literacy rate of Muslim women (64.77) is low compared to Muslim men (72.52) in the state. The results derived from the analysis of census of India (Figure 12.5) shows that in various levels of education the number of females is less compared to males for Muslims in the state. Figure 12.5 also exhibits that the rate of girls' school dropout is much high in the higher level of education (higher secondary and above). The field survey data analysis result shows a very interesting fact regarding the educational attainment level of married Muslim men and women. It is found that a large proportion of Muslim men (23.63 per cent) are uninterested in achieving higher educational attainment due to poor job opportunities. On the other hand, a large proportion of Muslim families (41.58 per cent) send their girls to achieve higher education reported to be for marriage purposes only. They have mentioned that higher educational attainment of their daughters will cause a low dowry for them. The data show that the average educational attainment of married Muslim women is comparatively higher where the rate of urbanization (based on the accessibility and availability of amenities and services) is also higher like Barasat – II and Englishbazar; these CD blocks have a comparatively low level of fertility (Figure 12.6). On the other hand, CD blocks Suti – II, Kaliachak – III, and Mograhat – II represent a higher fertility rate and their average educational level of married women is witnessed lower compared to men among Muslims. It is often seen that couples prefer low educational attainment of girls than boys. They believe that higher education is not necessary for women, rather they should concentrate on household work. Often women are forced to discontinue their higher educational aspiration and are forced to marry at an early age; this promotes higher school drop-out rate for

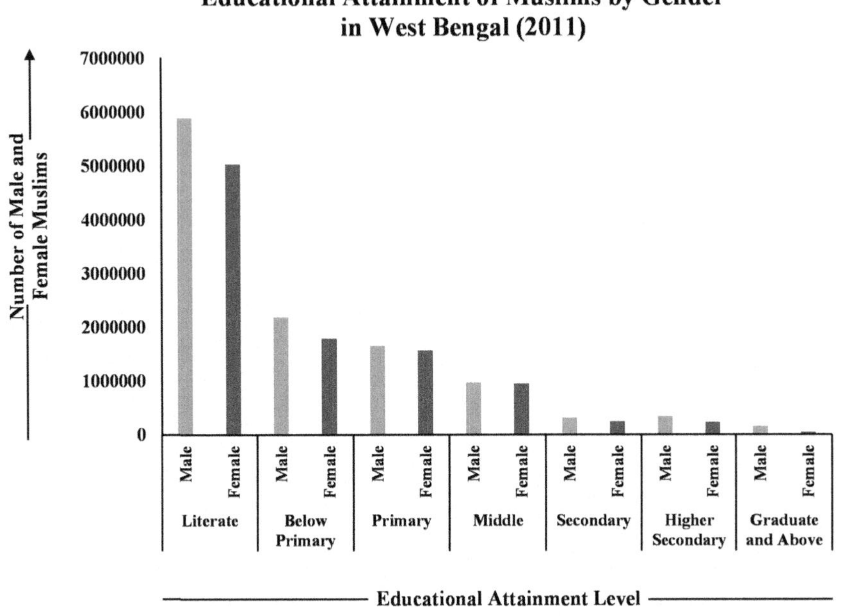

Figure 12.5 Educational attainment of Muslims by gender in West Bengal

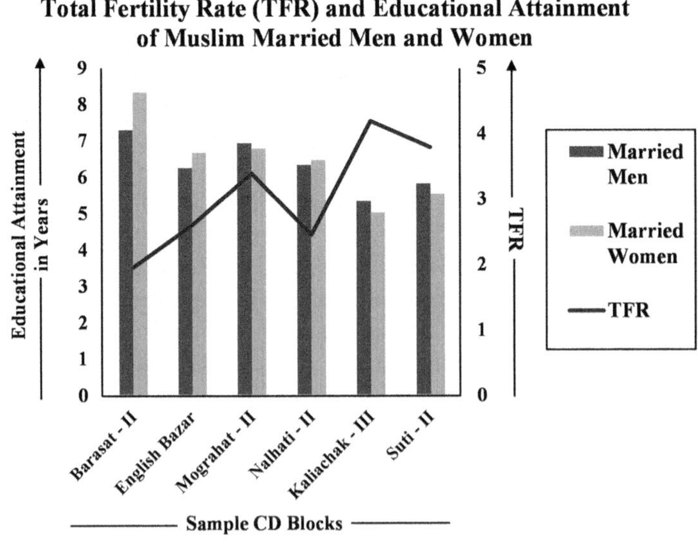

Figure 12.6 TFR and educational attainment of Muslim married men and women

females. It is derived that the CD block of Mograhat – II has the highest share of Muslim couples (28 per cent) having perception and preference for the support of lower educational attainment of girls than boys. Besides, the CD blocks of Kaliachak – III and Suti – II also have a comparatively higher share of Muslim couples considering higher educational attainment of girls is not as significant as it is in case of boys.

12.7 Financial Autonomy

The analysis of the status of the financial autonomy of Muslim women represents higher rate of deprivation of Muslim married women in terms of gender and shows very low work participation rate, engagement in mainly 'Bidi' rolling activities, a low percentage of women having their own bank account, and low rate of participation for outdoor work (minimum 3 km distance away from households). The CD blocks Barasat – II and Englishbazar represent comparatively better financial autonomy and low level of fertility (Figure 12.7). On the

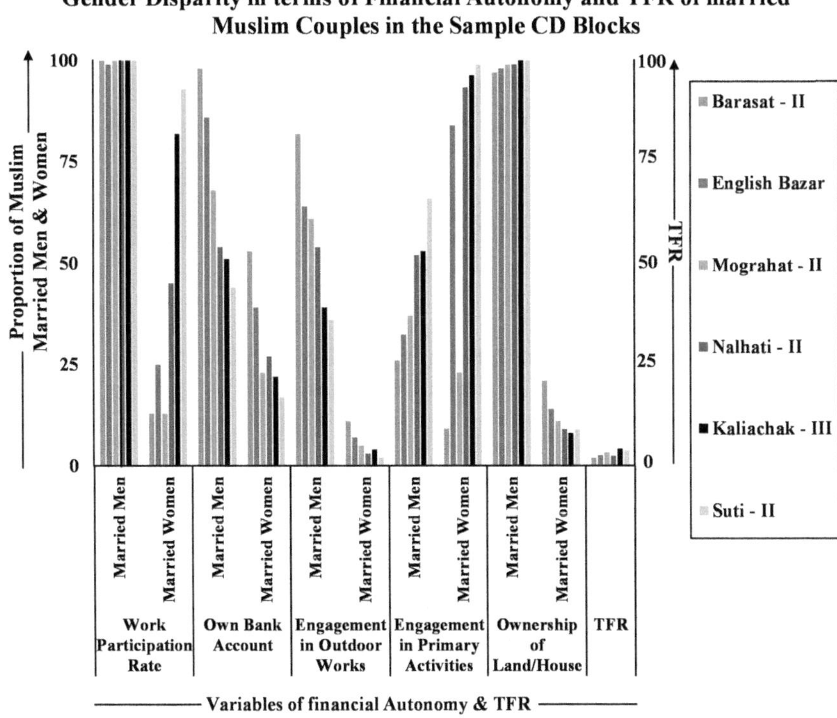

Figure 12.7 Gender disparity in terms of financial autonomy and TFR of married Muslim couples in the sample CD blocks

other hand, CD blocks Kaliachak – III and Suti – II have poor financial autonomy for married Muslim women with high rate of fertility. The condition of work participation shows that married Muslim women in Kaliachak – III and Suti – II are mainly engaged in 'Bidi' rolling activities which is not capable of reducing their fertility level.

12.8 Contraceptive Use and Non-consensual Sex

The analysis reflects that the practice of contraceptives is mainly adopted by Muslim women. Married men do not wish to practise the use of contraceptives, which is often related to their patriarchal view point. Figure 12.8 exhibits that around 75.66 per cent of couples in the study area, the wives bear the responsibility for practising modern family planning methods for Muslims. This indicates a status of low gender equality in terms of bearing the responsibility of practising modern contraceptives for limiting family size. Besides, the high prevalence of Non-consensual sex also indicates low gender equality among Muslims and this may have an association with their high fertility rate. The prevalence of non-consensual sex shows low women empowerment status and high patriarchal views of Muslim people in study area. The result derived from field survey data indicates a positive relationship between the prevalence of non-consensual sex and high fertility for Muslims exists. Figure 12.9 exhibits that in the study area around 29.95 per cent of respondents have experienced non-consensual sex in their married life. The CD block-wise analysis points out that the presence of non-consensual sex is very high among Muslim peoples in the CD blocks Suti – II and Kaliachak – III where the level of fertility is observed to be quite high. On the other hand, CD block Barasat – II has a low prevalence of non-consensual sex and the rate of fertility of Muslim people is low too.

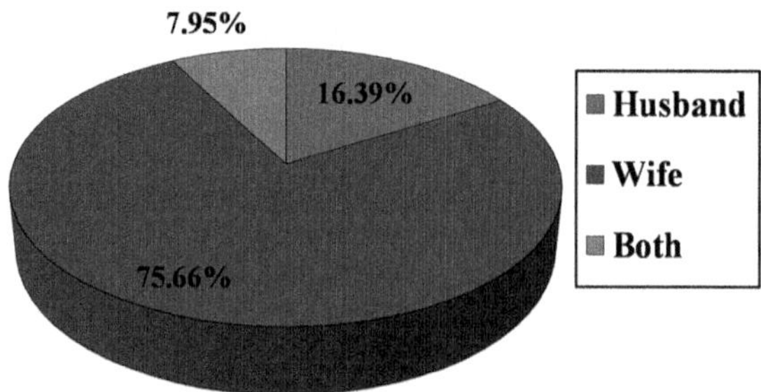

Figure 12.8 Use of modern contraceptives by married Muslim men/women in sample CD blocks

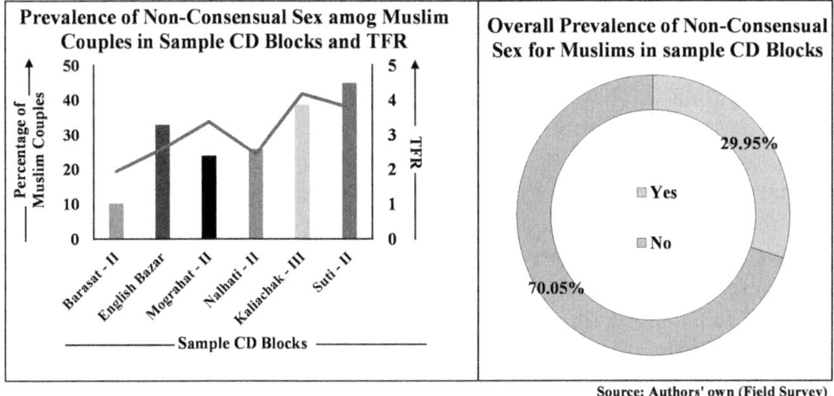

Figure 12.9 Prevalence of non-consensual sex among Muslim couples in sample CD blocks and TFR

12.9 Gender Preference at Birth and Practice of Patriarchy

The field survey data show that the preference for a male child is an important exhibition of gender disparity among Muslims and is also causing high fertility among Muslims. This finding is supported by a number of studies (Asgar et al. 01; Nasir & Kalla 276). The analysis reveals that the preference is for a male child. Figure 12.10 shows that around 72.92 per cent couples desire a male child and only in 27.08 per cent of cases do they desire a female child. This highlights the views of the couples towards the superiority of a male child. Patriarchy is a chronic social disease in the perception of people where the superiority and high preference for a male child is prevailing to a great extent. The causes of desire for a male child are social pride, lineage factor, patriarchal views, income source, etc. (Carranza 26; Nugent 75). The desire for a male

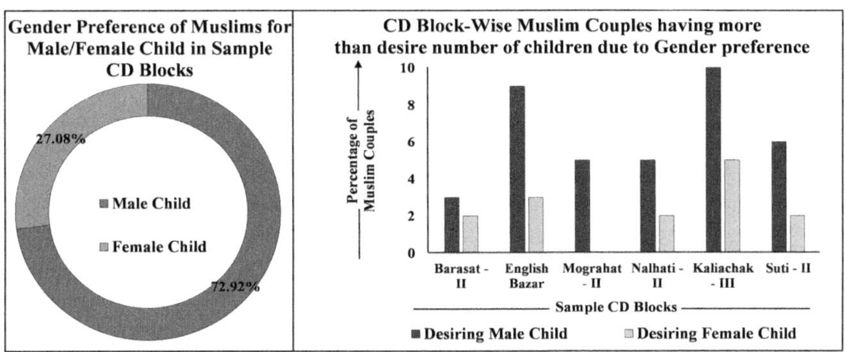

Figure 12.10 CD blocks of gender preference

child appears to be a determining factor in fertility analysis. The study shows that the desire for a male child often increases the rate of fertility of Muslim people. Figure 12.10 exhibits the CD block-wise percentage of couples who have more than the desired number of children due to gender preference as revealed in the field survey. It is observed that the CD blocks Kaliachak – III, Englishbazar, and Suti – II have a comparatively high percentage of couples who have more than the desired number of children due to the preference for a male child. The CD block of Barasat – II shows the lowest percentage of Muslim couples having more than the desired number of children due to gender preference.

Figure 12.11 shows the distribution of Muslim households in the study area who believe in the superiority of male child. It is derived through the result of field survey data analysis that CD blocks of Kaliachak – III, Suti – II and Mograhat – II have a comparatively higher percentage of Muslim households having the perception of superiority of male child. Besides, CD blocks of Barasat – II, Nalhati – II, and Englishbazar have comparatively low percentage of Muslim households having belief in male superiority; and there the level of fertility is also observed to be low. The prevalence of the superiority of a male child can be analysed in many ways. The analysis points out that the proportion of Muslim couples with the perception of the presence of at least one male child in the family is quite high in Suti – II, Kaliachak – III and Mograhat – II; this desire is responsible for an increase in their level of fertility. It is seen that although the level of fertility is the lowest in Barasat – II among Muslims at the same time the percentage of Muslim couples having a belief in the presence of at least one male child in the family is quite high. The CD block of Nalhati – II shows the lowest percentage of couples (21 per cent) having the perception of the presence of at least one male child in family. The percentage of Muslim households considering the birth of girl child to be undesirable also expresses the superiority of male child. Figure 12.11 exhibits that the CD block Mograhat – II has the highest percentage of Muslim households (20 per cent) considering the birth of girl child to be undesirable. On the other hand, the CD block of Nalhati – II has only 2.4 per cent of Muslim households considering the birth of girl child to be undesirable.

12.10 Domestic Violence

In sample CD blocks around 33.17 per cent of respondents have reported that they have experienced 'assault and torture' by the members of grooms' family; this is strongly associated with gender disparity (Kaur and Garg 74; Caragnano 51). The field survey observation brings out the fact that prevalence of 'assault and torture' by family members largely affect the empowerment status of women. The presence of 'assault and torture' results in break-down of the self-confidence of the respondents. This makes them mere dolls regarding decision making capacity at the hands of male members in our society. The respondents who have experienced 'assault and torture' by

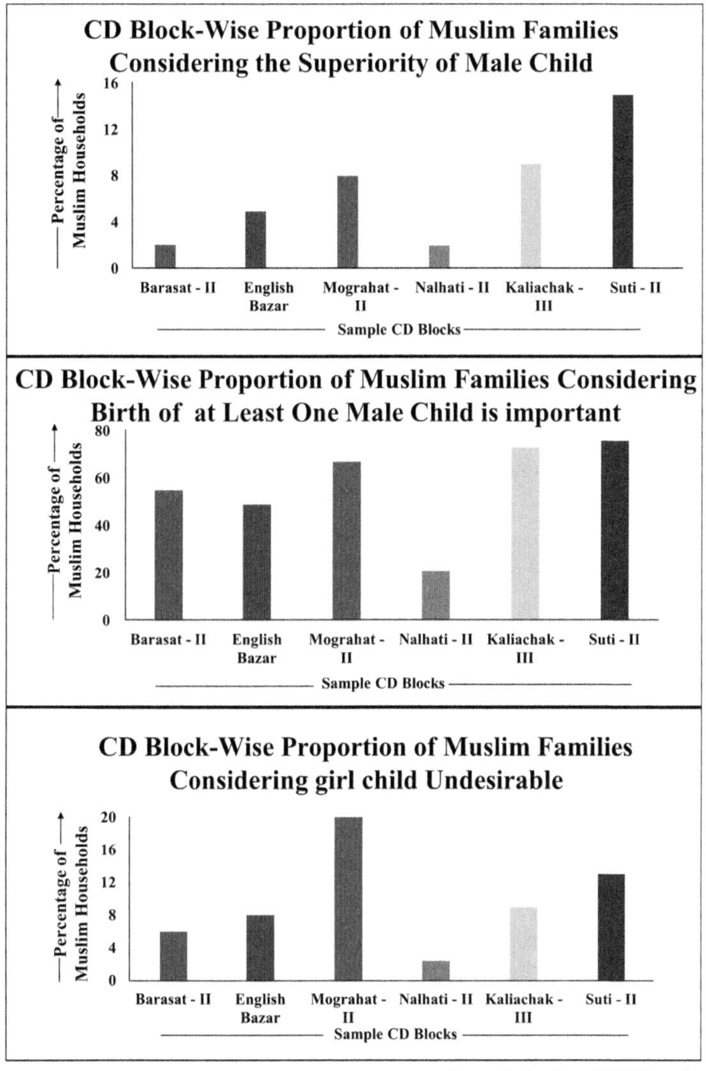

Source: Authors' own (Field Survey)

Figure 12.11 CD block-wise proportion data

the members of grooms' family show a very low level of women empowerment with high TFR (Figure 12.12). The percentage of respondents facing 'assault by husband' is relatively quite high in the CD blocks Kaliachak – III, Suti – II, and Mograhat – II. The presence of 'torture for the birth of girl child' is also observed to be significantly high in those CD blocks, except for Nalhati – II. In the CD block Nalhati – II, the overall attitude towards the birth of girl child is found to be good; although assault by husband is still

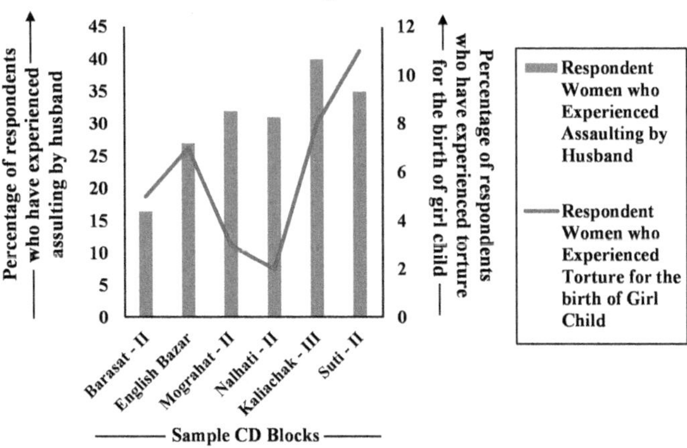

Figure 12.12 Percentage of respondent Muslim Women who have experienced assault and torture

prevalent. The CD block of Barasat – II shows low percentage of respondents who have experienced 'assault and torture' (16.5 percent for 'assault by husband' and 5 per cent for 'torture for the birth of girl child'), where the TFR is the lowest among all sample CD blocks (1.95). Therefore, a strong negative correlation can be ascertained between fertility level of women and presence of 'assault and torture' by grooms' family.

12.11 Conclusion

The above analysis indicates the poor level of gender equality among Muslims, which has a significant association with their high level of fertility. The CD blocks Barasat – II and English Bazar represent a comparatively high level of gender equality with a low level of fertility. On the other hand, CD blocks Suti – II and Kaliachak – III represent a comparatively higher level of deprivation of Muslim married women in terms of gender equality and witness high level of total fertility i.e. 3.79 and 4.19 respectively. CD blocks Mograhat – II and Nalhati – II show moderate level of fertility rate and gender equality of Muslims. Thus, through empowering Muslim women gender equality may be achieved that would lead to decreasing fertility among Muslims.

Bibliography

Arnaout, Nour, et al. "Gender Equity in Planning, Development and Management of Human Resources for Health: A Scoping Review." *Human Resources for Health*, vol. 17, no. 1, 2019, pp. 52–60, doi:10.1186/s12960-019-0391-3. Accessed 22 Feb. 2021.

Anderson, Thomas and Hans Peter Kohler. "Low Fertility, Socioeconomic Development, and Gender Equity." *Population and Development Review*, vol. 41, no. 3, 2015, pp. 381–407, www.ncbi.nlm.nih.gov/pmc/articles/PMC4623596/. Aceessed 24 Feb. 2021.

Anwaruzzaman, A. K. M. and Soleman Khan. "Child Marriage in West Bengal: Nature, Extent, Causes and Regional Dimensions." *Child Marriage The Root of Social Maladies*, edited by Gandhari Saha, Levant Books, Kolkata, 2017, pp. 20–34.

Asgar, Mohammad, et al. "Fertility Behaviour and Effect of Son Preference among the Muslims of Manipur, India." *Journal of Anthropology*, 2014, pp. 1–5, downloads. hindawi.com/archive/2014/108236.pdf. Accessed 26 Feb. 2021.

Caragnano, Roberta. "Violence against Women and Domestic Violence: General Framework and Overview of the Convention (Articles 1 and 2)." *Athens Journal of Law*, vol. 4, no. 1, 2018, pp. 51–66. www.athensjournals.gr/law/2018-4-1-3-Caragnano. pdf. Accessed 25 Feb. 2021.

Carranza, Eliana. *Islamic Inheritance Law, Son Preference and Fertility Behavior of Muslim Couples in Indonesia*. Washington, DC, World Bank, 2012. sites.bu.edu/ neudc/files/2014/10/paper_82.pdf. Accessed 28 Feb. 2021.

Chisamya, Grace, et al. "Gender and Education for All: Progress and problems in achieving gender equity." *International Journal of Educational Development*, vol. 32, no. 6, 2012, pp. 743–755, doi:10.1016/j.ijedudev.2011.10.004. Accessed 26 Feb. 2021.

Ghosh, Madhusudan. "Gender Equality, Growth and Human Development in India." *Asian Development Perspectives*, vol. 9, no.1, 2018, pp. 68–87. www.researchgate.net/ publication/326439355_Gender_Equality_Growth_and_Human_Development_in_ India. Accessed 23 Feb. 2021.

Hossain, Md. "Muslim Women of West Bengal: An Enquiry into their Minority Status." *IOSR Journal Of Humanities and Social Science*, vol. 4, no. 3, 2012, pp. 14–21. iosrjournals.org/iosr-jhss/papers/Vol4-issue3/C0431421.pdf. Accessed 20 Feb. 2021.

Hussain, Manzoor, et al. "Educational Status of Muslim Women in India: Issues and Challenges." *Scholars Journal of Arts, Humanities and Social Sciences*, vol. 6, no.2, 2018, pp. 311–316. www.researchgate.net/publication/330535000_Educational_ Status_of_Muslim_Women_in_India_Issues_and_Challenges. Accessed 18 Feb. 2021.

Islam, M., and Lubna Siddiqui. "A Geographical Analysis of Gender Inequality in Literacy among Muslims of West Bengal, India (2001–2011)." *GeoJournal*, vol. 85, no. 5, 2020, pp. 1325–1354. link.springer.com/article/10.1007/s10708-019-10025-1#:~:text=In%20fact%2C%20inter%2Dgenerational%20gender,Muslim%20 women%20than%20Muslim%20men.&text=In%202011%20census%2C%20the%20 gender,%25%20males%20and%2051.9%25%20females. Accessed 1 Mar. 2021.

Jahan, Yasmeen. "Intersectionality of Marginalization and Inequality: A Case Study of Muslims in India." *Journal of Political Sciences & Public Affairs*, vol. 4, no. 1, 2016, pp. 1–6. www.longdom.org/open-access/intersectionality-of-marginalization-and-inequalitya-case-study-of-muslims-in-india-2332-0761-1000187.pdf. Accessed 16 Feb. 2021.

Kanitkar, Tara and Asha A. Bhende. *Principal of Population Studies*. Nagpur: Himalaya Publishing House Pvt. Ltd, 2015.

Kaur, Ravneet and Suneela Garg. "Addressing Domestic Violence Against Women: An Unfinished Agenda." *Indian Journal of Community Medicine*, vol. 33, no. 2, 2008, pp. 73–76. www.ncbi.nlm.nih.gov/pmc/articles/PMC2784629/. Accessed 18 Feb. 2021.

Khan, Soleman and A. K. M. Anwaruzzaman. "Early Marriage of Girls in India: A Regional Perspective." *CHILD MARRIAGE The Root of Social Maladies*, edited by Gandhari Saha, Levant Books, Kolkata, 2017, pp. 85–97.

Mikkola, Anne. "Role of Gender Equality in Development - A Literature Review." *SSRN Electronic Journal*, 2005, pp. 1–44. https://papers.ssrn.com/sol3/papers. cfm?abstract_id=871461. Accessed 20 Feb. 2021.

Mollah, Kamruzzaman. "Status of Muslim Women in West Bengal." *International Journal of Research and Analytical Reviews*, vol. 5, no. 2, 2014, pp. 1009–1013. www.ijrar.com/upload_issue/ijrar_issue_936.pdf. Accessed 27 Feb. 2021.

Nasir, Rosina and A. K. Kalla. "Kinship System, Fertility and Son Preference among the Muslims: A Review." *The Anthoropologist*, vol. 8, no. 4, 2006, pp. 275–281. www.researchgate.net/publication/228342852_Kinship_System_Fertility_and_Son_Preference_among_the_Muslims_A_Review. Accessed 28 Feb. 2021.

Nugent, Jeffrey B. "The Old-Age Security Motive for Fertility." *Population and Development Review*, vol. 11, no. 1, 1985, pp. 75–97. www.jstor.org/stable/1973379?seq=1#metadata_info_tab_contents. Accessed 20 Jan. 2021.

Pasupuleti, Samba, et al. "Hindu–Muslim Fertility Differential in India: A Cohort Approach." *Journal of Biosocial Science*, vol. 49, no. 2, 2017, 147–172. www.researchgate.net/publication/303951929_HINDU-MUSLIM_FERTILITY_DIFFERENTIAL_IN_INDIA_A_COHORT_APPROACH. Accessed 28 Feb. 2021.

Sahin, Elmas. "Gender Equity in Education." *Open Journal of Social Sciences*, vol. 2, no. 1, 2014, pp. 59–63. www.researchgate.net/publication/275886441_Gender_Equity_in_Education#:~:text=Gender%20equality%20in%20education%20gives,%5B9%5D%20.%20...&text=Lilik%20Rita%20Handayani-,...,discrimination%20and%20exclusion%20%5B8%5D. Accessed 24 Feb. 2021.

Sengupta, Reshmi and Debasis Rooj. "Factors Affecting Gender Disparity in Muslim Education in India." *Journal of Development Policy and Practice*, vol. 3, no. 1 2018, pp. 87–113. doi: 10.1177/2455133317737936. Accessed 10 Feb. 2021.

Shehu, Fatmir, and Bukuri Zejno. "Gender Equality and the Participation of Muslim Women in Education and Work: A Critical Analysis." *IIUM Journal of Educational Studies*, vol. 3, no. 2, 2015, pp. 19–39. www.researchgate.net/publication/331741004_Gender_Equality_and_the_Participation_of_Muslim_Women_in_Education_and_Work_A_Critical_Analysis. Accessed 1 Mar. 2021.

Siddiqui, Shamsul, and Hasibur Rahaman. "Spatial Pattern of Muslim Women Empowerment in West Bengal." *International Journal of Applied Research*, vol. 2, no. 10, 2016, pp. 254–260. www.allresearchjournal.com/archives/2016/vol2issue10/PartD/2-10-4-976.pdf. Accessed 25 Feb. 2021.

Part III

Agriculture, Environment and Livelihoods

Mapping and Application

13 Trend Evaluation of Agriculture Production and Yield of Major Crops in West Bengal

A District-Level Time Series Data Analysis

Rukhsana

13.1 Introduction

Especially for the rural poor, agriculture is a vital source of income and employment in emerging nations, and it is regarded as one of the main engines of economic growth in the majority of them. In the 1960s, the Green Revolution began, and India quickly advanced in agricultural development. Today, the country is self-sufficient in food, having overcome the issue of food scarcity. The recent increase in crop production has, however, resulted in a number of environmental issues. Some of the issues that might threaten the future viability of agricultural systems at the local and regional levels include land degradation brought on by excessive cropping and irrigation, biodiversity loss, a drop in agricultural genetic variety, and climate change. Agricultural economists are quite interested in crop pattern change because of its notable impact on food production (Ranade, 1980; Rukhsana and Alam 2022). It is a well-known fact that increasing acreage and productivity are both necessary for increasing agricultural production. Increased yield and a shift in cropping patterns are two examples of productivity growth. While the latter analyzes the shift from crops with relatively low values of production per unit area to high-value crops, the former assesses the impact of a change in production per unit area (Boyce, 1987a; Rukhsana 2021). The World Food and Agriculture Organization (FAO) found that although 30 to 40% of people in the industrialized world experience overeating and obesity, over 500 million people worldwide experience famine and starvation. While developed industrial nations like the United States, Canada, the United Kingdom, West Germany, France, the Netherlands, and New Zealand experience food surpluses, developing nations experience food shortages and nutrient deficiencies. India produced 137 million tonnes of food grains in 1981–1982, up from 46 million tonnes in 1950–1951. In addition to the farm and the economy, several lakhs of small farmer families deserve praise for the outstanding achievement India has made in agriculture over the past three decades. During the past 30 years, there has been a significant increase

DOI: 10.4324/9781003275916-16

in food production and availability because to policy support, production techniques, public investment in infrastructure development, and the expansion of agricultural, livestock, and fisheries. A study in India shows that food grain production has increased from about 102 million tonnes in 1973 to about 203 million tonnes in 2000. The per capita availability of food grains has increased from 452 grams per person per day, while the population of the country has almost doubled from 548 million up to about 1000 million. Increased agricultural productivity and rapid industrial growth in recent years have contributed significantly to the reduction of poverty levels in India from 55% in 1973 to 26% in 1998 (Singh, P., 2003; Rukhsana and Alam 2021).

Over the past three decades, West Bengal's agricultural output has decreased. Both the growth rate and the per capita income have accelerated. Impressive outcomes from liberalization and regulation have been achieved, and the economy is quickly joining the global economy. Nevertheless, West Bengal's economy is based mostly on agriculture. Given that 72% of the state's population lives in rural areas and that agriculture is their primary industry, agriculture continues to be the most significant sector of the state's economy. However, with the structural transformation of the state's economy, the contribution of agriculture to the state domestic product (SDP) is observed to follow a declining trend. It contributes a significant share to SDP as compared to other sectors of the economy, even as the contribution of agriculture to total SDP (at constant prices) has declined from 41.16% in 1970–71 to 27.1% in 2000–01. The fact is that food grains dominate the cropping pattern in West Bengal (De, 2002; Ghosh and Kuri, 2005). Some 68% of the total area cultivated was made up of food crops like grains and pulses. West Bengal produced the most food grains in India, with a total production of 16,501.24 thousand tonnes. Despite having one of the highest rates of foodgrain production among the Indian states, there is evidence that the rate of growth in recent years has stagnated (Ghosh and Kuri, 2007).

The major objectives of the present study are as follows, first to analyze the regional trends and patterns of area, production and yield of Major crops in West Bengal at district level, and second to assess the spatio-temporal variations of levels of agricultural development to find out the nature and magnitude of disparity in agriculture in the study area, to make suggestions for furthering to increase area, production and yield of Major crop towards the sustainability of agriculture in the region.

Data related of various crops, agriculture during the study period at district in west Bengal have been collected for the years of 1981–82, 1991–92, 2001–02 and 2012–13 from 'Statistical Abstracts and District Handbooks' published by the Bureau of Applied Economics and Statistics and Office of the Directorate of Agriculture, Government of West Bengal. Socio-economic data have been taken from West Bengal Statistical Hand Book, Bureau of Applied Statistics and Economics, Kolkata.

13.2 Result and Discussion

13.2.1 Cropping Pattern and Trends in Area, Production and Yield of Rice

Rice is characterized as a cash crop which is one of the significant nourishment supplements and low on fat, cholesterol and therefore healthy. India is one of the main producers of rice. The rich soil of West Bengal is notable in delivering a portion of the best assortments of rice. Rice is an important component for meals and contains various measures of important vitamins and minerals. It is important to have an adjusted eating routine with rice being one of the fundamental dietary items in the Asian nations. The area occupied by rice cultivation decreased at a fast rate from 1981–82 to 2012–13, i.e. counted as 323.5 hectares to 302.5 thousand hectares respectively while production has augmented from 466.6 thousand quintal to 830.4 thousand quintal as per expanded demand of population with the percentage growth rate as −6.49 and 77.97 respectively during same period.

Table 13.1(a) shows the district wise distributional patterns of area, production and yield of rice in West Bengal. It has been found from the same table that the highest concentration in areas cultivating rice is recorded mainly in districts like Murshidabad, Hooghly, Birbhum, and Purulia during 1981–82 to 2012–13. It is due to the factors of low density of population and high productivity. The major concentration in growth of areas cultivating rice is confined to two districts, i.e., Birbhum, Murshidabad, and Hooghly. District wise analysis represents that the highest area growing rice has increased from 356.6 thousand hectares to 372.1 thousand hectares in Birbhum, exhibiting an increase of 4.35% during the last 30 years (1981–82 to 2012–13). This district occupies the highest area under rice due to large size of land holding. But the highest percentage growth rate of area under rice production was in Howrah (23.15%) during 1981–82 to 2012–13.

Table 13.1 illustrates that among all the districts of West Bengal, Cooch Behar district had a large-scale fluctuation in rice production which went up from 277.7 thousand tonnes in 1981–82 to 670.3 thousand tonnes in 2012–13 with a growth rate of 141.38%. During this period, the impact of new agricultural technology may have boosted the production rate and equally the situational change was observed that consumption increased in these last three decades, whereas the maximum decline in rice production has been found in Uttar Dinajpur with a decline rate of 17.97% (Table 13.1(b)) in 2012–13 compared to 1981–82. This district had a shortage in production of rice due to lack of area and small size of land holdings.

Yield is the main factor that affects the agricultural development. Table 13.1(a) illustrates that there is marked variation in per hectare yield of rice in the study area. The highest yield of rice has been noticed in Burdwan district, which has gone up from 1825.08 kg per hectares in 1981–82 to 3240 kg per hectares in 2012–13, but the highest percentage of growth has been seen in

Table 13.1 Distribution of area, production, percentage growth of area, production and yield of rice in West Bengal

(a): Distribution of area, production and yield of rice *in West Bengal (1981–82 to 2012–2013)*

Name of the district	1981–82			1991–92			2001–02			2012–13		
	Area	Production	Yield	Area	Production	Yield	Area	Production	Yield	Area	Production	Yield
Burdwan	549.4	1002.7	1825.08	560.6	1420.4	2533.71	582.7	1571.4	2696.76	593.4	1922.5	3240
Koch Behar	272.7	277.7	1018.34	305.7	401.9	1314.69	291.9	517.3	1772.18	277.8	670.3	2413
South 24 Parganas	374.3	479.2	1280.26	417.9	509.7	1219.67	425.9	866.9	2035.45	378.4	904.3	2390
North 24 Parganas	265.5	444.4	1673.82	287.4	646.9	2250.87	273.9	641	2340.27	217.5	649.2	2985
Bankura	380.9	592.5	1555.53	434.8	843.8	1940.66	395.6	993.4	2511.12	376.2	1033.7	2747
Dakshin Dinajpur	……	……	……	……	……	……	210.7	467.4	2218.32	181.4	510.7	2816
Darjeeling	43.1	52.8	2218.32	55.3	64.6	1168.17	34.1	52.9	1551.32	32.4	67.6	2084
Howrah	90.3	151.4	1676.63	133.1	232.5	1746.81	113.3	224.2	1978.82	111.2	286.8	2579
Malda	202.9	287.4	1416.46	263.8	493.3	1869.98	221.7	523.1	2359.495	207.6	644.6	3105
Hooghly	253.2	509.3	2011.45	275.3	610.7	2218.31	199.3	503.7	2527.35	273.5	793	2900
Jalpaiguri	264.7	294.7	1113.34	275.4	249.7	906.68	260.8	384.1	1472.78	228.1	492.9	2161
Murshidabad	302.6	421.8	1393.92	353.9	785.3	2218.99	224.1	548.2	2446.23	352.4	973.7	2763
Nadia	208.4	280.7	1346.93	285	695.1	2438.95	229.2	650	2835.95	232.8	668.3	2870
Paschim Medinipur	903.3	1255.5	1389.9	1024.9	1529.3	1492.15	1108.5	2584.4	2331.44	691.1	1880.2	2721
Purba Medinipur	…..	…..	…..	…..	…..	…..	…..	…..	…..	402.4	1060.3	2635
Uttar Dinajpur	457.3	486.4	1063.63	477.8	768.4	1608.2	283.2	628.6	2219.63	226.8	573.8	2530
Purulia	251	343.3	1367.73	285.3	372.8	1306.69	262.2	474.9	1811.21	289.1	730.7	2527
Birbhum	356.6	585.8	1642.74	376.7	813.1	2158.48	318.2	796.5	2503.14	372.1	1084.2	2914
Total	**323.5**	**466.6**	**1499.6**	**363.3**	**652.3**	**1774.6**	**319.7**	**731.1**	**2212.4**	**302.5**	**830.4**	**2687.8**

(b): Percentage growth of area, production and yield of rice in West Bengal (1981–82 to 2012–2013)

Name of the district	*1981–82 to 1991–92*			*1991–92 to 2001–02*			*2001–02 to 2012–13*			*1981–82 to 2012–13*		
	Area	*Production*	*Yield*	*Area*	*Production*	*Yield*	*Area*	*Production*	*Yield*	*Area*	*Production*	*Yield*
Burdwan	2.04	41.66	38.83	3.94	10.63	6.44	1.84	22.34	20.14	8.01	91.73	77.53
Koch Behar	12.10	44.72	29.10	−4.51	28.71	34.80	−4.83	29.58	36.16	1.87	141.38	136.95
South 24 Parganas	11.65	6.36	−4.73	1.91	70.08	66.89	−11.15	4.31	17.42	1.10	88.71	86.68
North 24 Parganas	8.25	45.57	34.48	−4.70	−0.91	3.97	−20.59	1.28	27.55	−18.08	46.08	78.33
Bankura	14.15	42.41	24.76	−9.02	17.73	29.40	−4.90	4.06	9.39	−1.23	74.46	76.60
Dakshin Dinajpur	.....	.....	.....	.....	.....	.....	−13.91	9.26	26.94			
Darjeeling	28.31	22.35	−47.34	−38.34	−18.11	32.80	−4.99	27.79	34.34	−24.83	28.03	−6.06
Howrah	47.40	53.57	4.19	−14.88	−3.57	13.28	−1.85	27.92	30.33	23.15	89.43	53.82
Maldah	30.01	71.64	32.02	−15.96	6.04	26.18	−6.36	23.23	31.60	2.32	124.29	119.21
Hoogly	8.73	19.91	10.28	−27.61	−17.52	13.93	37.23	57.43	14.74	8.02	55.70	44.17
Jalpaiguri	4.04	−15.27	−18.56	−5.30	53.82	62.44	−12.54	28.33	46.73	−13.83	67.25	94.10
Murshidabad	16.95	86.18	59.19	−36.68	−30.19	10.24	57.25	77.62	12.95	16.46	130.84	98.22
Nadia	36.76	147.63	81.07	−19.58	−6.49	16.28	1.57	2.82	1.20	11.71	138.08	113.08
Paschim Mednipur	13.46	21.81	7.36	8.16	68.99	56.25	−37.65	−27.25	16.71	−23.49	49.76	95.77
Purba Mednipur	.....	.....	.....	.....	.....	.....	.....	.....	.....			
Uttar Dinajpur	4.48	57.98	51.20	−40.73	−18.19	38.02	−19.92	−8.72	13.98	−50.40	17.97	137.86
Purulia	13.67	8.59	−4.46	−8.10	27.39	38.61	10.26	53.86	39.52	15.18	112.85	84.76
Birbhum	5.64	38.80	31.40	−15.53	−2.04	15.97	16.94	36.12	16.41	4.35	85.08	77.39
Total	**16.10**	**43.37**	**20.55**	**−14.18**	**11.65**	**29.09**	**−0.80**	**21.76**	**23.30**	**−6.49**	**77.97**	**79.23**

Source: Calculation is based on data from Statistical Handbook West Bengal (1981, 1991, 2001, 2011), Bureau of Applied Economics & Statistics Department, Kolkata, West Bengal

Notes: A: Area (in 1000 hectares) P: Production (in 1000 quintals) Y: Yield (in kg/hectare)
On 1 April 1992, the West Dinajpur district was divided into Uttar Dinajpur district (north) and Dakshin Dinajpur district (south)
On 1 January 2002, the Mednipure district was divided into Paschim Mednipur district and Purba Mednipur district

Koch Behar district, i.e. 136.95% during the same period. Whereas the maximum downturn in rice yield has been registered in Darjeeling with −6.06 kg per hectare in the last three decades (Table 13.1(b)). Farmers have been concentrating more on tobacco cultivation instead of rice cultivation because, the soil condition of Darjeeling is favorable for tobacco cultivation rather rice cultivation.

13.3 Cropping Pattern and Trends in Area, Production, and Yield of Wheat

Wheat is the most significant crop for the diet of the people. Table 13.2 represents the years of 1981, 1991, 2001, and 2011 in relation to area, production, and yield of wheat. Wheat is the second significant cash crop in India. Wheat has a critical part in the 'Green Revolution', as in West Bengal area under wheat cultivation is taking place slowly which increased from 17.7 thousand hectare to 17.9 thousand hectare during 1981–82 to 2012–13 respectively while production under wheat has augmented from 29.6 thousand quintal to 49.8 thousand quintal as per expanded demand of population with percentage growth rate of 1.13 and 68.24 respectively during same period.

District wise analysis represents that wheat cultivation has increased from 0.7 thousand hectares to 3.4 thousand hectares in South 24 Parganas, exhibiting an increase of 385.71 thousand hectares during the last 30 years (1981–82 to 2012–13). This district has the highest wheat cultivation due to the large size of land holding. But the highest percentage of decline was observed in Darjeeling district (−98.44%) during the said period. This is because the district is dominated by tobacco cultivation and farmers concentrated on the other profitable crops also. Besides these, Table 13.2 interprets that Birbhum, Paschim Medinipur has witnessed the comparatively increased rates in the area over the period of time (1981–82 to 2012–13). District wise analysis shows that Malda district secured the rank of highest production in West Bengal throughout the last 30 years, which has been reported to be augmented from 38.6 thousand tonnes in 1981–82, 88.5 in 1991–92,49.4 in 2001–02 and 125.1 in 2012–13.

The highest yield of wheat was reported in the district of Dakshin Dinajpur (3337 kg/hectare) during 2012–13 because of the assured availability of irrigation facilities, adequate amount of fertilizer and high yielding varieties of seeds. The percentage yield growth rate is high in Murshidabad district (110.70%). It has been found that those districts are low in wheat production like Darjeeling (1174 ke/hectare) which has a higher concentration of cash crop production.

13.3.1 Cropping Pattern and Trends in Area, Production, and Yield of Total Food Grain

Factors like urbanization, territorial varieties in utilization design, changes in food habits, and restrictions on energy requirements and changes in tastes and

Table 13.2 Distribution of area, production and yield of *wheat*, percentage growth of area in West Bengal (1981–82 to 2012–2013)

(a): Distribution of area, production and yield of wheat *in West Bengal (1981–82 to 2012–2013)*

Name of the district	*1981–82*			*1991–92*			*2001–02*			*2012–13*		
	Area	*Production*	*Yield*	*Area*	*Production*	*Yield*	*Area*	*Production*	*Yield*	*Area*	*Production*	*Yield*
Burdwan	10.4	17.5	1682.69	2.7	5.7	2111.11	6.3	15.3	2428.57	2	5.7	2864
Koch Behar	12.6	20.1	1595.24	13.6	22.5	1654.41	25.2	49.8	1976.19	10.6	25.9	2445
South 24 Parganas	0.7	1.6	2285.71	0.6	1	1666.67	3.2	5.7	1781.25	3.4	9.9	2920
North 24 Parganas	8.3	13.8	1662.65	8.9	18.7	2101.12	13.6	33.8	2485.29	7.2	19.8	2754
Bankura	10.2	15.5	1519.61	7.8	12.3	1576.92	8.8	21.1	2397.73	2.7	6.1	2272
Dakshin Dinajpur	.....	.....	.....	.....	.....	.....	10.2	25.2	2470.59	12.3	41	3337
Darjeeling	1.5	2.8	2470.59	3.7	3.8	1027.03	3.3	6.3	1909.09	2	2.3	1174
Hawrah	0.8	1.5	1875	0.1	0.2	2000	1	2.1	2100	0.1	0.3	2808
Maldah	24.6	38.6	1569.11	37.6	88.5	2353.72	49.4	49.4	1000	44.9	125.1	2784
Hoogly	6.4	12.1	1890.63	0.8	1.5	1875	1.6	3.7	2312.5	0.1	0.3	2545
Jalpaiguri	8.8	16.4	1863.64	10.6	16.4	1547.17	26.6	47	1766.92	17.4	38.9	2233
Murshidabad	86.7	118.3	1364.48	92.2	174.9	1896.96	135.5	373.5	2756.46	99.2	285.3	2875
Nadia	44	72.1	1638.64	41.5	81.5	1963.86	60.4	148.5	2458.61	41.7	138.5	3320
Paschim Mednipur	8.8	16.4	1863.64	5.6	8.8	1571.43	13.6	31.2	2294.12	5.9	11	1881
Purba Mednipur	.....	.....	.....	.....	.....	.....	.....	.....	.....	0.5	1.3	2564
Uttar Dinajpur	37	90.5	2445.95	28	67.2	2400	37.5	86.2	2298.67	37.5	89.3	2383
Purulia	1.7	2.5	1470.59	1.2	2.1	1750	2.9	7.5	2586.21	1.6	3.3	2008
Birbhum	20.5	33.5	1634.15	14.2	25.1	1767.61	26.9	77.7	2888.48	32.4	91.7	2832
Total	**17.7**	**29.6**	**1802.**	**16.8**	**33.1**	**1828.9**	**25.1**	**57.9**	**2230.0**	**17.9**	**49.8**	**2555.5**

Table 13.2 (Continued)

(b): Percentage growth otablef area, production and yield of wheat in West Bengal (1981–82 to 2012–2013)

Name of the district	*1981–82 to 1991–92*			*1991–92 to 2001–02*			*2001–02 to 2012–13*			*1981–82 to 2012–13*		
	Area	*Production*	*Yield*	*Area*	*Production*	*Yield*	*Area*	*Production*	*Yield*	*Area*	*Production*	*Yield*
Burdwan	−74.04	−67.43	25.46	133.33	168.42	15.04	−68.25	−62.75	17.93	−80.77	−67.43	70.20
Koch Behar	7.94	11.94	3.71	85.29	121.33	19.45	−57.94	−47.99	23.72	−15.87	28.86	53.27
South 24 Parganas	−14.29	−37.50	−27.08	433.33	470.00	6.87	6.25	73.68	63.93	385.71	518.75	27.75
North 24 Parganas	7.23	35.51	26.37	52.81	80.75	18.28	−47.06	−41.42	10.81	−13.25	43.48	65.64
Bankura	−23.53	−20.65	3.77	12.82	71.54	52.05	−69.32	−71.09	−5.24	−73.53	−60.65	49.51
Dakshin Dinajpur	.....	.....	.....	.....	.....	.....	20.59	62.70	35.07			
Darjeeling	146.67	35.71	−58.43	−10.81	65.79	85.88	−39.39	−63.49	−38.50	33.33	−17.86	−52.48
Howrah	−87.50	−86.67	6.67	900.00	950.00	5.00	−90.00	−85.71	33.71	−87.50	−80.00	49.76
Malda	52.85	129.27	50.00	31.38	−44.18	−57.51	−9.11	153.24	178.40	82.52	224.09	77.43
Hooghly	−87.50	−87.60	−0.83	100.00	146.67	23.33	−93.75	−91.89	10.05	−98.44	−97.52	34.61
Jalpaiguri	20.45	0.00	−16.98	150.94	186.59	14.20	−34.59	−17.23	26.38	97.73	137.20	19.82
Murshidabad	6.34	47.84	39.02	46.96	113.55	45.31	−26.79	−23.61	4.30	14.42	141.17	110.70
Nadia	−5.68	13.04	19.85	45.54	82.21	25.19	−30.96	−6.73	35.04	−5.23	92.09	102.61
Paschim Medinipur	−36.36	−46.34	−15.68	142.86	254.55	45.99	−56.62	−64.74	−18.01	−32.95	−32.93	0.93
Purba Medinipur	.....	.....	.....	.....	.....	.....	.....	.....	.....			
Uttar Dinajpur	−24.32	−25.75	−1.88	33.93	28.27	−4.22	.....	.....	.....	1.35	−1.33	−2.57
Purulia	−29.41	−16.00	19.00	141.67	257.14	47.78	−44.83	−56.00	−22.36	−5.88	32.00	36.54
Birbhum	−30.73	−25.07	8.17	89.44	209.56	63.41	20.45	18.02	−1.96	58.05	173.73	73.30
Total	**−4.91**	**−8.7**	**5.1**	**149.3**	**197.6**	**25.4**	**−38.8**	**−20.3**	**22.1**	**1.13**	**68.24**	**41.81**

Source: Calculation is based on data from Statistical Handbook West Bengal (1981,1991,2001,2011), Bureau of Applied Economics & Statistics Department, Kolkata, West Bengal

Notes: A: Area (in 1000 hectares) P: Production (in 1000 quintals) Y: Yield (in kg/hectare)
On 1 April 1992, the West Dinajpur district was divided into Uttar Dinajpur district (north) and Dakshin Dinajpur district (south).
On 1 January 2002, the Mednipure district was divided into Paschim Mednipur district and Purba Mednipur district.

inclinations in purchasing power play a major role in demand, trend, and pattern distribution of food grains. It has been noticed that the area growing food grains in West Bengal has found to be slightly increased from 1981–82 to 2012–13, i.e. counted as 6070.6 thousand hectares to 6121.5 thousand hectares respectively while production has just doubled as per demands of the population from 8281.4 thousand tonnes to 16521.5 thousand tonnes with a percentage growth rate of 084 and 99.50% respectively during the same period.

As per examined in Table 13.3(a) which depicts the regional disparity of change in areas growing food grains among the districts in West Bengal. District wise analysis represents that the cultivation of food grain has gone up from 279.50 to 308.2 thousand hectare in Purulia district, exhibiting an increase of 10.30% during the last 30 years (1981–82 to 2012–13) which means this district is trying to move towards advancement while it is one of the backward districts of West Bengal in terms of advancement or agriculture sector. On the other hand, the lowest percentage growth rate in the expansion of area growing food grains was reported in Uttar Dinajpur with the percentage growth rate −39.39% from 1981–82 to 2012–13.

A very high concentration in growth of area cultivating food grains has been recorded mainly in South 24 Parganas, Bankura, Murshidabad, and Paschim Medinipur. It is due to the factors of low density of population and high productivity. Major concentration in growth of area cultivating rice is confined to two districts, i.e. Murshidabad and Paschim Medinipur.

The highest production of food grains has been recorded in Burdwan district with 1931.9 thousand tonnes in 2012–13 which increased from 1028.4 in 1981–82 with a percentage growth rate of 87.85% during the same period. The lowest production in food grains was found in Darjilin with a growth of 26.06%, from 86.1 thousand tonnes in 1981–82 to 120.6 thousand tonnes in 2012–13. This district has a shortage in production of food grains due to lack of area and small size of land holdings. During this period, the impact of new agricultural technology was observed like fertilizer consumption. Per capita consumption of electricity rose. The average rate of growth hides the fact that there has been a large-scale fluctuation in food grain production year to year in the region. In the study region the production has increased due to increase in productivity.

The highest growth rate in yield of food grains has been reported in Malda, from 1248 kg per hectares to 2827 kg per hectares during 1981–82 to 2012–13, exhibiting an increased growth rate of 126.42% (Table 13.3(b)).

13.3.2 Cropping Pattern and Trends in Area, Production, and Yield of Pulses

From Table 13.4(a) we find that the area growing pulses in West Bengal has decreased at a fast rate from 1981–82 to 2012–13, which was 37.05 thousand hectares to 12.11 thousand hectares respectively. On the other hand, production also similarly decreased from 14.89 thousand tonnes to 11.19 thousand with a percentage growth rate of −67.31 and −24.85 respectively during the same study period.

Table 13.3 Distribution of area, production, and yield of *food grains* in West Bengal (1981–82 to 2012–2013)

(a): Distribution of area, production, and yield of food grains *in West Bengal (1981–82 to 2012–2013)*

Name of the district	*1981–82*			*1991–92*			*2001–02*			*2012–13*		
	Area	*Production*	*Yield*	*Area*	*Production*	*Yield*	*Area*	*Production*	*Yield*	*Area*	*Production*	*Yield*
Burdwan	577.9	1028.4	1779.55	566.1	1427.5	333.51	594.5	1592.9	2679.39	598.5	1931.9	3228
Koch Behar	295.3	303.8	1028.78	331.9	432.4	1302.8	328.3	574.6	1750.23	309.2	778.2	2517
South 24 Parganas	399.8	495.5	1239.37	430.6	513.9	1193.45	941	879.5	934.64	405.1	933.5	2304
North 24 Parganas	299.6	467.8	1561.42	311.9	674.2	2161.59	3000.4	685.2	228.37	236.9	679.1	2867
Bankura	401.8	613.8	1527.63	448.6	866.6	1931.79	406.9	1016.7	2498.65	379.6	1041.1	2743
Dakshin Dinajpur	-	-	-	-	-	-	225.1	495	2199.02	195.3	554.6	2840
Darjeeling	86.1	98.8	2199.02	105.1	121.1	1152.24	64	114.6	1790.63	62.3	120.6	1934
Hawrah	106.7	159.6	1495.78	137.7	234.9	1705.88	114.7	226.7	1976.46	113	288.8	2555
Maldah	309.4	386.3	1248.55	374.1	633.2	1692.6	313.7	313.7	1000	295.3	834.7	2827
Hoogly	268	524.7	1957.84	276.9	612.7	2212.71	202.4	508.4	2511.86	274.1	794	2897
Jalpaiguri	286.8	319.7	1114.71	295.8	275.2	930.36	296.6	441.5	1488.54	263.3	568.2	2158
Murshidabad	513.3	600	1168.91	503.9	995	1974.6	420.9	979.8	2327.87	501.7	1321.8	2635
Nadia	356.1	398.6	1119.35	405.3	830.1	2048.11	347.2	843.5	2429.44	333.9	867.8	2599
Paschim Mednipur	969.3	1289.2	1330.03	1054.2	1558.7	1478.56	1144.6	2636.7	2303.6	702.4	1897.9	2702
Purba Mednipur										410.1	1073.5	2618
Uttar Dinajpur	513.3	602.4	1173.58	529	846.8	1600.76	331.1	721.1	2177.89	311.1	890.7	2863
Purulia	279.5	357.9	1280.5	324.2	403.7	1245.22	298.8	516.2	1727.58	308.3	750.3	2436
Birbhum	407.7	634.9	1557.27	400.5	844.1	2107.62	365.9	892.1	2438.1	421.4	1194.8	2836
Total	**6070.6**	**8281.4**	**22782.29**	**6495.8**	**11270.1**	**25071.8**	**9396.1**	**13438.2**	**32462.27**	**6121.5**	**16521.5**	**47559**

(b): Percentage growth of area, production and yield of foodgrains in West Bengal (1981–82 to 2012–2013)

Name of the district	*1981–82 to 1991–92*			*1991–92 to 2001–02*			*2001–02 to 2012–13*			*1981–82 to 2012–13*		
	Area	*Production*	*Yield*	*Area*	*Production*	*Yield*	*Area*	*Production*	*Yield*	*Area*	*Production*	*Yield*
Burdwan	−2.04	38.81	−81.26	5.02	11.59	703.39	0.67	21.28	20.48	3.56	87.85	81.39
Koch Behar	12.39	42.33	26.64	−1.08	32.89	34.34	−5.82	35.43	43.81	4.71	156.16	144.66
South 24 Parganas	7.70	3.71	−3.71	118.53	71.14	−21.69	−56.95	6.14	146.51	1.33	88.40	85.90
North 24 Parganas	4.11	44.12	38.44	861.97	1.63	−89.44	−92.10	−0.89	1155.42	−20.93	45.17	83.61
Bankura	11.65	41.19	26.46	−9.30	17.32	29.34	−6.71	2.40	9.78	−5.53	69.62	79.56
Dakshin Dinajpur							−13.24	12.04	29.15			
Darjeeling	22.07	22.57	−47.60	−39.11	−5.37	55.40	−2.66	5.24	8.01	−27.64	22.06	−12.05
Hawrah	29.05	47.18	14.05	−16.70	−3.49	15.86	−1.48	27.39	29.27	5.90	80.95	70.81
Maldah	20.91	63.91	35.57	−16.15	−50.46	−40.92	−5.87	166.08	182.70	−4.56	116.08	126.42
Hoogly	3.32	16.77	13.02	−26.91	−17.02	13.52	35.42	56.18	15.33	2.28	51.32	47.97
Jalpaiguri	3.14	−13.92	−16.54	0.27	60.43	60.00	−11.23	28.70	44.97	−8.19	77.73	93.59
Murshidabad	−1.83	65.83	68.93	−16.47	−1.53	17.89	19.20	34.91	13.19	−2.26	120.30	125.42
Nadia	13.82	108.25	82.97	−14.34	1.61	18.62	−3.83	2.88	6.98	−6.23	117.71	132.19
Paschim Mednipur	8.76	20.90	11.17	8.58	69.16	55.80	−38.63	−28.02	17.29	−27.54	47.22	103.15
Purba Mednipur												
Uttar Dinajpur	3.06	40.57	36.40	−37.41	−14.84	36.05	−6.04	23.52	31.46	−39.39	47.86	143.95
Purulia	15.99	12.80	−2.76	−7.83	27.87	38.74	3.18	45.35	41.01	10.30	109.64	90.24
Birbhum	−1.77	32.95	35.34	−8.64	5.69	15.68	15.17	33.93	16.32	3.36	88.19	82.11
Total	**7.00**	**36.09**	**10.05**	**44.65**	**19.24**	**29.48**	**−34.85**	**22.94**	**46.51**	**0.84**	**99.50**	**108.75**

Source: Calculation is based on data from Statistical Handbook West Bengal (1981,1991,2001,2011), Bureau of Applied Economics & Statistics Department, Kolkata, West Bengal

Note: A: Area (in 1000 hectares) P: Production (in 1000 quintals) Y: Yield (in kg/hectare)
On 1 April 1992, the West Dinajpur district was divided into Uttar Dinajpur district (north) and Dakshin Dinajpur district (south)
On 1 January 2002, the Mednipure district was divided into Paschim Mednipur district and Purba Mednipur district

Table 13.4 Distribution of area, production and yield of *pulses* in West Bengal (1981–82 to 2012–2013)

(a): Distribution of area, production and yield of pulses *in West Bengal (1981–82 to 2012–2013)*

Name of the district	*1981–82*			*1991–92*			*2001–02*			*2012–13*		
	Area	*Production*	*Yield*	*Area*	*Production*	*Yield*	*Area*	*Production*	*Yield*	*Area*	*Production*	*Yield*
Burdwan	17.00	7.60	447.06	2.80	5.60	2000	5.40	6.00	1111.11	2.50	2.50	1027.00
Koch Behar	8.10	4.70	580.25	9.40	5.70	606.38	10.50	7.00	666.67	6.70	4.20	622.00
South 24 Parganas	24.20	13.50	557.85	12.10	3.20	264.46	5.30	6.90	1301.89	23.10	18.80	812.00
North 24 Parganas	25.80	9.60	372.09	15.60	8.60	551.28	12.90	10.40	806.2	12.00	9.60	798.00
Bankura	7.90	3.70	468.35	2.70	1.40	518.52	1.30	0.90	692.31	0.10	0.10	701.00
Dakshin Dinajpur	.....	.....	.....	.....	.....	.....	4.10	2.40	585.37	1.00	0.60	582.00
Darjeeling	1.40	0.90	585.37	1.00	0.60	600	1.90	1.20	631.58	1.20	0.90	751.00
Howrah	15.60	6.70	429.49	4.50	2.20	488.89	0.40	0.40	1000	1.60	1.60	968.00
Malda	56.10	37.30	664.88	59.20	38.50	650.34	36.10	36.10	1000	21.00	22.70	1083.00
Hooghly	8.40	3.30	392.86	0.80	0.50	625	1.50	1.00	666.67	0.30	0.30	1010.00
Jalpaiguri	7.90	4.50	569.62	4.90	2.90	591.84	6.60	4.40	666.67	4.80	2.90	612.00
Murshidabad	108.90	45.60	418.73	53.00	29.60	558.49	59.30	56.30	949.41	44.80	43.00	960.00
Nadia	101.00	43.60	431.68	73.80	49.70	673.44	57.00	44.20	775.44	56.30	53.40	950.00
Paschim Medinipur	55.60	15.90	285.97	22.20	19.10	860.36	20.80	20.10	966.35	4.30	3.70	860.00
Purba Medinipur	.....	.....	.....	.....	.....	.....	.....	.....	.....	7.10	11.80	1674.00
Uttar Dinajpur	108.90	19.80	181.82	22.60	10.90	482.3	9.10	4.60	505.49	3.50	2.70	761.00
Purulia	17.10	7.10	415.2	20.80	13.70	658.65	18.70	7.30	390.37	10.90	4.10	377.00
Birbhum	28.90	14.40	498.27	8.60	5.40	627.91	20.20	16.80	831.68	16.70	18.60	1117.00
Total	**37.05**	**14.89**	**456.22**	**19.63**	**12.35**	**672.37**	**15.95**	**13.29**	**796.89**	**12.11**	**11.19**	**870.28**

(b): Percentage growth of area, production and yield of pulses in West Bengal (1981–82 to 2012–2013)

Name of the district	*1981–82 to 1991–92*			*1991–92 to 2001–02*			*2001–02 to 2012–13*			*1981–82 to 2012–13*		
	Area	*Production*	*Yield*	*Area*	*Production*	*Yield*	*Area*	*Production*	*Yield*	*Area*	*Production*	*Yield*
Burdwan	–83.53	–26.32	347.37	92.86	7.14	–44.44	–53.70	–58.33	–7.57	–85.29	–67.11	129.72
Koch Behar	16.05	21.28	4.50	11.70	22.81	9.94	–36.19	–40.00	–6.70	–17.28	–10.64	7.20
South 24 Parganas	–50.00	–76.30	–52.59	–56.20	115.63	392.28	335.85	172.46	–37.63	–4.55	39.26	45.56
North 24 Parganas	–39.53	–10.42	48.16	–17.31	20.93	46.24	–6.98	–7.69	–1.02	–53.49	0.00	114.46
Bankura	–65.82	–62.16	10.71	–51.85	–35.71	33.52	–92.31	–88.89	1.26	–98.73	–97.30	49.67
Dakshin Dinajpur	.....	.....	.....	.....	.....	.....	–75.61	–75.00	–0.58			
Darjeeling	–28.57	–33.33	2.50	90.00	100.00	5.26	–36.84	–25.00	18.91	–14.29	0.00	28.29
Howrah	–71.15	–67.16	13.83	–91.11	–81.82	104.54	300.00	300.00	–3.20	–89.74	–76.12	125.38
Malda	5.53	3.22	–2.19	–39.02	–6.23	53.77	–41.83	–37.12	8.30	–62.57	–39.14	62.89
Hooghly	–90.48	–84.85	59.09	87.50	100.00	6.67	–80.00	–70.00	51.50	–96.43	–90.91	157.09
Jalpaiguri	–37.97	–35.56	3.90	34.69	51.72	12.64	–27.27	–34.09	–8.20	–39.24	–35.56	7.44
Murshidabad	–51.33	–35.09	33.38	11.89	90.20	70.00	–24.45	–23.62	1.12	–58.86	–5.70	129.26
Nadia	–26.93	13.99	56.00	–22.76	–11.07	15.15	–1.23	20.81	22.51	–44.26	22.48	120.07
Paschim Medinipur	–60.07	20.13	200.86	–6.31	5.24	12.32	–79.33	–81.59	–11.01	–92.27	–76.73	200.73
Purba Medinipur	.....	.....	.....	.....	.....	.....	.....	.....	.....			
Uttar Dinajpur	–79.25	–44.95	165.26	–59.73	–57.80	4.81	–61.54	–41.30	50.55	–96.79	–86.36	318.55
Purulia	21.64	92.96	58.63	–10.10	–46.72	–40.73	–41.71	–43.84	–3.42	–36.26	–42.25	–9.20
Birbhum	–70.24	–62.50	26.02	134.88	211.11	32.45	–17.33	10.71	34.31	–42.21	29.17	124.18
Total	**–44.48**	**–24.19**	**60.96**	**6.82**	**30.34**	**44.65**	**–2.38**	**–7.21**	**6.42**	**–67.31**	**–24.85**	**90.76**

Source: Calculation is based on data from Statistical Handbook West Bengal (1981, 1991, 2001, 2011), Bureau of Applied Economics & Statistics Department, Kolkata, West Bengal

Note: A: Area (in 1000 hectares) P: Production (in 1000 quintals) Y: Yield (in kg/hectare)
On 1 April 1992, the West Dinajpur district was divided into Uttar Dinajpur district (north) and Dakshin Dinajpur district (south)
On 1 January 2002, the Medinipur district was divided into Paschim Medinipur district and Purba Medinipur district

The highest decline in growth rate in area cultivating pulses is confined in Uttar Dinajpur with the growth margin of −96.79 thousand hectares from 1981–82 to 2012–13. But the lowest percentage of growth in area cultivating pulses was found in South 24 Parganas (4.55 thousand hectors, which declined from 24.20 thousand hectares in 1981–82 to 23.10 thousand hectares in 2012–13). Pulses have been replaced by other crops like wheat and rice.

The highest percentage of growth rate of pulses was reported in South 24 Parganas (39.26%) from 13.50 thousand tonnes to 18.80 thousand tonnes during 1981–82 to 2012–13, while the highest percentage of decline coincided in Hooghly district (−90.91%) due to increased food production like rice, potato, etc. (Table 13.4(b)). The maximum yield is reported in Uttar Dinajpur district with 318.55%, significantly improved from 181.82 kg per hectares to 782 kg per hectares during 1981–82 to 2012–13. But the highest decline in yield is observed in Purulia with −9.20% which decreased from 415.2 kg per hectare to in 1981–82 to 377 kg per hectare in 2012–13.

It has been concluded that including Birbhum there is a very high concentration of yield in pulses from these districts, namely, Hooghly and Burdwan, Howrah, Murshidabad, Nadia, Paschim Medinipur and Uttar Dinajpur. Few districts lie in the category of low yield of pulses cultivation which include Jalpaiguri and Koch Behar district.

13.4 Cropping Pattern and Trends in Area, Production, and Yield of Total Cereals

Cereals occupy a major and highly significant position on account of the lower percentage share of pulse crops. It has been noticed that the total area growing cereals in West Bengal has decreased slowly due to commercialization for increased earning, which shows that it declined its area from 348.44 thousand hectares to 327.98 during 1981–82 to 2012–13 while production has doubled as per demand of the population from 498.71 thousand tonnes to 906.67 thousand tonnes with a percentage growth rate of –5.87 and 81.80% respectively during the same period.

Table 13.5(a) shows that the maximum decline in area under cereals has been recorded in Uttar Dinajpur (–38.68%) which declined from 501.6 thousand hectares in 2001–02 to 307.6 thousand hectares in 2012–13. The area growing cereal crops has been reported to have decreased during the study period due to being replaced by commercial crops.

The production of cereals increased from 348.44 thousand tonnes to 906.67 thousand tonnes during 1981–82 to 2012–13 with the growth rate of 81.80%. The highest percentage of growth rate was observed in Uttar Dinajpur district (158.78%) during the same period. This was because of the increase in the use of fertilizer and irrigation facilities. On the other hand, the lowest percentage of growth rate was observed in Darjeeling district (22.27%). It is also reported in Table 13.4(a) that cereal production increased during

Table 13.5 Distribution of area, production and yield of *total cereals* in West Bengal (1981–82 to 2012–2013)

(a): Distribution of area, production and yield of total cereals *in West Bengal (1981–82 to 2012–2013)*

Name of the district	*1981–82*			*1991–92*			*2001–02*			*2012–13*		
	Area	*Production*	*Yield*	*Area*	*Production*	*Yield*	*Area*	*Production*	*Yield*	*Area*	*Production*	*Yield*
Burdwan	560.9	1020.8	1819.93	563.3	1426.1	2531.69	589.1	1586.9	2693.77016	596	1929.4	3237
Koch Behar	287.2	299.1	1041.43	322.5	426.7	1323.10	317.8	567.6	1786.03	302.5	774	2559
South 24 Parganas	375.6	418.2	1113.42	418.5	510.7	1220.31	935.7	872.6	932.56	382	914.7	2395
North 24 Parganas	273.8	458.2	1673.48	296.3	665.6	2246.37	287.5	674.8	2347.13	224.9	669.5	2977
Bankura	393.9	610.1	1548.87	445.9	865.2	1940.35	405.6	1015.8	2504.44	379.5	1041	2743
Dakshin Dinajpur	.....	.....	.....	.....	.....	.....	221	492.6	2228.96	194.3	554	2851
Darjeeling	84.7	97.9	2228.96	104.1	120.5	1157.54	62.1	113.4	1826.09	61.1	119.7	1958
Howrah	91.1	152.9	1678.38	133.2	232.7	1747	114.3	226.3	1979.88	111.4	287.2	2578
Malda	253.3	349	1377.81	314.9	594.7	1888.536	277.6	277.6	1000.00	274.3	812	2960
Hooghly	259.6	521.4	2008.47	276.1	612.2	2217.3126	200.9	507.4	2525.63	273.8	793.7	2899
Jalpaiguri	278.9	315.2	1130.15	290.9	272.3	936.0605	290	437.1	1507.24	258.5	565.3	2187
Murshidabad	404.4	554.4	1370.92	450.9	965.4	2141.0512	361.6	923.5	2553.93	456.9	1278.8	2799
Nadia	255.1	355	1391.61	331.5	780.4	2354.1478	290.2	799.3	2754.31	277.6	814.4	2934
Paschim Medinipur	913.7	1273.3	1393.56	1032	1539.6	1491.8605	1123.8	2618.6	2330.13	698.1	1894.2	2713
Purba Medinipur	.....	.....	.....	.....	.....	.....	.....	.....	.....	403	1061.7	2634
Uttar Dinajpur	501.6	582.6	1161.48	506.4	835.9	1650.6714	322	716.5	2225.16	307.6	888	2887
Purulia	262.4	350.8	1336.89	303.4	390	1285.4318	280.1	508.9	1816.85	297.4	746.2	2512
Birbhum	378.8	620.5	1638.07	391.9	838.7	2140.0868	345.7	875.3	2531.96	404.7	1176.2	2907
Total	**348.44**	**498.71**	**1494.59**	**386.36**	**692.29**	**1766.97**	**377.94**	**777.31**	**2090.83**	**327.98**	**906.67**	**2707.22**

(Continued)

Table 13.5 (Continued)

(b): Percentage growth of area, production and yield of total cereals in West Bengal (1981–82 to 2012–2013)

Name of the district	*1981–82 to 1991–92*			*1991–92 to 2001–02*			*2001–02 to 2012–13*			*1981–82 to 2012–13*		
	Area	*Production*	*Yield*	*Area*	*Production*	*Yield*	*Area*	*Production*	*Yield*	*Area*	*Production*	*Yield*
Burdwan	0.43	39.70	39.11	4.58	11.28	6.40	1.17	21.58	20.17	6.26	89.01	77.86
Koch Behar	12.29	42.66	27.05	–1.46	33.02	34.99	–4.81	36.36	43.28	5.33	158.78	145.72
South 24 Parganas	11.42	22.12	9.60	123.58	70.86	–23.58	–59.17	4.82	156.82	1.70	118.72	115.10
North 24 Parganas	8.22	45.26	34.23	–2.97	1.38	4.49	–21.77	–0.79	26.84	–17.86	46.12	77.89
Bankura	13.20	41.81	25.27	–9.04	17.41	29.07	–6.43	2.48	9.53	–3.66	70.63	77.10
Dakshin Dinajpur	.....	.....	.....	.....	.....	.....	–12.08	12.46	27.91			
Darjeeling	22.90	23.08	–48.07	–40.35	–5.89	57.76	–1.61	5.56	7.22	–27.86	22.27	–12.16
Howrah	46.21	52.19	4.09	–14.19	–2.75	13.33	–2.54	26.91	30.21	22.28	87.84	53.60
Malda	24.32	70.40	37.07	–11.85	–53.32	–47.05	–1.19	192.51	196.00	8.29	132.66	114.83
Hooghly	6.36	17.41	10.40	–27.24	–17.12	13.91	36.29	56.42	14.78	5.47	52.22	44.34
Jalpaiguri	4.30	–13.61	–17.17	–0.31	60.52	61.02	–10.86	29.33	45.10	–7.31	79.35	93.51
Murshidabad	11.50	74.13	56.18	–19.80	–4.34	19.28	26.36	38.47	9.60	12.98	130.66	104.17
Nadia	29.95	119.83	69.17	–12.46	2.42	17.00	–4.34	1.89	6.52	8.82	129.41	110.83
Paschim Medinipur	12.95	20.91	7.05	8.90	70.08	56.19	–37.88	–27.66	16.43	–23.60	48.76	94.68
Purba Medinipur	.....	.....	.....	.....	.....	.....	.....	.....	.....			
Uttar Dinajpur	0.96	43.48	42.12	–36.41	–14.28	34.80	–4.47	23.94	29.74	–38.68	52.42	148.56
Purulia	15.63	11.17	–3.85	–7.68	30.49	41.34	6.18	46.63	38.26	13.34	112.71	87.90
Birbhum	3.46	35.17	30.65	–11.79	4.36	18.31	17.07	34.38	14.81	6.84	89.56	77.46
Total	**14.01**	**40.36**	**20.18**	**–3.65**	**12.76**	**21.08**	**–4.71**	**29.72**	**40.78**	**–5.87**	**81.80**	**81.13**

Source: Calculation is based on data from Statistical Handbook West Bengal (1981,1991,2001,2011), Bureau of Applied Economics & Statistics Department, Kolkata, West Bengal

Notes: A: Area (in 1000 hectares) P: Production (in 1000 quintals) Y: Yield (in kg/hectare)
On 1 April 1992, the West Dinajpur district was divided into Uttar Dinajpur district (north) and Dakshin Dinajpur district (south).
On 1 January 2002, the Mednipure district was divided into Paschim Mednipur district and Purba Mednipur district.

1981–82 to 2012–13 all over the district. The growth rate was also high in South 24 Parganas, Malda, Murshidabad, Nadia district compared to the others districts in the study area.

Table 13.5(a) displays spatial variation in yield of cereals which indicates that Murshidabad and Uttar Dinajpur districts come under very high concentration in yield of cereals, though those are scattered throughout the study area. There is a low concentration in yield of cereals in these districts, namely, Malda and Darjeeling.

13.4.1 Cropping Pattern and Trends in Area, Production and Yield of Oilseed

Table 13.6(b) demonstrates the spatial variation in area under oilseed. A high concentration in growth rate of areas growing oilseed has been reported in several of the study areas including the highest Nadia district in growth margin, namely, Paschim Medinipur and Murshidabad. A low concentration of areas growing oilseed was reported in these districts, namely, Birbhum, Bankura, Koch Behar, Uttar Dinajpur, and Purulia.

The amount of oilseed production increased in North 24 Parganas (814.12%) from 8.50 thousand tonnes in 1981–82 to 77.70 thousand tonnes in 2012–13, while the Burdwan district has the lowest growth rate of 49.47% thousand tonnes in growth margin. Table 13.6(a) and (b) shows the spatial distribution in oilseed production which has been mapped into various districts of the study area of West Bengal. Oilseed production was concentrated in the districts of Paschim Medinipur, Murshidabad and North 24 Parganas, Hooghly. The maximum yield of oilseed is reported in Howrah district with 270.67%, significantly improved from 615.38 kg per hectares to 2281 kg per hectares during 1981–82 to 2012–13. But The greatest decline in yield is observed in Darjeeling with –38.55% which decreased from 929.2 kg per hectare to in 1981–82 to 571 kg per hectare in 2012–13 (Table 13.6(a and b)).

13.4.2 Cropping Pattern and Trends in Area, Production, and Yield of Jute Crop

Jute is a critical characteristic fiber crop in India alongside cotton. It assumes a critical part in the nation's economy. Jute was initially considered as a wellspring of raw material for bundling ventures as it were. Table 13.7(a) discloses that the area with the highest growth rate of jute is Murshidabad with a growth margin of 91.25 thousand hectares in 1981–82 to 2012–13. In Burdwan cultivation of jute reduced (−94.80%), declining from 173 thousand hectares in 1981–82 to 9 thousand hectares in 2012–13. Jute has been replaced by other crops like wheat and rice.

The maximum growth rate in production has been recorded in Murshidabad, significantly increasing from 486.2 thousand tonnes in 1981 to 2176.8 thousand tonnes in 2012–13. In the last three decades, the lowest growth rate in

Table 13.6 Distribution of area, production and yield of *oilseed* in West Bengal (1981–82 to 2012–2013)

(a): Distribution of area, production and yield of oilseed *in West Bengal (1981–82 to 2012–2013)*

Name of the district	*1981–82*			*1991–92*			*2001–02*			*2012–13*		
	Area	*Production*	*Yield*	*Area*	*Production*	*Yield*	*Area*	*Production*	*Yield*	*Area*	*Production*	*Yield*
Burdwan	41.70	28.10	673.86	64.90	101.70	1567.03	56.20	38.40	683.27	36.10	42.00	1163.00
Koch Behar	10.00	2.90	290.00	10.30	5.30	514.56	12.60	7.70	611.11	14.50	6.60	450.00
South 24 Parganas	.....	.....	.....	4.20	4.90	1166.67	5.50	4.00	727.27	13.10	17.00	1297.00
North 24 Parganas	15.80	8.50	537.97	34.80	35.10	1008.62	48.50	45.70	942.27	57.20	77.70	1359.00
Bankura	23.80	11.90	500.00	38.30	33.60	877.28	27.50	21.20	770.91	32.10	30.40	950.00
Dakshin Dinajpur	.....	.....	.....	.....	.....	.....	22.60	21.00	929.20	25.10	29.30	1167.00
Darjeeling	1.00	0.40	929.20	1.80	1.30	722.22	0.40	0.20	500.00	0.60	0.40	571.00
Hawrah	6.50	4.00	615.38	3.80	2.30	605.26	5.30	4.30	811.32	7.90	18.00	2281.00
Maldah	18.30	7.20	393.44	28.70	21.70	756.10	35.70	35.70	1000.00	34.30	36.30	1059.00
Hoogly	21.50	12.00	558.14	34.30	34.10	994.17	33.60	31.50	937.50	56.10	71.10	1266.00
Jalpaiguri	7.40	2.80	378.38	12.30	7.60	617.89	12.40	7.60	612.90	17.60	17.50	994.00
Murshidabad	45.20	20.40	451.33	55.60	60.40	1086.33	71.10	72.10	1014.06	106.40	109.40	1029.00
Nadia	33.00	16.70	506.06	59.30	63.80	1075.89	108.90	100.00	918.27	123.50	152.30	1233.00
Paschim Mednipur	21.50	9.40	437.21	52.00	44.90	863.46	72.50	82.90	1143.45	93.90	95.70	1019.00
Purba Mednipur	.....	.....	.....	.....	.....	.....	.....	.....	.....	19.30	39.20	2030.00
Uttar Dinajpur	44.00	11.20	254.55	55.00	49.90	907.27	45.40	32.40	713.66	50.30	43.00	855.00
Purulia	3.40	1.00	294.12	7.60	5.60	736.84	2.80	1.50	535.71	2.80	1.90	658.00
Birbhum	24.20	13.80	570.25	50.30	39.40	783.30	37.40	42.10	1125.67	36.70	33.20	904.00
Total	**21.15**	**10.02**	**492.66**	**32.08**	**31.98**	**892.68**	**35.20**	**32.25**	**822.15**	**40.42**	**45.61**	**1126.94**

(b): Percentage growth of area, production and yield of oilseed in West Bengal (1981–82 to 2012–2013)

Name of the district	*1981–82 to 1991–92*			*1991–92 to 2001–02*			*2001–02 to 2012–13*			*1981–82 to 2002–13*		
	Area	*Production*	*Yield*	*Area*	*Production*	*Yield*	*Area*	*Production*	*Yield*	*Area*	*Production*	*Yield*
Burdwan	55.64	261.92	132.54	–13.41	–62.24	–56.40	–35.77	9.38	70.21	–13.43	49.47	72.59
Koch Behar	3.00	82.76	77.44	22.33	45.28	18.76	15.08	–14.29	–26.36	45.00	127.59	55.17
South 24 Parganas	.....	.....	.....	30.95	–18.37	–37.66	138.18	325.00	78.34			
North 24 Parganas	120.25	312.94	87.49	39.37	30.20	–6.58	17.94	70.02	44.23	262.03	814.12	152.62
Bankura	60.92	182.35	75.46	–28.20	–36.90	–12.13	16.73	43.40	23.23	34.87	155.46	90.00
Dakshin Dinajpur	.....	.....	.....	.....	.....	.....	11.06	39.52	25.59			
Darjeeling	80.00	225.00	–22.27	–77.78	–84.62	–30.77	50.00	100.00	14.20	–40.00	0.00	–38.55
Howrah	–41.54	–42.50	–1.64	39.47	86.96	34.04	49.06	318.60	181.15	21.54	350.00	270.67
Malda	56.83	201.39	92.17	24.39	64.52	32.26	–3.92	1.68	5.90	87.43	404.17	169.16
Hooghly	59.53	184.17	78.12	–2.04	–7.62	–5.70	66.96	125.71	35.04	160.93	492.50	126.82
Jalpaiguri	66.22	171.43	63.30	0.81	0.00	–0.81	41.94	130.26	62.18	137.84	525.00	162.70
Murshidabad	23.01	196.08	140.70	27.88	19.37	–6.65	49.65	51.73	1.47	135.40	436.27	127.99
Nadia	79.70	282.04	112.60	83.64	56.74	–14.65	13.41	52.30	34.27	274.24	811.98	143.65
Paschim Medinipur	141.86	377.66	97.49	39.42	84.63	32.43	29.52	15.44	–10.88	336.74	918.09	133.07
Purba Medinipur	.....	.....	.....	.....	.....	.....	.....	.....	.....			
Uttar Dinajpur	25.00	345.54	256.43	–17.45	–35.07	–21.34	10.79	32.72	19.81	14.32	283.93	235.89
Purulia	123.53	460.00	150.53	–63.16	–73.21	–27.30	0.00	26.67	22.83	–17.65	90.00	123.72
Birbhum	107.85	185.51	37.36	–25.65	6.85	43.71	–1.87	–21.14	–19.69	51.65	140.58	58.53
Total	**64.12**	**228.42**	**91.85**	**5.04**	**4.78**	**–3.67**	**27.57**	**76.88**	**33.03**	**91.11**	**355.19**	**128.75**

Source: Calculation is based on data from Statistical Handbook West Bengal (1981, 1991, 2001, 2011), Bureau of Applied Economics & Statistics Department, Kolkata, West Bengal

Notes: A: Area (in 1000 hectares) P: Production (in 1000 quintals) Y: Yield (in kg/hectare)
On 1 April 1992, the West Dinajpur district was divided into Uttar Dinajpur district (north) and Dakshin Dinajpur district (south)
On 1 January 2002, the Mednipure district was divided into Paschim Mednipur district and Purba Mednipur district

Table 13.7 Distribution of area, production and yield of *jute* in West Bengal (1981–82 to 2012–2013)

(a): Distribution of area, production and yield of jute *in West Bengal (1981–82 to 2012–2013)*

Name of the district	*1981–82*			*1991–92*			*2001–02*			*2012–13*		
	Area	*Production*	*Yield*	*Area*	*Production*	*Yield*	*Area*	*Production*	*Yield*	*Area*	*Production*	*Yield*
Burdwan	173.00	208.80	1206.94	.....	.....	.....	10.50	190.40	1813.33	9.00	139.10	15455.56
Koch Behar	75.40	541.80	7185.68	61.90	478.70	7733.44	84.10	782.60	9305.59	75.90	987.40	13009.22
South 24 Parganas	.....	.....	.....	1.70	20.90	12294.12	7.70	18.10	2350.65	2.60	49.40	19000.00
North 24 Parganas	74.60	651.00	8726.54	37.30	436.00	11689.01	.....	.....	.....	47.40	754.30	15913.50
Bankura	1.90	20.80	10947.37	0.40	5.60	14000.00	0.40	6.90	17250.00	.....	.....	.....
Dakshin Dinajpur	.....	.....	.....	.....	.....	.....	15.10	146.10	9675.50	18.30	272.80	14907.10
Darjeeling	4.70	26.90	9675.50	3.70	22.40	6054.05	2.20	18.10	8227.27	3.00	36.90	12300.00
Hawrah	6.20	53.20	8580.65	3.50	46.60	13314.29	6.80	101.30	14897.06	2.40	43.20	18000.00
Maldah	31.50	163.10	5177.78	27.50	218.70	7952.73	24.00	24.00	1000.00	23.80	394.60	16579.83
Hoogly	35.30	363.00	10283.29	31.40	447.80	14261.15	29.90	532.00	17792.64	26.70	526.40	19715.36
Jalpaiguri	55.50	388.10	6992.79	40.60	266.70	6568.97	44.30	404.50	9130.93	35.10	437.30	12458.69
Murshidabad	83.40	486.20	5829.74	105.80	1379.50	13038.75	141.00	1901.50	13485.82	159.50	2176.80	13647.65
Nadia	117.50	816.40	6948.09	116.60	1542.50	13228.99	130.30	1702.80	13068.30	125.10	1775.30	14191.05
Paschim Mednipur	21.60	257.10	11902.78	10.00	140.10	14010.00	7.70	142.60	18519.48	2.70	42.70	15814.81
Purba Mednipur	0.00	0.00	0.00	0.00	0.00	0.00	0.00	0.00	0.00	0.50	7.60	15200.00
Uttar Dinajpur	85.20	463.70	5442.49	45.90	299.10	6516.34	59.00	457.50	7754.24	44.50	578.10	12991.01
Purulia	.....	.....	.....	.....	.....	.....	.....	.....	.....	.....	.....	.....
Birbhum	0.30	2.60	8666.67	0.20	2.70	13500.00	0.10	2.00	20000.00	0.30	6.10	20333.33
Total	**51.07**	**296.18**	**7171.09**	**32.43**	**353.82**	**10277.46**	**35.19**	**401.90**	**10266.93**	**36.05**	**514.25**	**14264.91**

(b): Percentage growth of area, production and yield of jute in West Bengal (1981–82 to 2012–2013)

Name of the district	*1981–82 to 1991–92*			*1991–92 to 2001–02*			*2001–02 to 2012–13*			*1981–82 to 2012–13*		
	Area	*Production*	*Yield*	*Area*	*Production*	*Yield*	*Area*	*Production*	*Yield*	*Area*	*Production*	*Yield*
Burdwan	.....	.....	.....	.....	.....	.....	−14.29	−26.94	752.33	−94.80	−33.38	1180.56
Koch Behar	−17.90	−11.65	7.62	35.86	63.48	20.33	−9.75	26.17	39.80	0.66	82.24	81.04
South 24 Parganas	0.00	0.00	0.00	352.94	−13.40	−80.88	−66.23	172.93	708.29			
North 24 Parganas	−50.00	−33.03	0.00	.....	.....	.....	.....	.....	.....	−36.46	15.87	82.36
Bankura	−78.95	−73.08	27.88	0.00	23.21	23.21	0.00	0.00	0.00			
Dakshin Dinajpur	.....	.....	.....	.....	.....	.....	21.19	86.72	54.07			
Darjeeling	−21.28	−16.73	−37.43	−40.54	−19.20	0.00	36.36	103.87	49.50	−36.17	37.17	27.13
Hawrah	−43.55	−12.41	55.17	94.29	117.38	11.89	−64.71	−57.35	20.83	−61.29	−18.80	109.77
Maldah	−12.70	34.09	53.59	−12.73	−89.03	−87.43	−0.83	1544.17	1557.98	−24.44	141.94	220.21
Hoogly	−11.05	23.36	38.68	−4.78	18.80	24.76	−10.70	−1.05	10.81	−24.36	45.01	91.72
Jalpaiguri	−26.85	−31.28	−6.06	9.11	51.67	39.00	−20.77	8.11	36.44	−36.76	12.68	78.16
Murshidabad	26.86	183.73	123.66	33.27	37.84	3.43	13.12	14.48	1.20	91.25	347.72	134.10
Nadia	−0.77	88.94	90.40	11.75	10.39	−1.21	−3.99	4.26	8.59	6.47	117.45	104.24
Paschim Mednipur	−53.70	−45.51	17.70	−23.00	1.78	32.19	−64.94	−70.06	−14.60	−87.50	−83.39	32.87
Purba Mednipur	.....	.....	.....	.....	.....	.....	.....	.....	.....			
Uttar Dinajpur	−46.13	−35.50	19.73	28.54	52.96	19.00	−24.58	26.36	67.53	−47.77	24.67	138.70
Purulia	0.00	0.00	0.00	0.00	0.00	0.00	0.00	0.00	0.00			
Birbhum	−33.33	3.85	55.77	−50.00	−25.93	48.15	200.00	205.00	1.67	0.00	134.62	134.62
Total	**−24.62**	**4.99**	**29.78**	**31.05**	**16.43**	**3.75**	**−0.63**	**127.29**	**205.90**	**−29.41**	**73.63**	**98.92**

Source: Calculation is based on data from Statistical Handbook West Bengal (1981, 1991, 2001, 2011), Bureau of Applied Economics & Statistics Department, Kolkata, West Bengal

Notes: A: Area (in 1000 hectares) P: Production (in 1000 quintals) Y: Yield (in kg/hectare)
On 1 April 1992, the West Dinajpur district was divided into Uttar Dinajpur district (north) and Dakshin Dinajpur district (south).
On 1 January 2002, the Mednipure district was divided into Paschim Mednipur district and Purba Mednipur district.

production has been recorded in Paschim Medinipur, significantly falling with a margin of −83.39 thousand tonnes. The highest yield is reported in Burdwan district, significantly improving from 1206.94 kg per hectares to 15455.56 kg per hectares during 1991–92 with a growth rate of 1180.56%. But lowest growth rate in yield is observed in Paschim Medinipur (32.87%) which increased from 11902.78 kg per hectare in 1981 to 15814.81 kg per hectare in 2012–13 (Table 13.7(b)).

13.4.3 Cropping Pattern and Trends in Area, Production, and Yield of Potato

Potato is the fourth most important crop after corn, rice and wheat. This product is developed all through the world. Potato is substantial source of nutrition. Its root framework is shallow and stringy, thus fertilization is prescribed for higher supplement accessibility and utilize productivity. The point of the fertilization program is to cover the contrast between trim request and supply. The supplement necessities of potato are generally high. Other important factors are irrigation, disease, harvesting and post-gathering tasks to limit problems.

The highest growth rate in areas growing potato is in Jalpaiguri with a growth rate of 2700% from 1981–82 to 2012–13. In Jalpaiguri district the growth rate of potato drastically has changed due demand of potato and use of new technology for maximum production. But the lowest growth in areas growing potato was found in North 24 Parganas which declined from 4.80 thousand hectares in 1981–82 to 6.40 thousand hectares in 2012–13. Purulia district has been falling behind in production of potato. This is because the soil condition of the district is not suitable for potato cultivation due to the impact of extreme climate condition.

Table 13.8(a) and (b) demonstrates that the spatial variation in areas growing potatoes. A high concentration in growth rate of areas growing potato has been reported in several of the study areas including the highest in margin, namely, Paschim Medinipur, Burdwan, and Jalpaiguri. A low concentration of areas growing potato was reported in these districts, namely, North 24 Parganas, Darjeeling, Howrah, Malda, Murshidabad, Nadia and Uttar Dinajpur.

The highest percentage of growth rate observed in Jalpaiguri district (16569.84%) from 6.30 thousand quintal in 1981–82 to 1050.20 thousand tonnes in 2012–13. On the other hand, the lowest percentage of growth of production reported in North 24 Parganas (153.96%) with an increase of 80.80 thousand tonnes to 205.20 thousand tonnes. In North 24 Parganas production of potato decreased day by day due because farmer concentrated more on other vegetables rather than potato production. Lack of storage facilities is another important reason for the decrease in potato production in the said district.

Table 13.8(a) and (b) also depicts the spatial variation in production of potato among districts of the study area. Including the highest in margin, a very high concentration in production of potato is mainly observed in some

Table 13.8 Distribution of area, production and yield of *potato* in West Bengal (1981–82 to 2012–2013)

(a): Distribution of area, production and yield of potato *in West Bengal (1981–82 to 2012–2013)*

Name of the district	*1981–82*			*1991–92*			*2001–02*			*2012–13*		
	Area	*Production*	*Yield*	*Area*	*Production*	*Yield*	*Area*	*Production*	*Yield*	*Area*	*Production*	*Yield*
Burdwan	24.50	479.00	19551.02	.....	.....	.....	42.10	1115.00	26484.56	57.00	1854.20	32578.00
Koch Behar	1.20	7.30	6083.33	5.30	57.20	10792.45	12.20	281.70	23090.16	24.60	744.30	30291.00
South 24 Parganas	.....	.....	.....	1.10	15.60	14181.82	92.90	78.60	846.07	3.30	80.00	23931.00
North 24 Parganas	4.80	80.80	16833.33	3.40	58.80	17294.12	8.50	189.00	22235.29	6.40	205.20	31851.00
Bankura	6.80	115.90	17044.12	19.40	510.50	26314.43	21.50	571.30	26572.09	26.30	691.90	26328.00
Dakshin Dinajpur	.....	.....	.....	.....	.....	.....	4.80	87.00	18125.00	5.60	163.00	29108.00
Darjeeling	4.60	33.50	18125.00	4.20	39.70	9452.38	7.10	97.20	13690.14	7.40	119.30	16124.00
Hawrah	1.90	39.50	20789.47	3.60	101.60	28222.22	9.20	246.30	26771.74	7.80	247.90	31866.00
Maldah	1.10	6.00	5454.55	1.50	20.40	13600.00	2.40	2.40	1000.00	4.90	176.80	36077.00
Hoogly	33.00	762.00	23090.91	58.30	1486.10	25490.57	80.60	2337.00	28995.04	100.20	3246.40	32414.00
Jalpaiguri	1.20	6.30	5250.00	4.10	49.20	12000.00	14.30	308.90	21601.40	33.60	1050.20	31214.00
Murshidabad	6.90	72.30	10478.26	6.30	107.60	17079.37	8.90	213.70	24011.24	13.10	364.70	27898.00
Nadia	2.00	25.90	12950.00	1.90	34.40	18105.26	3.60	93.10	25861.11	5.60	177.60	31905.00
Paschim Mednipur	15.60	252.30	16173.08	33.60	763.60	22726.19	60.40	1596.60	26433.77	58.60	1463.60	24973.00
Purba Mednipur	.....	.....	.....	.....	.....	.....	.....	.....	.....	4.30	108.80	25222.00
Uttar Dinajpur	4.90	27.10	5530.61	4.30	31.70	7372.09	6.70	133.40	19910.45	11.70	357.70	30592.00
Purulia	0.30	4.80	16000.00	1.40	16.50	11785.71	0.90	12.80	14222.22	1.00	17.20	18842.00
Birbhum	6.80	58.30	8573.53	8.10	171.90	21222.22	12.80	266.60	20828.13	17.90	551.80	30906.00
Total	**7.71**	**131.40**	**13461.81**	**10.43**	**230.99**	**17042.59**	**22.88**	**448.86**	**20039.91**	**21.63**	**645.59**	**28451.11**

(*Continued*)

Table 13.8 (Continued)

(b): Percentage growth of area, production and yield of potato in West Bengal (1981–82 to 2012–13)

Name of the district	*1981–82 to 1991–92*			*1991–92 to 2001–02*			*2001–02 to 2012–13*			*1981–82 to 2012–13*		
	Area	*Production*	*Yield*	*Area*	*Production*	*Yield*	*Area*	*Production*	*Yield*	*Area*	*Production*	*Yield*
Burdwan	.....	.....	.....	.....	.....	.....	35.39	66.30	23.01	132.65	287.10	66.63
Koch Behar	341.67	683.56	77.41	130.19	392.48	113.95	101.64	164.22	31.19	1950.00	10095.89	397.93
South 24 Parganas	0.00	0.00	0.00	8345.45	403.85	−94.03	−96.45	1.78	2728.49			
North 24 Parganas	−29.17	−27.23	2.74	150.00	221.43	28.57	−24.71	8.57	43.25	33.33	153.96	89.21
Bankura	185.29	340.47	54.39	10.82	11.91	0.98	22.33	21.11	−0.92	286.76	496.98	54.47
Dakshin Dinajpur	.....	.....	.....	.....	.....	.....	16.67	87.36	60.60			
Darjeeling	−8.70	18.51	−47.85	69.05	144.84	44.83	4.23	22.74	17.78	60.87	256.12	−11.04
Hawrah	89.47	157.22	35.75	155.56	142.42	−5.14	−15.22	0.65	19.03	310.53	527.59	53.28
Maldah	36.36	240.00	149.33	60.00	−88.24	−92.65	104.17	7266.67	3507.70	345.45	2846.67	561.41
Hoogly	76.67	95.03	10.39	38.25	57.26	13.75	24.32	38.91	11.79	203.64	326.04	40.38
Jalpaiguri	241.67	680.95	128.57	248.78	527.85	80.01	134.97	239.98	44.50	2700.00	16569.84	494.55
Murshidabad	−8.70	48.82	63.00	41.27	98.61	40.59	47.19	70.66	16.19	89.86	404.43	166.25
Nadia	−5.00	32.82	39.81	89.47	170.64	42.84	55.56	90.76	23.37	180.00	585.71	146.37
Paschim Mednipur	115.38	202.66	40.52	79.76	109.09	16.31	−2.98	−8.33	−5.53	275.64	480.10	54.41
Purba Mednipur	.....	.....	.....	.....	.....	.....	.....	.....	.....			
Uttar Dinajpur	−12.24	16.97	33.30	55.81	320.82	170.08	74.63	168.14	53.65	138.78	1219.93	453.14
Purulia	366.67	243.75	−26.34	−35.71	−22.42	20.67	11.11	34.38	32.48	233.33	258.33	17.76
Birbhum	19.12	194.85	147.53	58.02	55.09	−1.86	39.84	106.98	48.39	163.24	846.48	260.48
Total	**93.90**	**195.23**	**47.24**	**633.12**	**169.71**	**25.26**	**31.33**	**492.99**	**391.47**	**180.54**	**391.32**	**111.35**

Source. Calculation is based on data from Statistical Handbook West Bengal (1981, 1991, 2001, 2011), Bureau of Applied Economics & Statistics Department, Kolkata, West Bengal

Notes: A: Area (in 1000 hectares) P: Production (in 1000 quintals) Y: Yield (in kg/hectare)
On 1 April 1992, the West Dinajpur district was divided into Uttar Dinajpur district (north) and Dakshin Dinajpur district (south).
On 1 January 2002, the Mednipure district was divided into Paschim Mednipur district and Purba Mednipur district.

districts including Burdwan, Jalpaiguri, and Paschim Medinipur. Districts like North 24 Parganas, Darjeeling, Malda, and Nadia come under the grade of low concentration in production growth. This is because said districts are highly dominated in various types of vegetables rather than potato.

It has been concluded that highest concentration of yield in potato is from these districts, namely, Burdwan, Koch Behar, South 24 Parganas, North 24 Parganas, Dakshin Dinajpur, Howrah, Jalpaiguri, Murshidabad, Nadia, Purba Medinipur, Uttar Dinajpur, and Birbhum. Darjeeling district lies in the category of low yield concentration of potato in the study area, −30.07% growth margin (1981–82 to 2012–13) (Table 13.8(b)).

13.5 Summary

Sustainable development of any country specially a developing country like India depends on agriculture advancement. Over the years, irrigation has played an significant role in the conversion of semi-agricultural areas from unstable agriculture to productive agriculture. The success of high yielding programs mainly depends on the availability of adequate irrigation facilities as high doses of chemical fertilizers and more water supplies to the plants are required. It is clear that food security based on food production is one of the key aspects of agricultural development. The present chapter has examined the nature of crop disparity in terms of the changes in cropping pattern with respect to acreage and production distribution. The area and production over the time span in West Bengal and among the districts can be observed to be increasingly dominated by boro, paddy, oilseeds (including, rapeseed and mustard) and potato. The growth in the area under above-mentioned crops originated from the expansion of area under irrigation. Crops like boro rice and potatoes are highly dependent on irrigation, as they are grown in off-monsoon seasons. From the above analysis it is also seen that for the expansion of the new commercial crops like boro rice, oilseeds and potatoes, the exchange effect played a major role which resulted in a shifting of resources from the lower value crops to higher value crops. The farmers of the sample study area prefer to grow cash crops mostly because of profit, but it is particularly beneficial for large farmers as compared to small and marginal farmers due to lack of purchasing power. If farmers can be provided with fertilizers and other inputs at affordable rates, regional disparities between food crops can be reduced to some extent and more benefits can be obtained for increasing the income of poor and small farmers.

Bibliography

Abid, S., Shah, N. A., Hassan, A., Farooq, A., & Masood, M. A. (2014). Growth and Trend in Area, Production and Yield of Major Crops of Khyber Pakhtunkhwa, Pakistan,*Asian Journal of Agriculture and Rural Development*, 04, 149–155.

Ali, H., Ali, H., Faridi, Z., & Ali, H. (2013). Production and Forecasting Trends of Cotton in Pakistan: An Analytical View. *Journal of Basic and Applied Sciences, III* (12), 97–101.

Boyce, J. K. (1987a). *Agrarian Impasse in Bengal: Institutional Constraints to Technological Change*. New York: Oxford University Press.

Boyce, J. K. (1987b). *Agrarian Impasse in Bengal: Institutional Constraints to Technological Change*. New York: Oxford University Press.

Chand, R., Raju, S. S., & Pandey, L. M. (2007). Growth Crisis in Agriculture: Severity and Options at National and State Levels. *Economic and Political Weekly, XXXXII* (26), 2528–2533.

Chattopadhyay, A. K., & Das, P. S. (2000). Estimation of Growth Rate: A Critical Analysis with Reference to West Bengal Agriculture. *LV, Indian Journal of Agricultural Economics*, 116–135.

De, U. K. (2002). *Economics of Crop Diversification*. New Delhi: Akansha Publishing House.

Ghosh, B. K., & Kuri, P. K. (2007). Agricultural Growth in West Bengal from 1970–71 to 2003–04: A Decomposition Analysis. *ICFAI J. Agric. Econ*, *IV*, 30–46.

Ghosh, B. K., & Kuri, P. K. (2005). Changes in Cropping Pattern in West Bengal during 1970–71 to 2000–01. *Journal of Development and Agricultural Economics, IASSI, XXIV*, 39–56.

Harris, J. (1992). What Is Happening in Rural West Bengal: Agrarian Reforms, Growth and Distribution. *Econ Polit Week*, *XXVIII*, 1237–1247.

Kumar, P., & Mitta, S. (2006). Agricultural Productivity Trends in India: Sustainability Issues. *Agricultural Economics Research Review*, *XIX*, 71–88.

Mamoria, C. B. (1953). *Agricultural Problems in India* (p. 33). Allahbad: Kitab Mahal.

Mukherji, B., & Mukhopadhyay, S. (1995). Impact of Institutional Change on Productivity in a Small-Farm Economy: Case of Rural West Bengal. *Econ Polit Week*, *XXX*, 2134–2137.

Ranade, C. G. (1980). Impact of Cropping Pattern on Agricultural Production. *Indian Journal of Agricultural Economics, XXXV* (2). DOI: 10.22004/ag.econ.269129

Ravanera, R. (2003). Hunger and Poverty in Asia, Focus on Rural Household. In P. Chaturvedi (Ed.), *Food Security in South Asia*. New Delhi: Indian Association for the Advancement of Science Concept Publishing Company.

Rukhsana, Alam A. (2021). Agriculture, Food, and Nutritional Security: An Overview. In: Rukhsana, Alam, A. (eds) *Agriculture, Food and Nutrition Security*. Cham: Springer. https://doi.org/10.1007/978-3-030-69333-6_1

Rukhsana, Alam A. (2022). Agriculture, Environment and Sustainable Development: An Overview. In: Alam A. Rukhsana (Eds.), *Agriculture, Environment and Sustainable Development*. Cham: Springer. https://doi.org/10.1007/978-3-031-10406-0_1

Rukhsana (2021). Levels of Agriculture Development and Estimated Growth and Pattern of Area, Production and Yield of Major Crops in Selected Region of Uttar Pradesh. In: Alam A., Rukhsana (Eds.), *Agriculture, Food and Nutrition Security*. Cham: Springer. https://doi.org/10.1007/978-3-030-69333-6_1

Saha, A., & Swaminathan, M. (1994). Agricultural Growth in West Bengal in the 1980s: A Disaggregation by Districts and Crops. *Econ Polit Week, XXXIX*, A-8.

Sanyal, M. K., Biswas, P. K., & Bardhan, S. (1998). Institutional Change and Output Growth in West Bengal Agriculture: End of Impasse. *Econ Polit Week*, *XXXIII*, 2979–2986.

Shafi, M.(2006). *Agricultural Geography*. Licenses of Pearson Education. South Asia: Dorling Kindersley, India, Pvt. Ltd.

Singh, P. (2003). *Indian Agriculture*. South Asia: New Dimensions.

Singh, R. L. (2002). *Food Security Issues in South Asia. In P. Chaturvedi, Food Security in South Asia, Pub: Indian Association for the Advancement of Science*. New Delhi: Concept Publishing Company.

Vyas, V. S. (1996). Diversification in Agriculture: Concept, Rationale and Approaches. *Indian Journal of Agricultural Economics, LI* (4), 636–643.

14 Mapping and Analyzing Air Pollution Using Google Earth Engine of Thiruvananthapuram City

S. Bala Subramaniyam and Sabirul Sk

14.1 Introduction

Air pollution is a complex and multifaceted problem affecting the health and well-being of people worldwide (Zhao et al., 2021). It occurs when harmful substances are released into the atmosphere, leading to a range of negative impacts on human health and the environment (Ghasempour et al., 2021). It is important to understand the causes and consequences of air pollution, as well as the strategies that can be employed to reduce its negative impacts.

There are many sources of air pollution, including industrial processes, transportation, and household energy use (Holman, 1999). Industrial facilities, such as power plants and factories, emit a range of pollutants, including carbon dioxide, sulfur dioxide, and nitrogen oxide. These substances can have harmful effects on human health, including respiratory problems and an increased risk of heart disease, stroke, and lung cancer (Bălă et al., 2021). Transportation is another significant source of air pollution, as burning fossil fuels by cars, buses, and airplanes release harmful substances into the air. In addition, the use of wood, coal, and other fossil fuels for household energy can also contribute to air pollution (Colvile et al., 2001). Air pollution also has negative consequences for the environment. Pollutants released into the air can damage crops and forests, and contribute to climate change by increasing the amount of greenhouse gases in the atmosphere (Manisalidis et al., 2020). The effects of climate change, including rising temperatures, more frequent extreme weather events, and sea level rise, have the potential to have significant impacts on human communities and ecosystems around the world (Clarke et al., 2022). In recent years, Thiruvananthapuram has faced challenges related to air pollution and environmental degradation (Bency et al., 2003). The city's growing population and rapid urbanization have contributed to increased levels of air pollution, which can have negative impacts on public health and the environment. As a result, there have been efforts to address these issues through measures such as the implementation of stricter emissions standards and the promotion of sustainable transportation options.

One way to understand and mitigate the impacts of air pollution is through mapping and analysis (Tainio et al., 2021). By visualizing the distribution and

DOI: 10.4324/9781003275916-17

intensity of pollutants in the atmosphere, we can identify sources of pollution, assess the risk to human health, and develop strategies to reduce emissions. Air pollution is a significant global environmental and health problem that affects both urban and rural areas. It is caused by the emission of harmful substances such as particulate matter, ozone, and greenhouse gases into the atmosphere. These pollutants can have serious impacts on human health, including respiratory and cardiovascular diseases and the environment, such as climate change and acid rain (Manisalidis et al., 2020).

Mapping and analyzing air pollution is important for understanding its spatial and temporal distribution, as well as for identifying sources and trends (Zhao et al., 2021). Traditional methods of air pollution monitoring, such as ground-based monitoring stations, are limited in their spatial coverage and may not provide a complete picture of the distribution of pollutants (Fuentes et al., 2020). Remote sensing offers an alternative approach to monitoring air pollution at a large scale and over time (Martin, 2008).

Google Earth Engine (GEE) is a cloud-based platform for geospatial data analysis and visualization that combines satellite and aerial imagery with other data sources, such as ground-based measurements and climate models (Tamiminia et al., 2020). GEE provides access to a vast archive of satellite imagery and makes it possible to perform complex analyses on large datasets in a matter of minutes. There are several ways in which GEE can be used to map and analyze air pollution. One approach is to use satellite data to measure the concentration of pollutants in the atmosphere. This can be done using instruments on board satellites that measure the intensity of certain wavelengths of light, which can be used to infer the concentration of certain pollutants. Another approach is to use satellite data to track the movement of pollutants through the atmosphere. This can be done by analyzing the trajectory of pollutants over time and identifying patterns in their movement. This information can be used to understand the sources and sinks of pollutants and to develop strategies to reduce emissions.

GEE can also be used to analyze the impacts of air pollution on human health (Ghasempour et al., 2021). By overlaying data on population density and health outcomes, it is possible to identify areas where air pollution may be having the greatest impact on public health (Jerrett et al., 2001). This information can be used to prioritize interventions and policy efforts to reduce pollution in these areas. In addition to mapping and analysis, GEE can also be used to monitor the effectiveness of efforts to reduce air pollution. By comparing data on emissions and air quality before and after interventions, it is possible to assess the impact of these efforts and determine whether they are effective in reducing pollution.

The problem addressed in this study is the mapping and analysis of air pollution using GEE. The ability to map and analyze air pollution is crucial for understanding its distribution and identifying sources of pollution. GEE is a powerful tool for mapping and analyzing large-scale environmental data, but its application to air pollution has not been fully explored. The research aims to develop a GEE application that can be used to map and analyze air pollution data to better understand its distribution and identify sources.

14.2 Study Area

Thiruvananthapuram is the capital city of the Indian state of Kerala (Tiwari et al., 2020). Located on the western coast of India, Thiruvananthapuram has a population of over 9.7 million people and is known for its cultural and historical importance, as well as its natural beauty. Thiruvananthapuram is situated on the Arabian Sea, and the city is known for its long sandy beaches and coastal backwaters. The city is also home to several landmarks and tourist attractions, including the Padmanabhaswamy Temple, the Napier Museum, and the Kanakakkunnu Palace. Thiruvananthapuram is an important economic and cultural hub in southern India, and the city is known for its strong education and healthcare sectors. The city is home to several universities, including the University of Kerala and the Indian Institute of Technology. Overall, Thiruvananthapuram is an interesting and diverse city with a rich history and culture. Its coastal location, cultural attractions, and strong economy make it an important study area for a range of fields, including environmental science. The study of air pollution monitoring is important in Thiruvananthapuram because it can help to identify and address issues related to public health, environmental degradation, economic losses, and legal and policy requirements. By understanding the causes and impacts of air pollution in the city, it is possible to take action to protect public health and the environment and to promote sustainable economic development (Figure 14.1).

14.3 Datasets

14.3.1 Sentinal-5P NO_2

Sentinel-5P is a satellite operated by the European Space Agency (ESA) as part of the Copernicus Earth Observation Program. The satellite is equipped with a range of instruments, including the Tropospheric Monitoring Instrument (TROPOMI), which is designed to measure a variety of atmospheric gases, including nitrogen dioxide (NO_2) (Wang et al., 2022).

NO_2 is a gas that is produced by the burning of fossil fuels, such as coal and gasoline. It is a key contributor to air pollution and has a range of negative impacts on human health, including respiratory problems and an increased risk of heart disease and stroke. NO_2 is also a major contributor to the formation of ozone, which can have harmful effects on plants and ecosystems (Ghasempour et al., 2021).

Sentinel-5P is designed to measure the concentration of NO_2 in the Earth's atmosphere with high spatial and temporal resolution. The satellite uses a spectrometer, which is an instrument that measures the intensity of light at different wavelengths, to infer the concentration of NO_2 in the atmosphere.

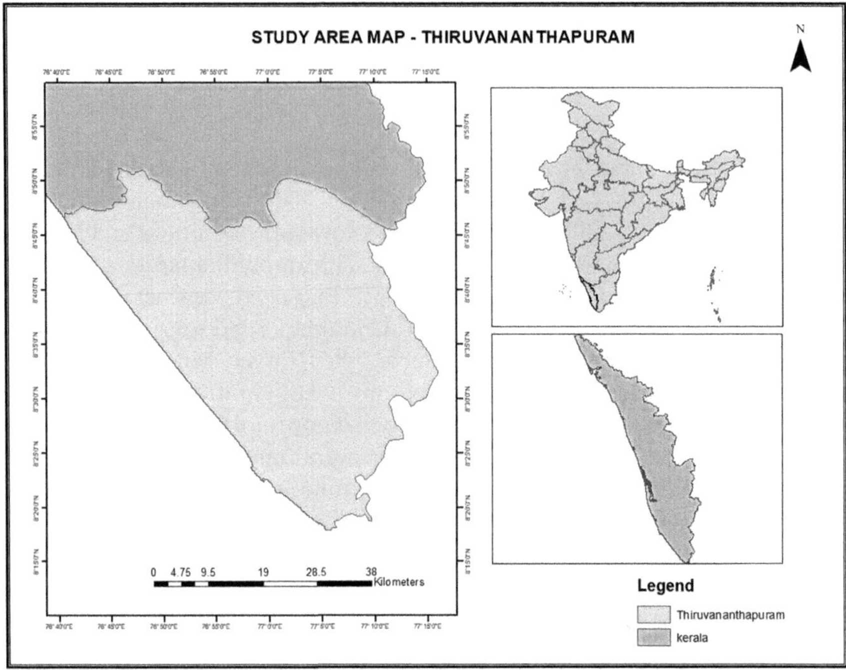

Figure 14.1 Study area map

The data collected by Sentinel-5P is used to understand the sources and sinks of NO_2, as well as the spatial and temporal patterns of its distribution in the atmosphere.

The data collected by Sentinel-5P is used by a range of stakeholders, including governments, researchers, and environmental organizations, to inform policy decisions and improve our understanding of the impacts of air pollution on human health and the environment. The satellite's high-resolution data is also used to monitor and track the effectiveness of interventions designed to reduce NO_2 emissions, such as the implementation of emissions controls on vehicles and industrial processes (Faisal et al., 2021).

14.4 The Population Count Data from Gridded Population of the World Version 4

The Gridded Population of the World (GPW) is a dataset that provides estimates of population counts and densities at a global scale. The dataset is produced by the Socioeconomic Data and Applications Center (SEDAC) at the Earth Institute at Columbia University. The GPW is based on data from a variety of sources, including censuses, surveys, and satellite imagery, and is designed to provide a consistent and accurate representation of global population patterns (Rabiei-Dastjerdi et al., 2022).

The GPW is available in several versions, with the latest version being GPWv4. GPWv4 provides population estimates for the years 2000, 2005, 2010, 2015, and 2020. It is available at a resolution of approximately 30 arc-seconds (approximately 1 kilometer at the Equator), which allows for detailed analysis at the regional and local scales.

The GPWv4 dataset is organized into several layers, including population counts, population densities, and urban extents. The population count layer provides estimates of the total number of people living in each grid cell. The population density layer provides estimates of the number of people per square kilometer in each grid cell. The urban extents layer provides information on the extent and location of urban areas, based on the presence of built-up land and other indicators of urbanization.

The GPWv4 dataset is a valuable resource for researchers, policymakers, and others interested in understanding global population patterns and trends. It can be used to inform a range of applications, including the analysis of population-environment interactions, the identification of areas with high population densities, and the assessment of the impacts of urbanization (Rabiei-Dastjerdi et al., 2022).

14.5 Methodology

Mapping and analyzing air pollution using GEE involves several steps. These steps involve accessing and preparing data, applying algorithms and techniques for analysis, and visualizing and interpreting the results. In this section, we will outline a methodology for using GEE to map and analyze air pollution (Figure 14.2).

14.5.1 Data Importing and Cleaning

The following satellite-based datasets are available for analyzing air pollution in the Earth engine data catalog: the moderate resolution imaging spectroradiometer and advanced very-high-resolution radiometer for monitoring aerosol optical depth (a proxy for pm2.5), the total ozone mapping spectrometer ozone monitoring instrument for monitoring ozone, and the tropospheric monitoring instrument (TROPOMI) on board the Sentinel-5 precursor (Sentinel-5p) for monitoring a range of air pollutants. In this study, we will use Sentinel-5p data, specifically the offline (offl) NO_2 product.

The TROPOMI sensor on Sentinel-5P measures ultraviolet, visible, near-infrared, and shortwave infrared wavelengths to monitor NO_2, ozone, aerosols, methane, formaldehyde, carbon monoxide, and sulfur dioxide in the atmosphere (Rabiei-Dastjerdi et al., 2022). The satellite's swath width is approximately 2,600 km on the ground, resulting in global daily coverage with a spatial resolution of 7 × 7 km. The Sentinel-5P data is available from July 2018 onward and includes Near Real-Time (NRTI) and Offline (OFFL) versions for all pollutants except methane, which is only available as OFFL. The NRTI assets

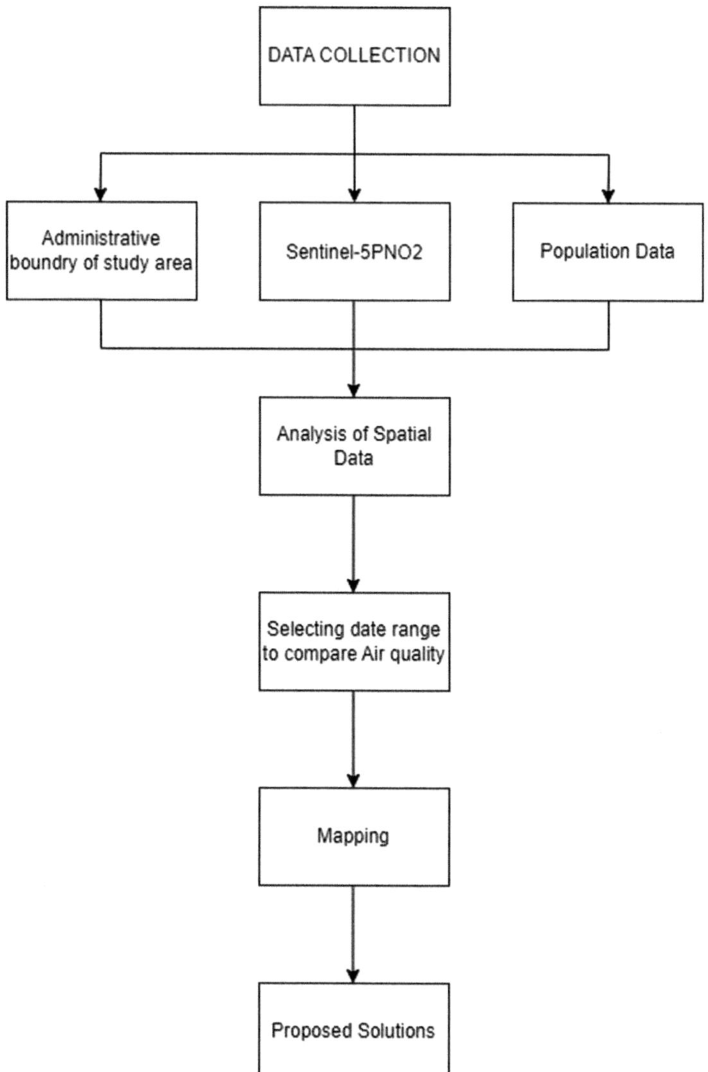

Figure 14.2 Flow chart of work carried out for the project

have faster availability but cover a smaller area, while the OFFL assets have a delayed availability but contain data from an entire orbit and are easier to use for retrospective analyses.

To import the NO_2 data, we created a function to mask out pixels with a cloud fraction above a certain threshold (e.g., 30% cloud cover). We can test different masking thresholds to determine the optimal value for our specific use case. After masking out cloudy pixels, we created a median composite from images taken in March 2021. It is important to note that we have used the band that gives measurements for the tropospheric vertical column of NO_2, not the stratospheric or total vertical column. The troposphere is the closest we can get to ground-level measurements with Sentinel-5P (Rabiei-Dastjerdi et al., 2022).

14.5.2 Quantifying and Visualizing Changes

To investigate changes in NO_2 concentrations during the COVID-19 lockdowns in 2020, we compared the median NO_2 concentration during March 2020 (when Hubei Province was in lockdown) with the median value during March 2019. Weather can significantly affect air pollutant concentrations, and therefore any differences between the two years could be due to weather conditions. By comparing the same month in different years, we partially control for the influence of seasonal weather patterns, but not completely. To more thoroughly control for weather effects, we can refer to the methods described by (Abbass et al., 2022). To visualize the comparison, we will calculate and create median composite images for March 2019 and March 2020, and use Earth Engine's user-interface widgets, specifically the UI.SplitPanel widget, to view the side of the image by side. The UI.SplitPanel widget can also be set to have a wiping effect where the maps are overlaid on top of each other.

14.5.3 Calculating Population-Weighted Concentrations

In a previous section, we used the Erath Engine Reducer which reduces the region function to calculate the average NO_2 concentration over the study area. However, when aggregating pollutant concentrations to define population exposure, we need to take a different approach. For example, if there was a high concentration of NO_2 in a study area with a small population, calculating the average of all pixels would not accurately represent the population's exposure (Dey et al., 2020). To address this issue, we can use a population dataset and calculate the population-weighted exposure, aggregated across pixels in the study area, using the equation (1).

$$\text{Population weighted exposure} = (\text{NO}_2 \text{ concentration} * \text{subpopulation in pixel})/\text{total population in the area of interest.} \quad (1)$$

We mapped a function to calculate the population-weighted exposure over all the images in the NO_2 Image Collection (Dey et al., 2020). Since we masked out pixels from images with more than 30% cloud cover in a previous section, we will also need to calculate the percentage of available Sentinel-5P pixels within Hubei Province per image. We will choose a threshold of 25% pixel coverage to calculate a representative average for the province, but this value may vary depending on the research question. By comparing the simple average and population-weighted average, we can see the difference that accounting for population density can make in the calculation.

14.6 Results and Analysis

Air pollution refers to any substance or agent that changes the natural composition of the atmosphere and can be harmful to human health. Examples of such pollutants include particulate matter with a diameter of less than 2.5 micrometers (PM2.5), carbon monoxide, ozone, NO_2, and sulfur dioxide. Chronic exposure to air pollution is a major global health threat, causing more deaths than HIV/AIDS, malaria, and tuberculosis combined, and significantly more fatalities than all forms of violence. It is estimated that PM2.5 and ozone exposure contribute to approximately 4.7 million excess deaths worldwide each year. However, these estimates can vary depending on the disease categories considered and the exposure-response function used, with some estimates ranging from 3 to 10 million excess deaths per year. NO_2 exposure may also lead to the development of 4 million new cases of pediatric asthma annually.

Over the past decade, there has been a significant increase in our understanding of the global distribution and sources of air pollutants due to the expansion of ground-based monitoring networks, the development of satellite products, and the advancement of atmospheric chemistry models (Dey et al., 2020). Research has shown that more than 70% of the global health burden from air pollution is caused by human-generated emissions (Fuller et al., 2022). The main sources of anthropogenic air pollution are industries, vehicles, power plants, agricultural activities, and household combustion, while non-anthropogenic sources include desert dust, biogenic emissions, forest fires, and volcanoes. The reduction in transportation and industrial activity during the COVID-19 lockdowns resulted in a significant decrease in global air pollution levels, highlighting the impact of human-generated emissions (Venter et al., 2020). It is estimated that the decline in air pollution in the first five months of 2020 led to 49,900 avoided deaths and 89,000 fewer pediatric asthma emergency department visits (Venter et al., 2020)

Despite the expansion of monitoring networks, most regions of the world do not have sufficient air monitoring, which limits air quality management. As a result, alternative monitoring methods such as satellite remote sensing are becoming more popular and accurate. In recent decades, we have gained access to a range of satellite sensors that monitor the contents of the Earth's atmosphere. However, it is important to note that satellites measure pollutant concentrations in the troposphere and stratosphere, which are many kilometers above the Earth's surface. Therefore, satellite measurements may not accurately represent the concentrations humans are exposed to on the ground, and relying solely on satellite data for human health applications is not recommended. However, more advanced techniques that combine information from satellite remote sensing data, atmospheric chemistry models, and ground-based monitors can provide reliable estimates of ground-level pollutant concentrations (Figure 14.3).

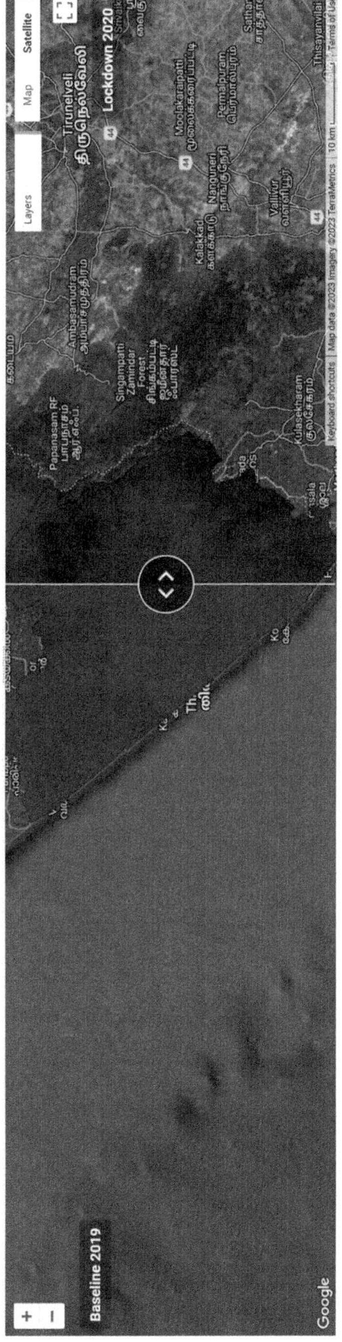

Figure 14.3 Screenshot of application showing air pollution of Kerala for two different years simultaneously (2019 on left, 2020 on right)

The line graph provided illustrates the mean concentration of NO_2 in the air before and during a lockdown period. The data represented by the red line represents the mean NO_2 content in March 2019, before the lockdown, while the data represented by the blue line represents the mean NO_2 content during the lockdown period in March 2020. From the graph, it is apparent that the air quality was significantly better during the lockdown period as compared to before the lockdown. This can be attributed to a decrease in human activity, specifically a decrease in commuter traffic and private vehicle usage. As many people were working remotely during the lockdown, there was a reduction in the number of vehicles on the road, leading to a decrease in emissions of pollutants, including NO_2. NO_2 is a key pollutant that is emitted primarily by the burning of fossil fuels, such as in vehicles and power plants (Figures 14.4 and 14.5). It is one of the gases that comprise smog and can have detrimental effects on human health and the environment. A reduction in the concentration of NO_2 in the air, as seen during the lockdown period, would therefore have a positive impact on air quality (Figures 14.6 and 14.7).

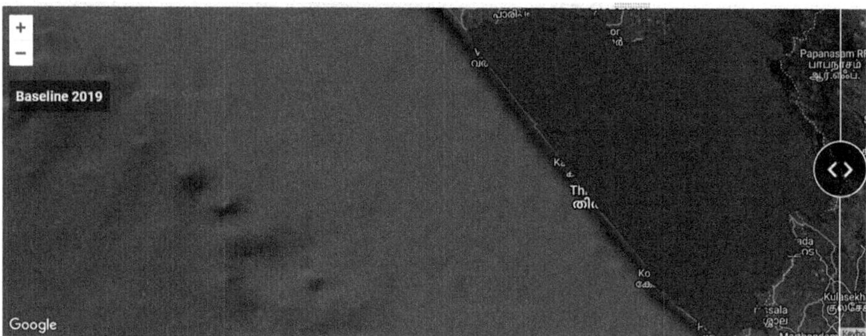

Figure 14.4 Application showing air pollution of Kerala for 2019 March

Figure 14.5 Application showing air pollution of Kerala for 2020 March

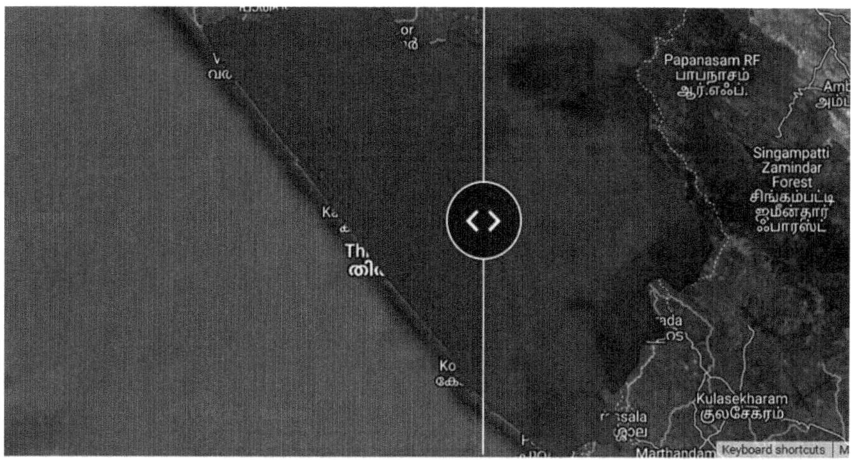

Figure 14.6 Screenshot of application showing air pollution of Kerala for two different years simultaneously (2019 on left, 2020 on right)

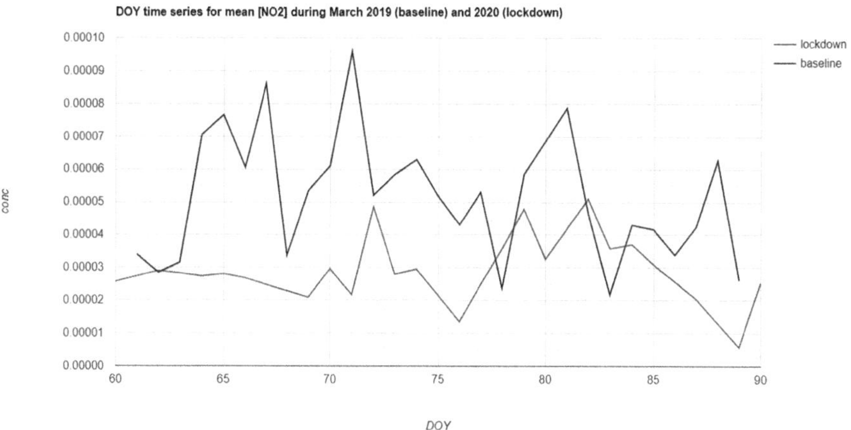

Figure 14.7 Time series for mean NO_2 during March 2019 and 2020

The chart provided shows a time series of mean NO_2 concentrations in the air, represented by the red line, and population-weighted NO_2 concentrations, represented by the blue line, for the month of March 2020.

NO_2 is a key air pollutant that is emitted primarily by the burning of fossil fuels, such as in vehicles and power plants. It is one of the gases that comprise smog and can have detrimental effects on human health and the environment. The mean NO_2 concentration represents the average concentration of NO_2 in the air over a given period, while the population-weighted NO_2 concentration represents the average concentration of NO_2 experienced by the population, taking into account the population distribution (Figure 14.8).

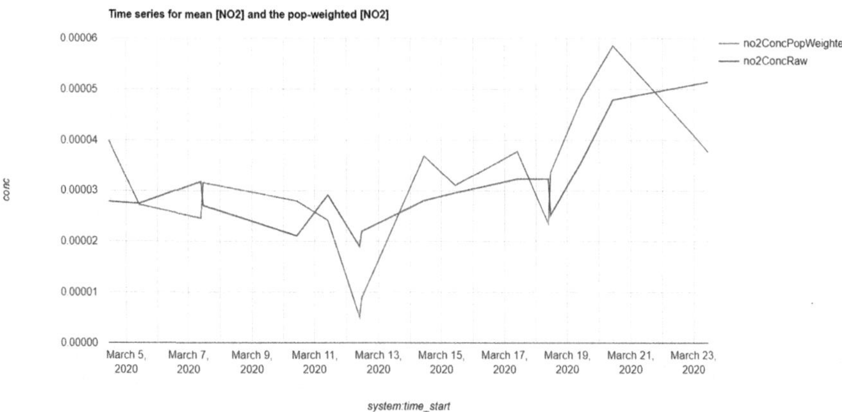

Figure 14.8 Time series for mean NO_2 and population-weighted NO_2

From the chart, it is clear that the population-weighted NO_2 concentration (blue line) is consistently higher than the mean NO_2 concentration (red line) throughout the month of March 2020. This indicates that certain areas or populations are experiencing higher concentrations of NO_2 than the overall average. This could be due to a higher density of population in areas with high NO_2 emission sources, such as near major roads or industrial areas (Figure 14.8).

It is important to consider population-weighted concentrations when evaluating the impacts of air pollution on public health, as certain populations may be disproportionately affected by higher levels of pollution. Understanding the differences in mean and population-weighted concentrations can aid in identifying and targeting interventions to reduce exposure and improve air quality for affected populations.

14.7 Conclusion

Mapping and analyzing air pollution using GEE is a powerful way to understand and address the problem of air pollution. GEE is a cloud-based platform that provides access to a wide range of satellite data and advanced analytical capabilities, making it easier for scientists, policymakers, and the general public to understand and address environmental issues.

There are several ways in which GEE can be used to map and analyze air pollution. One approach is to use satellite data to measure the concentration of pollutants in the atmosphere. This is done using instruments on board satellites that measure the intensity of certain wavelengths of light, which can be used to infer the concentration of certain pollutants. GEE can also be used to analyze the impacts of air pollution on human health. By overlaying data on population density and health outcomes, it is possible to identify areas where air pollution may be having the greatest impact on public health. This information can be used to prioritize interventions and policy efforts to reduce pollution in these areas. In addition to mapping and analysis, GEE can also be used to monitor the effectiveness of efforts to reduce air pollution. By comparing data on emissions and air quality before and after interventions, it is possible to assess the impact of these efforts and determine whether they are effective in reducing pollution.

Overall, GEE is a valuable tool for understanding and addressing the problem of air pollution. Providing access to data and advanced analytical capabilities, enables scientists, policymakers, and the general public to better understand the sources and impacts of air pollution and develop strategies to reduce emissions and improve air quality.

Bibliography

Abbass, K., Qasim, M. Z., Song, H., Murshed, M., Mahmood, H., & Younis, I. (2022). A Review of the Global Climate Change Impacts, Adaptation, and Sustainable Mitigation Measures. *Environmental Science and Pollution Research*, 29(28), 42539–42559. https://doi.org/10.1007/s11356-022-19718-6

Bălă, G.-P., Râjnoveanu, R.-M., Tudorache, E., Motișan, R., & Oancea, C. (2021). Air Pollution Exposure—The (In)visible Risk Factor for Respiratory Diseases. *Environmental Science and Pollution Research*, 28(16), 19615–19628. https://doi.org/10.1007/s11356-021-13208-x

Bency, K. T., Jansy, J., Thakappan, B., Kumar, B., Sreelekha, T. T., Hareendran, N. K., Nair, P. K. K., & Nair, M. K. (2003) A Study On The Air Pollution Related Human Diseases In Thiruvananthapuram City, Kerala. In Martin J. Bunch, V. Madha Suresh and T. Vasantha Kumaran, eds., *Proceedings of the Third International Conference on Environment and Health, Chennai, India, 15-17 December, 2003*. Chennai: Department of Geography, University of Madras and Faculty of Environmental Studies, York University. pp 15–22.

Clarke, B., Otto, F., Stuart-Smith, R., & Harrington, L. (2022). Extreme Weather Impacts of Climate Change: An Attribution Perspective. *Environmental Research: Climate*, 1(1), 012001. https://doi.org/10.1088/2752-5295/ac6e7d

Colvile, R. N., Hutchinson, E. J., Mindell, J. S., & Warren, R. F. (2001). The Transport Sector is a Source of Air Pollution. *Atmospheric Environment*, 35(9), 1537–1565. https://doi.org/10.1016/S1352-2310(00)00551-3

Dey, S., Purohit, B., Balyan, P., Dixit, K., Bali, K., Kumar, A., Imam, F., Chowdhury, S., Ganguly, D., Gargava, P., & Shukla, V. K. (2020). A Satellite-Based High-Resolution (1-km) Ambient PM2.5 Database for India over Two Decades (2000–2019): Applications for Air Quality Management. *Remote Sensing*, 12(23), 3872. https://doi.org/10.3390/rs12233872

Faisal, Moh., Adi Prakoso, K., Sanjaya, H., & Rohmaneo Darminto, M. (2021). Spatio-Temporal Analysis of Air Pollutants Changes During The COVID-19 Using Sentinel-5P in Google Earth Engine (Case Study: Java Island). 2021 IEEE Asia-Pacific Conference on Geoscience, Electronics and Remote Sensing Technology (AGERS), 102–108. https://doi.org/10.1109/AGERS53903.2021.9617331

Fuentes, M., Millard, K., & Laurin, E. (2020). Big Geospatial Data Analysis for Canada's Air Pollutant Emissions Inventory (APEI): Using Google Earth Engine to Estimate Particulate Matter from Exposed Mine Disturbance Areas. *GIScience & Remote Sensing*, 57(2), 245–257. https://doi.org/10.1080/15481603.2019.1695407

Fuller, R., Landrigan, P. J., Balakrishnan, K., Bathan, G., Bose-O'Reilly, S., Brauer, M., Caravanos, J., Chiles, T., Cohen, A., Corra, L., Cropper, M., Ferraro, G., Hanna, J., Hanrahan, D., Hu, H., Hunter, D., Janata, G., Kupka, R., Lanphear, B., … Yan, C. (2022). Pollution and Health: A Progress Update. *The Lancet Planetary Health*, 6(6), e535–e547. https://doi.org/10.1016/S2542-5196(22)00090-0

Ghasempour, F., Sekertekin, A., & Kutoglu, S. H. (2021). Google Earth Engine based Spatio-Temporal Analysis of Air Pollutants Before and During the First Wave COVID-19 Outbreak Over Turkey via Remote Sensing. *Journal of Cleaner Production*, 319, 128599. https://doi.org/10.1016/j.jclepro.2021.128599

Holman, C. (1999). Sources of Air Pollution. In *Air Pollution and Health* (pp. 115–148). Elsevier. https://doi.org/10.1016/B978-012352335-8/50083-1

Jerrett, M., Burnett, R. T., Kanaroglou, P., Eyles, J., Finkelstein, N., Giovis, C., & Brook, J. R. (2001). A GIS–Environmental Justice Analysis of Particulate Air Pollution in Hamilton, Canada. *Environment and Planning A: Economy and Space*, 33(6), 955–973. https://doi.org/10.1068/a33137

Manisalidis, I., Stavropoulou, E., Stavropoulos, A., & Bezirtzoglou, E. (2020). Environmental and Health Impacts of Air Pollution: A Review. *Frontiers in Public Health*, 8, 14. https://doi.org/10.3389/fpubh.2020.00014

Martin, R. V. (2008). Satellite Remote Sensing of Surface Air Quality. *Atmospheric Environment*, 42(34), 7823–7843. https://doi.org/10.1016/j.atmosenv.2008.07.018

Rabiei-Dastjerdi, H., Mohammadi, S., Saber, M., Amini, S., & McArdle, G. (2022). Spatiotemporal Analysis of NO_2 Production Using TROPOMI Time-Series Images and Google Earth Engine in a Middle Eastern Country. *Remote Sensing*, 14(7), 1725. https://doi.org/10.3390/rs14071725

Tainio, M., Jovanovic Andersen, Z., Nieuwenhuijsen, M. J., Hu, L., De Nazelle, A., An, R., Garcia, L. M. T., Goenka, S., Zapata-Diomedi, B., Bull, F., & Sá, T. H. D. (2021). Air Pollution, Physical Activity and Health: A Mapping Review of the Evidence. *Environment International*, 147, 105954. https://doi.org/10.1016/j.envint.2020.105954

Tamiminia, H., Salehi, B., Mahdianpari, M., Quackenbush, L., Adeli, S., & Brisco, B. (2020). Google Earth Engine for Geo-Big Data Applications: A Meta-Analysis and Systematic Review. *ISPRS Journal of Photogrammetry and Remote Sensing*, 164, 152–170. https://doi.org/10.1016/j.isprsjprs.2020.04.001

Tiwari, V., Kumar, V., Matin, M. A., Thapa, A., Ellenburg, W. L., Gupta, N., & Thapa, S. (2020). Flood Inundation Mapping- Kerala 2018; Harnessing the Power of SAR, Automatic Threshold Detection Method and Google Earth Engine. *PLOS ONE*, 15(8), e0237324. https://doi.org/10.1371/journal.pone.0237324

Venter, Z. S., Aunan, K., Chowdhury, S., & Lelieveld, J. (2020). COVID-19 Lockdowns Cause Global Air Pollution Declines. *Proceedings of the National Academy of Sciences*, 117(32), 18984–18990. https://doi.org/10.1073/pnas.2006853117

Wang, S., Chu, H., Gong, C., Wang, P., Wu, F., & Zhao, C. (2022). The Effects of COVID-19 Lockdown on Air Pollutant Concentrations across China: A Google Earth Engine-Based Analysis. *International Journal of Environmental Research and Public Health*, 19(24), 17056. https://doi.org/10.3390/ijerph192417056

Zhao, Q., Yu, L., Li, X., Peng, D., Zhang, Y., & Gong, P. (2021). Progress and Trends in the Application of Google Earth and Google Earth Engine. *Remote Sensing*, 13(18), 3778. https://doi.org/10.3390/rs13183778

15 Mapping and Development of Agriculture Diversity in West Bengal

An Evaluation at District Level

Rukhsana

15.1 Introduction

There are several benefits to having diversified agricultural systems for both farmers and the environment. These benefits include improved agricultural losses due to pests and other risks, improved ecosystem performance, considerable services provided for agriculture (Bommarco et al., 2013; Chaplin Kramer, et al., 2011; Gardiner et al., 2009; Rukhsana, 2021), and augmentation of healthy soil (McDaniel et al., 2014; Tiemann et al., 2015). A number of studies demonstrate how diversified agricultural practices boost a region's ability to adapt to climate change (Altieri, 1999; Makate et al., 2016; McCord et al., 2015; Smit and Skinner, 2002; Smit and Wandel, 2006) and reduce the risks of farmers associated with climatic factors (Birthal et al., 2015; Bradshaw et al., 2004; Lin, 2011; McCord et al., 2015). Crop diversification as an agricultural approach can improve crop production stability for a variety of crops and raise household income (Abson et al., 2013; Barrett et al., 2001; Bigsten and Tengstam, 2011; Demissie and Legesse, 2013; Makate et al., 2016; Mhango et al., 2013; Njeru, 2016; Smith et al., 2008).

Crop diversity is a prerequisite for an economy reliant on agriculture. Farmers have had to diversify their agricultural endeavors in order to satisfy their families' monetary demands throughout time and to protect water tables, mitigate the effects of climate change, and decrease hazards like shrinking the area under cultivation (Sinha and Ahmad Nasim, 2016; Rukhsana, 2020). Crop diversification is also receiving a lot of attention due to market infrastructure, resource availability, public participation (pricing and credit policies, research and development), and globalization of agriculture (Kumar et al., 2012 and Singh et al., 2013). Crop diversification has a number of advantages, including increased income, a decline in poverty, a secure supply of food and nutrition, the creation of jobs, efficient use of land and water resources, sustainable farming methods, and environmental improvement. Additionally, diversification is frequently required to repair or replace a depleted natural resource base. In order to preserve or raise the value of natural resources, it is typically claimed that cropping systems have varied or new cropping systems have been implemented. Additionally, diversification has been seen to stabilize

DOI: 10.4324/9781003275916-18

agricultural income at higher levels by allowing for the cultivation of increasingly profitable crops. This is crucial for small farmers who work hard to maintain their farms (Saleth and Maria, 1995; Rukhsana, 2021). Crop diversification is a crucial scientific technique that connects the spatial variance of different crops to one another, directly defining the number of crops that can be grown. It emphasizes a shift from low value to high value agriculture and is a significant approach to boost agricultural output (Dutta, 2012). It is fundamentally regulated by both the physical and socioeconomic conditions of the region and where there will be a higher level of agricultural technology, and a lower degree of diversification (Raju, 2012).

West Bengal, the economic center of the nation, is one of the least developed areas in the nation with the highest concentration of economically depressed districts and 32.10% of the population living below the poverty line. A population density of 1.91, greater than the national average in the eastern states, is found in the region, which comprises around 21.85% of the country's land area and sustains 34% of its population. In a region of 29.17 million hectares, the net sowing area has been measured at a 150% intensity. The average annual rainfall in this region ranges from 1091 to 2477 mm (average: 1526 mm), which is adequate for the cultivation of a variety of crops. While 45% of the country is on average irrigated, 67% of tenant farmers are from the marginal group, and more than 75% of their income is spent on ensuring food security, thus the irrigated area is only approximately 39% in the eastern region. Natural resources have been given to eastern India. The rate of agricultural growth is extremely slow in spite of the abundance of natural resources (fertile land, plentiful ground water) (Ahmad *et al.*, 2017; Rukhsana, 2021).

The following objectives have been taken in to consideration for the study: first is to explore the nature of the changes in the crop diversification or specialization at the district level in West Bengal. Second is to find out the changes in cropping pattern and agricultural development and to make suggestions for furthering the diversification towards the sustainability of agriculture in the region.

15.2 Geo-environmental Setting of the Study Area and Research Methodology

The state of West Bengal is located between latitudes 21°31′ and 27°13′14″ north and latitudes 85°45′20″ and 89°long53′ east. West Bengal makes up 88,752 square kilometers, or roughly 2.7%, of the nation's total land area (Kolkata Gazette, 2005). According to the 2011 Census, West Bengal is divided into 19 districts and 66 sub-divisions, totaling 341 development blocks. It is a state in eastern India that shares borders with Bangladesh and Nepal and has a wide range of physical diversity. Aside from the tea case of the crop harvest, the variance in crop pattern in West Bengal has just recently become a phenomenon. It is primarily a rice-producing state that became the biggest rice-producing nation in the 1980s and 1990s thanks to exceptionally high paddy agriculture productivity. However, diversification toward larger proportion crops is being regarded as a means to boost the contribution to the production ratio of non-rice crops in order to attain higher agricultural growth rates in the future. Diversification is thought to support growth while also improving nutrition, reducing poverty, creating jobs, and managing natural resources in a sustainable way.

The present study is based on secondary data, collected from various issues of the *Statistical Abstract and Economic Review of West Bengal*, published by the Bureau of Applied Economics and Statistics, Government of West Bengal, for the time periods of 1981, 1991, 2001, and 2011, based on census data. The nature of crop diversification has been investigated through changes in cultivation of various crops grown in different seasons over the years. There are various methods for studying the extent of diversification which are available in the literature. Several statistical measures has been applied for crop diversity like the Simpson index, Ogive index, modified entropy index and composite entropy index etc. (Chand 1996; Pandey and Sharma, 1996).

In this study the crop diversification (CDI) index was calculated to determine the crop diversification for particular crops of interest which is obtained by subtracting the Herfindalh index (HI) from 01. The crop diversification index is a measure of concentration and has a direct relationship with diversification where a zero value represents specialization and greater than zero indicated crop diversification. Thus, it becomes easier to identify farmers who are practicing crop diversification. The Herfindalh index formula has been applied to refine the crop diversification area in the study area. The Herfindalh index given below is computed by taking sum of squares of acreage proportion of each crop in the total cropped area. Mathematically, the index is given as below.

$$\textbf{Herfindalh index (HI)} = \sum_{i-1}^{N} P_i^2$$

where N represents the total number of crops and Pi is the area proportion of the i-th crop in total cropped area. With the increase in diversification, the Herfindalh index would decrease. This index takes a value of one when there is complete concentration and approaches zero when diversification is perfect. Thus the Herfindalh index is bounded by zero and one.

15.3 Result and Discussions

15.3.1 Agriculture

In West Bengal, productivity growth in agriculture since the early 1980s contributed significantly to the overall economic development of states, particularly in food production, with agricultural development having a significant impact on poverty reduction (Rukhsana, 2021). It has been observed that after a long period of stagnation, the development of agriculture in West Bengal began in the early 1980s, using technology based on high-yielding seeds (HYVs) and chemicals within a high-yielding distribution frame. Tenancy reforms in the shape of Operation Barga, which were put into effect in the state in the late 1970s, have been granted the authority to register tenants and legal rights to high crop shares. There has been increasing concern in recent years about the decline in agricultural production in most of the agricultural states of India since the early 1990s. Adoption of HYVs technology without considering the soil and moisture conditions, inadequate rural infrastructure, and weak network of agricultural marketing, sharply tilted land distribution and tenancy against the tenants in most parts of the country are the major obstruction to agricultural growth in India (Rukhsana, 2021). West Bengal (after Uttar Pradesh) is the second largest producer of potatoes and ranks first in terms of average yield. Jute production also increased to 5.4 per year in the 1990s, after growing at a rate of only 2.2% per year in the 1980s. Some of the more important crop diversifications are related to horticulture. West Bengal is now a major producer of vegetables, accounting for about 17% of the total vegetable production in the country. It is worth repeating that all this is the result of the actions of small to medium farmers, and there is a lot of scope for further expansion of horticultural production.

15.3.2 Changes in Cropping Pattern

Change in cropped area through the net area sown or change in intensity of cropping pattern is brought about multiple cropping supported by irrigation. The introduction of multiple cropping of non-traditional varieties of seeds with suitable irrigation led to in gross cropped area. Cropping intensity has been seen to have grown significantly throughout the 1990s and subsequently, which has contributed to the state's increase in gross cultivated area. This increase in cropping intensity is mostly connected to an extraordinary rise in the area planted to rice (boro rice). There has been a substantial diversification of crops under cultivation, as shown by the index of crop pattern, in West Bengal during the past three decades. The index of cropping pattern in 1990–1991 with 1980–81 as the base period is estimated at around 114.2 (Table 15.1). Beginning in the 1990s, there was noticeable variation of crop patterns in West Bengal away from food grains. The share of cropped area

Table 15.1 Indices of cropping pattern and cropping intensity in West Bengal

Year	*Cropping pattern*	*Cropping intensity*
1980–81	100	100
1990–91	114.2	111.8
2000–01	128.8	113.4

Source: Government of West Bengal. Economic Review 2007 08

Table 15.2 Area of principal crops in West Bengal

Year/Crops	*1980–81*	*1990–91*	*2000–01*
Rice	72.13	75.21	70.26
Aus	8.57	7.9	5.09
Aman	58.73	55.72	47.04
Boro	4.83	11.59	18.12
Wheat	3.94	3.48	5.51
Pulses	7.31	4.06	3.55
Total foodgrains	85	84.05	80.05
Oilseeds	4.42	6.47	7.74
Jute	8.51	6.47	7.92
Potato	1.61	2.52	3.87
Non-food grains	15	15.95	19.95

Source: Government of West Bengal, Economic Review 2007–08

under non-food grains increased considerably in the last two and a half decades. During 1980–2001, the percentage of oilseeds, particularly mustard, was found to be almost double. The potato area also grew spectacularly during the same period (Table 15.1). But the cultivation of jute decreased during this period, although there was a slight increase in 1990. The cropping pattern varies widely across the district. Table 15.2 shows the share of average for major crops among the various districts of the state which shows that most of the districts are producing rice, and Aman is the leading variety. Howrah, Medinipur (East), Bardhaman, 24 Parganas (North) and Hooghly are the major borough producing districts of West Bengal. Murshidabad, Malda, Nadia are the leaders in the production of wheat and pulses. The production of oilseeds is mainly concentrated in Nadia, 24 Parganas (North), Murshidabad and Malda. On the other hand, jute is cultivated in Nadia, Cooch Behar, Murshidabad, Dinajpur (North) and 24 Parganas (North). Hooghly has taken the lead in potato production.

15.3.3 Regional Trend of Crops Diversification in West Bengal from 1981–82 to 2012–13

The last three decades have witnessed significant changes in agricultural practice and the pattern of state agriculture. Changing agricultural patterns and practices is the result of successful implementation of the "Green Revolution Program (GRP)" and policies. In addition, the Indian agricultural sector has seen significant changes from traditional food grain farming to diversification and cultivation of commercial horticultural crops since the 1990s. Furthermore, similar results have been observed in the state of West Bengal in relation to agricultural practices. West Bengal is mainly divided into five categories (very high, high, medium, low and very low crop diversified regions) based on the Herfindahl index of crop diversification (Figure 15.1).

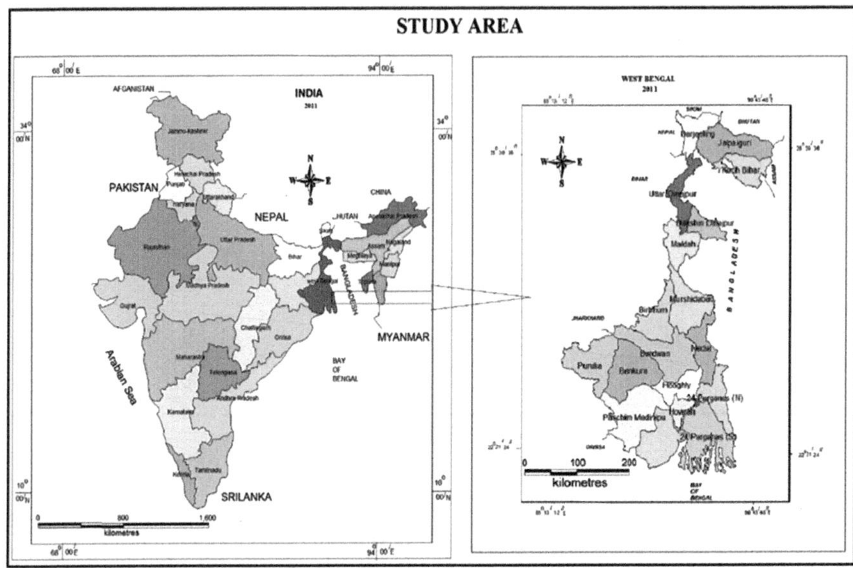

Figure 15.1 Location map of study area

Figure 15.2 and Table 15.3 illustrate the spatial pattern of crops diversification at district level in West Bengal which discloses that only three districts have very high diversity compared to others districts in terms of agricultural practices during 1981–82 including South 24 Parganas (0.01), Darjeeling (0.01) and Purulia (0.01). Figure 15.2 shows that four districts including Howrah (0.002), South 24 Parganas (0.016), North 24 Parganas (0.021) and Jalpaiguri (0.018), have shown very high diversity in term of agricultural practices during 1991–1992. It has been shown with analysis of crop diversification that low diversity was found in these districts such as Murshidabad (0.221) and Darjeeling (0.206) diversify in terms of agricultural practice which has been observed steadily moving towards crops specialization, medium diversified districts devoted their agricultural land to certain crops specialization in state of West Bengal during 1991–92. But the districts like Burdwan showed much less diversification in term of agriculture practices during 1991–92. It is mainly because of high specialization of jute and rice cultivation in the district of Murshidabad and on the other hand rice and tea cultivation in the district of Darjeeling and Burdwan. The same result has been observed during 1981–82.

Crop diversification has been also noticed significantly during 2001–02 in respect of 1991–92, Figure 15.2 and the same table demonstrates that five districts such as Hawrah (0.001), Cooch Behar (0.017), Bankura (0.02), Jalpaiguri (0.014) and Dakshin Dinajpur (0.006) disaplyed high diversity in terms of agricultural practices during 2001–02. Therefore, districts including Uttar Dinajpur (0.023), Burdwan (0.040), Maldah (0.052) and Hoogly (0.027) have made tremendous improvement in terms of crops diversification from 1991 to 2001,

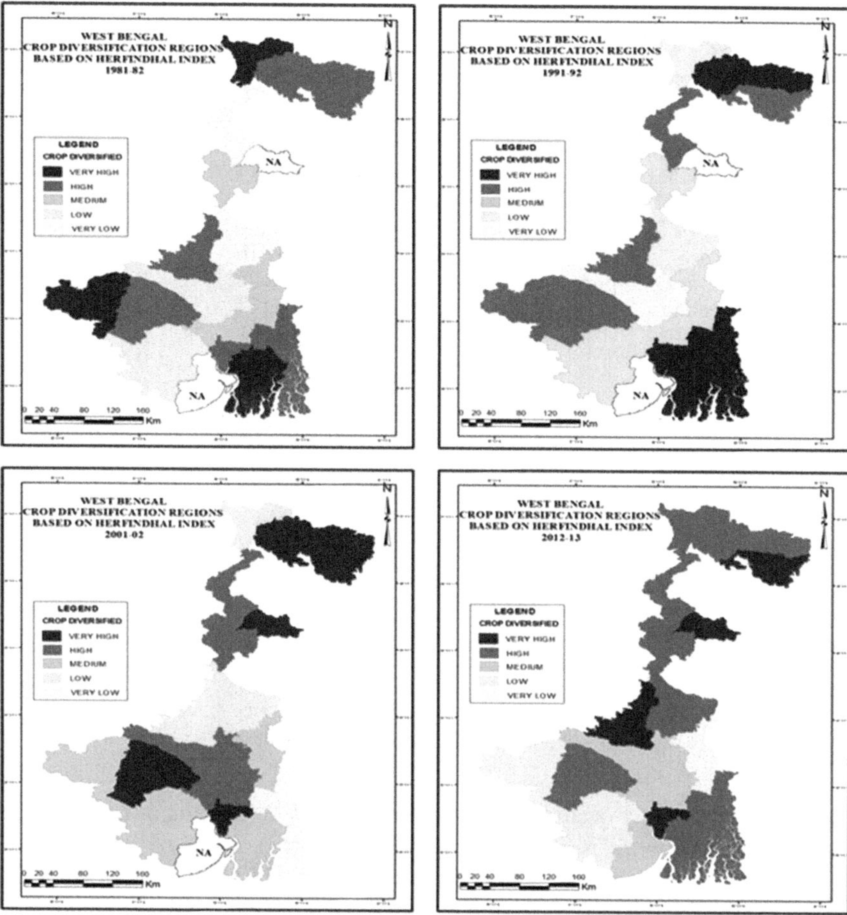

Figure 15.2 The spatial pattern of crops diversification at district level in West Bengal

which demonstrates high diversification and Birbhum (0.220) has shown low diversity in agricultural cropping practices during 2001–02. Therefore, these three districts have made significant improvements in the agriculture field in general and crops diversification in particular due to improvement of irrigation facilities and changes in institutional land holding size. North 24 Parganas (0.491) is categorized as having very low crop diversification in 2001–02. The Birbhum district is steadily moving towards crops specialization especially towards rice cultivation in the state of West Bengal during 2001–02.

Table 15.3 shows that the present scenario of crop diversification in West Bengal during 2012–13 shows significantly improvement compared to previous study periods (Table 15.3). Figure 15.2 describes the annual crop diversification index by regions in West Bengal. The sample mean of the Herfindahl index is 0.057 which is the highest Herfindalh index. Conversely crop

Table 15.3 Extent and changes of crops diversity in West Bengal at district level (1981–82 to 2012–13)

Sl. No.	*Name of the district*	*1981–82*	*1991–92*	*2001–02*	*2012–13*
1	Burdwan	0.13	0.8	0.04	0.06
2	Koch Behar	0.02	0.029	0.017	0.01
3	South 24 Parganas	0.01	0.016	0.069	0.03
4	North 24 Parganas	0.02	0.021	0.491	0.02
5	Bankura	0.02	0.033	0.016	0.02
6	Dakshin Dinajpur	NA	NA	0.006	0.01
7	Darjeeling	0.01	0.206	0.187	0.04
8	Hawrah	0.04	0.002	0.001	0
9	Maldah	0.06	0.088	0.052	0.03
10	Hoogly	0.08	0.104	0.027	0.08
11	Jalpaiguri	0.02	0.018	0.014	0.03
12	Murshidabad	0.18	0.221	0.176	0.03
13	Nadia	0.09	0.158	0.112	0.25
14	Paschim Mednipur	0.11	0.131	0.124	0.17
15	Purba Mednipur	NA	NA	NA	0.08
16	Uttar Dinajpur	0.22	0.057	0.023	0.02
17	Purulia	0.01	0.041	0.079	0.13
18	Birbhum	0.03	0.027	0.22	0.01
Total Average		0.066	0.122	0.097	0.057

Source: Compiled by Author, calculation is based on data from *Statistical Handbook West Bengal* (1988, 1992, 2002, 2012) published by Bureau of Applied Economics & Statistics Department, Kolkata, West Bengal Office of the Directorate of Agriculture, Government of West Bengal

diversification mainly occurs in northern regions of West Bengal. Figure 15.2 explains that Howrah is the most diversified district followed by Dakshin Dinajpur (0.01), Birbhum (0.01) and Cooch Behar (0.01). Therefore, it is observed that districts like Birbhum along with Uttar Dinajpur have made tremendous improvement in terms of crops diversification during the said period in the state of West Bengal. Moderate diversification has been noticed in the districts of Hoogly (0.08), Burdwan (0.08) and Purba Mednipur (0.08). These districts have made significant improvement in agriculture field in general and crops diversification in particular during 2012–13 due to improvement of irrigation facilities and changes in institutional holding size and accessibility to market.

15.4 Summary and Conclusion

It can be summarized that West Bengal is one of the important agriculturally developed states of India with specialization in a few crops like rice and jute. The chapter has shown that the state has recorded a major shift in crop diversity during the last three decades. The chapter shows that in the northern portion of West Bengal including districts such as Darjiling, Jalpaiguri,

Cooch Behar, Dakhin Dinajpur, Uttar Dinajpur, Burdwan and Malda etc. tremendous progress has been made in term of crop diversity in West Bengal during the study period. Cooch Behar district which accounted for very high diversification in terms of agriculture due to district's population is dominated by the agriculture occupation only. Cooch Behar district agriculturally is very much developed; here most of the 80–90% population is engaged in agriculture. It has been observed that the district Nadia was found to have a very low grade of crop diversification during 2012–13, which has been severely reduced from the previous study period to the post study period, due to agricultural land encroached by the urban area. Similarly, the district is located in the southwest of Purulia State, which was Not improving agricultural practice due to low economic conditions and favorable climatic conditions, faced shortage of irrigation because of dry area among other districts of West Bengal. Bardhaman district, a rice-dominated area in the state, is also making efforts to improve diversification of crops. Districts in the southern part of the state such as the North 24 Parganas and the South 24 Parganas have reduced crop diversification due to natural disasters, low economic conditions and increasing salinity of the soil which is not conducive for cultivation. Apart from this, other districts have also made progress in the field of crop diversification in the state of West Bengal in the last three decades. West Bengal is one of the important agriculturally developed states of India, specializing in some crops like rice and jute. The study showed that the state has made major changes in crop diversity during the last three decades. The study shows that the northern parts of West Bengal such as Darjeeling, Jalpaiguri, Cooch Behar, Dakhin Dinajpur, Uttar Dinajpur, Burdwan and Malda etc. have made tremendous progress in the period of crop diversity in the state of West Bengal during the study period. Cooch Behar district, which shown much diversification in terms of agriculture due to the population of the district, is dominated only by agribusiness. Agriculture is very developed in Cooch Behar district; most of the 80–90% population here is engaged in agriculture. Therefore, it has been safely declared that, most of the districts of West Bengal are trying to move towards diversification of crops and some of them have made significant progress in this area, and some are trying to do the same. The results of the study show that West Bengal's agricultural sector is gradually moving towards high value commodities. It is ascertained that most of the diversification comes with little support from the government through individual efforts of small farms. The reason for this is that the issues of food security in the state as well as in the country are still serious and the government policy is still self-sufficient in terms of food grains. Also the degree of diversification is not evenly distributed across districts. While some districts are increasingly picking up diversification, others are lagging far behind. This may be due to the fact that even though the state has achieved self-sufficiency in staple food, the emphasis is still focused on increasing rice production.

Acknowledgement

I am sincerely indebted to Indian Council of Social Science Research (Ministry of Human Resource Development, New Delhi, 110067, India, for sponsoring my Major Research Project.

Bibliography

Abson DJ, Fraser ED, & Benton TG, (2013) Landscape diversity and the resilience of agricultural returns: a portfolio analysis of land-use patterns and economic returns from lowland agriculture. *Agriculture & Food Security* 2 (1), 2.

Ahmad N., et al. (2017) Determinants of crop diversification in Bihar agriculture–An economic analysis, 35 (4E). *Environment and Ecology* XXXV, 3683–3687.

Altieri MA, (1999) The ecological role of biodiversity in agro ecosystems. *Agriculture, Ecosystems and Environment* 74 (1), 19–31. https://doi.org/101016/S0167-8809 (99)00028-6

Anderson, SA (1990) Core indicators of nutritional state for difficult-to-sample populations. *The Journal of Nutrition*, 120 (11), 1555–1598, https://doi.org/10.1093/jn/120.suppl_11.1555

Barrett CB, Clark MB, Clay DC, & Reardon T, (2001) Heterogeneous constraints, incentives and income diversification strategies in rural *Africa. Quarterly Journal of International Agriculture* 44 (1), 37–60. https://doi.org/102139/ssrn258371

Benin S, Smale M, Pender J, Gebremedhin B, & Ehui S, (2004) The economic determinants of cereal crop diversity on farm in Ethiopian highlands. *Agricultural Economics* III, 197–208.

Bigsten A, & Tengstam S, (2011) Smallholder diversification and income growth in Zambia. *Journal of African Economy* 20 (5), 781–822. https://doi.org/101093/jae/ejr017

Birthal P, Joshi, PK, & Minot NW, (2007) *Sources of agricultural growth in India: Role of diversification towards high value crops*. New Delhi: Markets, Trade and Institutions Division, IFPRI.

Birthal PS, Roy D, & Negi DS, (2015) Assessing the impact of crop diversification on farm poverty in India. *World Development* 72, 70–92. https://doi.org/101016/jworlddev201502015

Bommarco R, Kleijn D, & Potts SG, 2013 Ecological intensification: Harnessing ecosystem services for food security. *Trends in Ecology & Evolution* 28 (4), 230–238. https://doi.org/101016/jtree201210012

Bradshaw B, Dolan H, & Smit B, (2004) Farm-level adaptation to climatic variability and change: Crop diversification in the Canadian prairies. *Climatic Change* 67 (1), 119–141.

Chand R, (1996) Diversification through high value crops in Western Himalayan region: Evidence from Himachal Pradesh. *Indian Journal of Agricultural Economics* XXXXI (4), 652–663.

Chaplin Kramer R, O'Rourke ME, Blitzer EJ, & Kremen C, (2011) A meta-analysis of crop pest and natural enemy response to landscape complexity: Pest and natural enemy response to landscape complexity. *Ecology Letters* 14 (9), 922–932. https://doi.org/101111/j1461-0248201101642x

Dalsgaard JPT, & Oficial RT, (1997) A quantitative approach for assessing the productive performance and ecological contributions of smallholder farms. *Agricultural Systems* 55(4), 503–533. https://doi.org/101016/S0308-521X(97)00022-X

Demissie A, & Legesse B, (2013) Determinants of income diversification among rural households: the case of smallholder farmers in Fedis district, Eastern hararghe zone, Ethiopia. *Journal of Development and Agricultural Economics* 5 (3), 120–128. https://doi.org/105897/JDAE12104

Department of Economic Affair, (2012–13) Economic Division, Government of India, Economic Survey, New Delhi, Ministry of Finance, various issues. https://dea.gov.in/divisionbranch/economic-division

Dutta S, (2012) *A spatio temporal analysis of crop diversification in Hugli District, West Bengal*. Kolkata: Geo-Analyst.

FAO, (1996) *World food food summit*. Rome: FAO.

Field CB, Barros VR, Mastrandrea MD, Mach KJ, Abdrab MK, & Adger N, (2014) Summary for policymakers. In *Climate change impacts adaptation, and vulnerability* (pp. 1–32). Cambridge University Press. Retrieved from http://epicawide/37531/

Gardiner MM, Landis DA, Gratton C, DiFonzo CD, O'Neal M, Chacon JM, & Heimpel GE, (2009) Landscape diversity enhances biological control of an introduced crop pest in the north-central USA. *Ecological Applications* 19 (1), 143–154. https://doi.org/101890/07-12651

Gollin D, Morris M, & Byerlee D, (2005) Technology adoption in intensive post-Green Revolution systems. *American Journal of Agricultural Economics* 05, 1310–1316.

Gopalan C, Sastri Rama BV, & Balasubramanian SC, (1999) *Nutritive value of Indian foods*. Hyderabad: National Institute of Nutrition, ICMR.

Goyal S, & Singh JP, (2012) Demand versus supply of food grains in India: Implications to food security, Paper presented at International Farm Management Association's 13th Congress, Wageningen, Netherlands, July 7–12.

Hoddinott J, & Yohannes Y, (2002) *Dietary as a food security indicator*. Washington DC: Food Consumption and Nutrition Division, International Food Policy Research Institute.

Hopper GR, (1999) Changing food production and quality of diet in India, 1947–98. *Population and Development Review* 25, 443–477.

India Go, (2006) *National guidelines on infant and young child feeding*. New Delhi: Ministry of Women and Child Development.

Jha AE, (2000) Growth and instability in agriculture associated with new technology: District level evidences. *Agricultural Situation in India* XXXXIX (4), 517–524.

Kolkata Gazette (2005), Extraordinary, Government Of West Bengal. Finance Department. Audit Branch, June 14, chrome-https://extension://efaidnbmnnnibpcajpcglclefindmkaj/sabangcollege.ac.in/wp-content/uploads/GOs/dir_list/2005-JuneTreasury-Rules.pdf

Kumar A, Pramod K, & Alakh NS, (2012) Crop diversification in Eastern India: Status and determinants. *Indian Journal Of Agricultural Economics* LXVII (4), 600–616.

Lin B, (2011) Resilience in agriculture through crop diversification: Adaptive management for environmental change. *Bioscience* 61 (3), 183–193. https://doi.org/101525/bio20116134

Makate C, Wang R, Makate M, & Mango N, (2016) Crop diversification and livelihoods of smallholder farmers in Zimbabwe: Adaptive management for environmental change. *Springer Plus* 5 (1), 1135. https://doi.org/101186/s40064-016-2802-4

McCord PF, Cox M, Schmitt Harsh M, & Evans T, (2015) Crop diversification as a smallholder livelihood strategy within semi-arid agricultural systems near Mount Kenya. *Land Use Policy* 42, 738–750. https://doi.org/101016/jlandusepol201410012

McDaniel MD, Tiemann LK, & Grandy AS, (2014) Does agricultural crop diversity enhance soil microbial biomass and organic matter dynamics? A meta-analysis. *Ecological Applications* 24 (3), 560–570.

Meenakshi JV, & Vishwanathan B, (2003) Calorie deprivation in rural India. *Economic and Political Weekly* XXXVIII (4), 369–375.

Mhango WG, Snap SS, & Phiri GYK, (2013) Opportunities and constraints to legume diversification for sustainable maize production on smallholder farms in Malawi. *Renewable Agriculture and Food Systems* 28 (3), 234–244. https://doi.org/101017/S1742170512000178

Njeru EM, (2016) Crop diversification: A potential strategy to mitigate food insecurity by smallholders in sub Sahara African. *Journal of Agriculture, Food System, and Community Development* 3 (4), 63–69. https://doi.org/105304/jafscd2013034006

Organization WH, (1995) *Physical status: The use and interpretation of anthropometry*. Geneva: World Health Organization.

Pandey VK, & Sharma KC, (1996) Crop diversification and self-sufficiency in food grains. *Indian Journal of Agricultural Economics* LI (4), 644–651.

Petit M, & Barghouti S, (1992) Diversification: Challenges and opportunities. In: *Trends in agricultural diversification: Regional perspectives*, LG Shawki Barghouti, Washington D.C.: World Bank Technical Paper Number.

Radhakrishna R, (2005) Food and nutrition security of poor: Emerging perspectives and policy issues. *Economic and Political Weekly* XXXX (18), 1817–1823.

Raju A, (2012) Patterns of crop concentration and diversification in Vizianagaram District of Andhra Pradesh. *Transactions* II, 34.

Rao SE, (2001) Intake of micronutrient-rich foods in rural Indian mothers is associated with the size of their babies at birth: Pune maternal nutrition study. *Journal of Nutrition* 131, 1217–1224.

Rukhsana, (2020) Micro and macro-level analysis of crop diversification: Evidence from an agrarian state West Bengal. *Indian Journal of Economics and Development* 16 (4), 1. ISSN: 2277-5412.

Rukhsana, (2021) An assessment of disparities in diversity of crop at block level in West Bengal. *Acta Scientific Agriculture* 5 (12). ISSN: 2581-365X. https://actascientific.com/ASAG/pdf/ASAG-05-1078.pdf

Saleth, & Maria R, (1995) *Prospects, agricultural diversification in Tamil Nadu*. Delhi: Institute of Economic Growth.

Singh J, Yadav S, & Singh N, (2013) Crop diversification in Punjab agriculture: A temporal analysis. *Journal of Environmental Science, Computer Science and Engineering and Technology* II (2), 200–205.

Sinha DK, & Ahmad Nasim SK, (2016) Shrinking net sown area: An analysis of changing land use pattern in Bihar. *Journal of Agrisearch* III (4), 238–243.

Smit B, & Skinner M, (2002) Adaptation options in agriculture to climate change: A typology. *Mitigation and Adaptation Strategies for Global Change* 7, 85–114.

Smit B, & Wandel J, (2006) Adaptation, adaptive capacity and vulnerability. *Global Environmental Change* 16 (3), 282–292. https://doi.org/101016/jgloenvcha200603008

Smith, Bruce W, Dalen, Jeanne, Wiggins, Kathryn, Tooley, Erin, Christopher, Paulette, & Bernard, Jennifer (2008). The brief resilience scale: Assessing the ability to bounce back. *International Journal of Behavioral Medicine*, 15(3), 194–200. https://doi.org/10.1080/10705500802222972

Soora NK, Aggarwal PK, Saxena R, Rani S, Jain S, & Chauhan N, (2013) An assessment of regional vulnerability of rice to climate change in India. *Climatic Change* 118 (3–4), 683–699. 3

Swati G, Choudhary H, & Bisht A, (2017) Factors influencing crop diversification as Tool to two fold farmers' earnings in Uttarakhand. *Indian Journal of Economics and Development* XIII (2), 228–233.

Tao F, Hayashi Y, Zhang Z, Sakamoto T, & Yokozawa M, (2008) Global warming, rice production, and water use in China: Developing a probabilistic assessment. *Agricultural and Forest Meteorology* 148(1), 94–110. https://doi.org/101016/jagrformet200709012

Tiemann LK, Grandy AS, Atkinson EE, Marin Spiotta E, & McDaniel MD, (2015) Crop rotational diversity enhances belowground communities and functions in an agro ecosystem. *Ecology Letters* 18 (8), 761–771. https://doi.org/101111/ele12453

Vyas SV, (1996) Diversification of agriculture: Concept, rationale and approaches. *Indian Journal Of Agricultural Economics* XXXXXI (4), 636–646.

16 Does Agricultural Cooperative Membership Help to Reduce the Overuse of Agrochemicals Pesticides (CPs)?

Evidence from a Rural Area in West Region Cameroon

Kaldjob Mbeh Christian Bernard, Douya Emmanuel, Tata Ngome Precillia, Nso Ngang Andre, and Bamou Tankoua Lydie

16.1 Introduction

Pesticides spraying in agriculture is an important condition that guarantees small-scale African in general and specifically Cameroonian producers' effective protection of their crops. It ensures productivity growth that aligns with the food supply objectives of the population. Over the last 50 years, these chemicals pesticides (insecticides, fungicides, herbicides, and nematicides) have made it possible to increase the productivity of agricultural factors such as land and labor (Djoumessi 2020).

As with many African countries, Cameroonian producers are very dependent on pesticides, which is reflected in the constant increase in the quantities consumed (Wilson and Tisdell 2001). According to the National Institute of Statistics (NIS), in 2018, more than 1373 tons of pesticides, were used by about 40% of market garden and cash crops producers, while 74% used chemical fertilizers on their farms. Although in general, small-scale farmers often lack financial resources to purchase these pesticides, Okolle et al. (2016), Tarla et al. (2015) and Tarla et al. (2013) consider those specific producers among the greater consumers of chemicals pesticides in Cameroon.

Overall, with the economic crisis of the 1980s, and the severity of its effects in most rural Cameroonian agrarian zones, the government decided to reduce the level of imports and limit the supply of subsidized pesticides to producers. During this critical period, the distribution of fungicides, for example, through official outlets dropped from 30 million bags in the mid-1980s to less than 3 million in 1993 (Varlet and Berry 1997). This adjustment in macroeconomic policy has unfortunately led to unproductive pest and disease management for Cameroonian agriculture (Coulibaly et al. 2002).

However, it is the desire to modernize the production system through second-generation agriculture which requires a strong mechanization and the

DOI: 10.4324/9781003275916-19

massive adoption of agricultural inputs that has the led the Cameroonian government since the end of 2010 to produce the growth and employment strategy document (GESD) which since this year has been reviewed and corrected to become the document for the national development strategy NSD30. Technically, the main objective was to improve smallholder yields. This implied an increase in the level of chemical pesticide imports as illustrated in Figure 16.1.

It is in that context, characterized by several campaigns to supply pesticides to producers through agricultural cooperatives, that a significant increase in the quantities of pesticides was subsequently observed. In addition, the development of common pest management strategies within those agricultural cooperatives evolved in response to an increasing demand for pesticides initially supplied by the government. However, Tarla et al. (2013) observe that these strategies have proven to be unsustainable, marked by numerous pesticide abuses with devastating effects on food quality and ecological damage.

As the main rural institution for supervising agricultural activities, rural cooperatives play a crucial role in organizing smallholder farmers to increase their knowledge and technical capacities (Ma and Zhu 2020; Liu et al. 2019; Ma et al. 2018; Ito, et al. 2012). In the same line, agricultural cooperatives in Cameroon have multiple objectives including training of members, marketing of products, provision of inputs and various services. Training and marketing of products are equally crucial for the farmers who are not members of those rural associations. Furthermore, most of those rural cooperatives in Cameroon contract with supermarkets to sell their members' products in a "farmer-cooperative-supermarket" supply chain model. Under this model, some

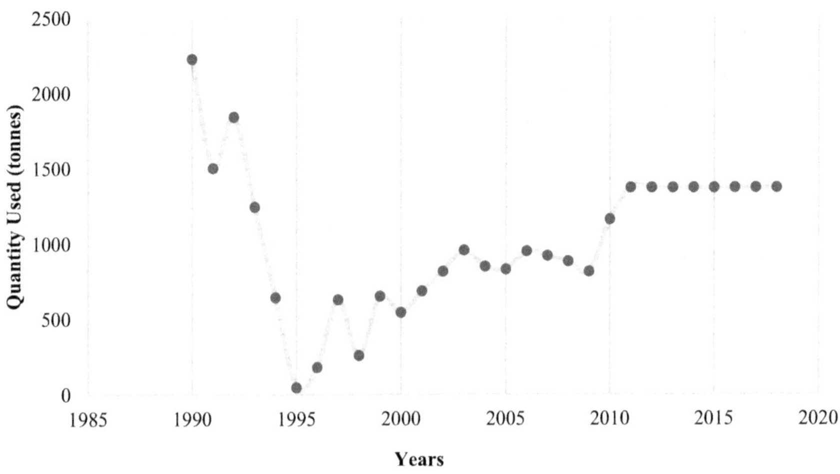

Figure 16.1 Trends in pesticide use in Cameroon from 1990 to 2020

Source: Authors' calculations from Statistical Database, Food and Agriculture Organization of the United Nations, Rome.

agricultural cooperatives established strict quality standards of their products, including residual levels of pesticides, to meet their customers 'quality demand. Therefore they play an important role in the management of agricultural inputs such as pesticides and other agricultural technologies which, in turn, increases crop productivity and farmers' income.

In the literature, several studies have then exposed the impact of cooperative membership on the adoption of agricultural technologies (Manda et al. 2020; Zhang et al. 2020; Ma and Abdulai 2019; Kolade and Harpham 2014; Abebaw and Haile 2013). For example, Manda et al. (2020) examined empirical evidence of the link between cooperative membership and the adoption of improved maize, inorganic fertilizers, and crop rotation agricultural technology in Zambia. Their results showed that a membership in an agricultural cooperative increased the probability of technology adoption by 11 to 24 percentage points. By assessing the impact of cooperative membership on the intensity of adoption of 13 technologies in 413 farm households in China a study by Zhang et al. (2020) indicates that membership in a cooperative does not have a significant impact on the number of production technologies adopted, but it does significantly increase the number of post-harvest technologies adopted. Their results are in line with the new theories of institutional economics.

Ma and Abdulai (2019) assess the impact of agricultural cooperative membership on apple farmers' decisions to adopt integrated pest management (IPM) technology and estimate the impact of IPM adoption on farms' economic performance in China. The results show that cooperative membership exerts a positive and significant impact on the adoption of IPM technology. IPM technology adoption in turn has a positive and statistical impact on apple yields and agricultural income. They conclude that agricultural cooperatives would be an opportunity for dissemination and vulgarization of the adoption of IPM technologies. The research of Abebaw and Haile (2013) uses a cross-sectional data and a propensity score matching technique to investigate the impact of cooperatives on adoption of agricultural technologies in Ethiopia. Their estimation results show that cooperative membership has a strong positive impact on fertilizer adoption. Further analysis help them to conclude that cooperative membership has a heterogeneous impact on fertilizer adoption among its members.

This previous literature exposes empirical evidence of the significant impact of rural cooperative membership in adoption of agricultural technologies. In the same line, this chapter fills the gaps and contributes to the literature by investigating the impacts of cooperative memberships on reducing the overuse of CPs for large-scale specialized agrarian family in rural area in Cameroon. The chapter also aims to highlight the scope for incorporating Cameroonian cooperative societies into income-generating policies and priority settings of policy makers. Since income generation is an integral part of the poverty reduction strategy, adequate understanding of the effect of cooperative membership on income generation also remains crucial.

16.2 Materials and Method

16.2.1 Study Area

This study was carried out in the Noun Division, West region of Cameroon. It is located roughly 5° N by 10° 30 E. The Noun river separates two main plateaus, one of which is 1200 m high, the other 1500 m high. These two plateaus have been beveled during the past geological times and covered by volcanic materials which have altered relief and disturbed the hydrography pattern. The local population cultivate their land with great care in the sprawling part of the plateaus but without any care on the escarpments. They are grouped around the seven main localities (Magba, Djinoun, Koutaba, Foumban, Kouoptamo, Foumbot, and Massagam) that made up our survey area.

The primary data used for our empirical analysis were obtained from a survey conducted during the period of May to July 2020 in seven localities of Noun Division which is known as an important basket of "market garden products" due to its fertile land. Table 16.1 presents the distribution of sub-samples by localities. The Noun Division was selected for two reasons: first, because of the high production of market garden product which is approximately 30–40 % of the whole vegetable production of the country (Minader 2017). Second, because of the significant consumption of chemical pesticides in this Division (Sonchieu et al. 2019). But also, as a result of the presence and organization of rural cooperatives, the socio-economic importance of agriculture for rural households, the intensity of agricultural activities, and finally the soil and climate characteristics are favorable to agricultural production (Figure 16.2).

16.2.2 Survey Design and Data Collection

The multistage sampling technique was applied for the present study. In the first step, the stratified random sampling technique was used, and Noun Division was geographically divided into three main zones the North, the West, and the Centre. In the second step, one district was selected from each

Table 16.1 Distribution of producers by survey location

Localities	*Number of producers (N)*	*Percentages (%)*	*Total*
Foumbot	130	23,21	130
Koutaba	80	14,28	80
Djinoun	80	14,28	80
Kouoptamo	70	12,5	70
Massagam	100	17,85	100
Magba	50	8,92	50
Foumban	50	8,92	50
Total	**560**	**100**	**560**

Source: Authors' construction 2021

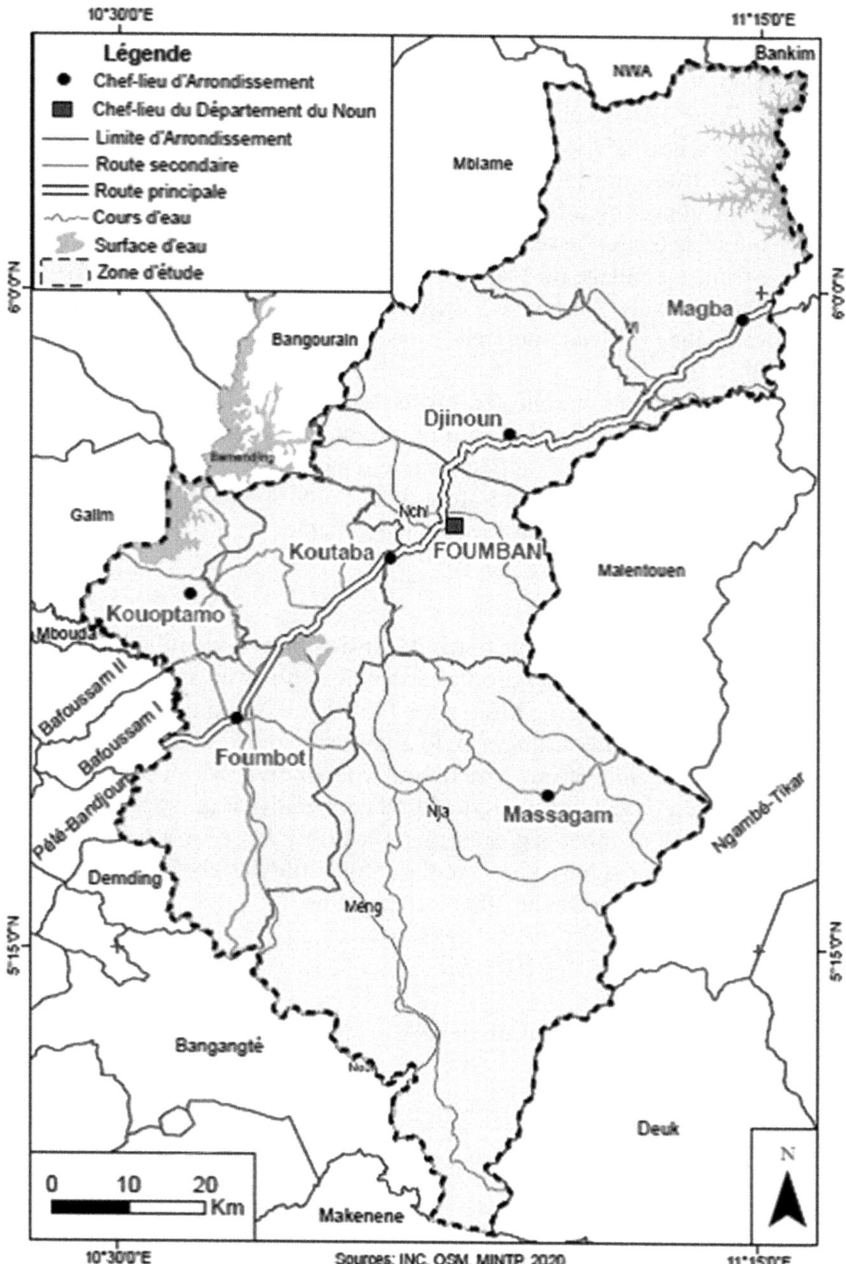

Figure 16.2 Study area

Source: Prepared by authors from collected survey data.

geographical stratum, followed by the selection of villages per district and finally market garden producers per village (Table 16.1). In the third stage, a random sample technique was used, and a sample of 560 market garden was obtained. Prior to the field survey, verbal approval was obtained from the chief and his senior advisors in the respective villages to conduct face-to-face interviews with the producers. Face-to-face interviews are a more sophisticated way of collecting adequate information, although they are expensive. Finally, out of 560 interviewed producers, 130 were from the Foumbot village, 100 were from the village of Massagam, and 80 were from the Koutaba and Djinoun, respectively. All farmers included in the sample list were obtained from seven villages. The interview process was conducted from May to July 2020.

The main information collected on the surveys was on household sociodemographic characteristics, agricultural production practices, application of agricultural pesticides and agricultural cooperatives' services. After data screening, 560 valid observations were finally used for the study. Table 16.2 presents descriptive statistics of the selected variables.

16.2.3 Econometric Model

The econometric analysis of our research consists first of identifying the main determinants of Cameroonian producers' membership in an agricultural cooperative. Based on the work of Mojo et al. (2017) and Ma and Abdulai (2019), we use the random utility framework to analyze the decision of Cameroonian producers' membership in an agricultural cooperative. We postulate that a rational producer i will join an agricultural cooperative if and only if the utility U_{im} of being a member is greater than the utility U_{in} of not being a member. Since it is not possible to observe the actual utility level of a producer, we can define surplus utility as the difference given by:

$$\Delta_{i*} = U_{im} - U_{in} \tag{16.1}$$

with Δ_{i*} as a vector function of observable explanatory variables Y_i in the following latent variable model:

$$P_i^* = \beta Y_i + \delta_i \tag{16.2}$$

where

$$P_i^* = \begin{cases} 1 & if\ P_i^* > 0 \\ 0 & if\ P_i^* < 0 \end{cases}$$

P_i^* is the probability for a Cameroonian producer to register in an agricultural cooperative or not, which is defined by: $P_i = 1$ if a producer has obtained cooperative membership $P_i = 0$ otherwise. Y_i is a vector of socioeconomic and

Table 16.2 Distribution of descriptive statistic of variables

Variables	*Description*	*Code*	*Value*	*S.D*
Dependent variables				
Pesticides reduction	Whether the producer has lowered the usage of pesticides per ha more than non-members of rural cooperatives?	1 = Yes 2 = No	0.716	.4513057
Net return per XAF	Net return per XAF cost of CPs	Amount in XAF	154623.7	120461.2
Treatment variables				
Membership	Whether the producer is an agricultural cooperative member	1 = Yes 2 = No	0.617	0.486
Production services	The producer adopts the technology service or machinery service provided by cooperatives in the last year?	1 = Yes 2 = No	0.216	0.411
Marketing services	Has the producer used the agri-products sales services provided by the cooperatives in the last year?	1 = Yes 2 = No	0.337	.473
Independents variables				
Age	Age of producer (head of farm)	Year	42.587	12.009
Sex	Gender of producer	1 = Male 2 = Female	1.381	0.486
Education	Level of education in years of the producer (farm head's)	0 = illiterate 1 = Primary 2 = Secondary 3 = Tertiary	1.442	0.800
Year	Producer experience in years of large-scale farm management	Year	15.180	9.877
Area	Area of land operated by farms	hectare	2.683	1.877
Subsidy	Whether the farm has received a rental subsidy from the government?	1 = Yes 2 = No	1.256	1.886
Rent	The average rent of farmland	XAF/hectare	–	–

(*Continued*)

Table 16.2 (Continued)

Variables	*Description*	*Code*	*Value*	*S.D*
Size	The proportion of family labors in the total farm labor force	Number of people in the house	7.235	3.626
Soil	Whether the farm tested soil for formulated fertilization in 2020?	0 = no; 1 = yes	–	–

Source: Authors' construction 2021

agricultural determinants that influence producers' decision to join an agricultural cooperative. β is a vector of parameter to be determined and δ_i is a disturbance term assumed to be normally distributed. The probability of a Cameroonian producer registering an agricultural cooperative is given as follows:

$$C_r\left(P_i=1\right)=C_r\left(\delta_i \geq \beta Y_i\right)=1-\Pi\left(-\beta Y_i\right) \tag{16.3}$$

where Π is the cumulative distribution function for δ_i. Thus, to create a statistical relation between an agricultural cooperative membership and chemical pesticides (CPs) use, we hypothesize that chemical Pesticide use is a linear function of a vector of covariates X_i and the dichotomous agricultural cooperative membership variable P_i, expressed as follows:

$$Z_i=\alpha X_i+\varphi P_i+\upsilon_i \tag{16.4}$$

where Z_i is a vector of outcome variables of producer i such as chemical pesticides reduction, and application efficacy of CPs, defined as the net return per FCFA[1] cost of CPs. α and φ are parameters to be estimated, υ_i is the disturbance term. In the previous equation (16.4) above, the agricultural cooperatives membership variable P_i is theoretically endogenous since the producer's choices to be cooperative members or not are affected by observed socioeconomic parameters such as sex, age, and education, and unobserved parameters as the inherent skills and motivations of producers. Furthermore, the perturbation term δ_i in equation (16.2) and the perturbation term υ_i in equation (16.4) can be correlated, if the correlation coefficient is different from 0 corr(δ_i, υ_i)υ ≠ 0, if not, there is a selection bias. Faced with this methodological constraint, standard regression techniques, such as the Probit model or ordinary least squares (OLS) do not account for this selection bias and tend to produce inconsistent estimation results.

1 The FCFA is the unit of currency in Cameroon. In July of this year 2021, one FCFA is about US $0.0018.

The propensity score matching (PSM) is a method that addresses selection bias. However, according to Ma and Abdulai (2016), it does not account for the unobservable selection bias issue. In contrast, the two-stage residual inclusion approach (2SRI) resolves selection biases resulting from both observed and unobserved parameters (Ying et al. 2019). Furthermore, this 2SRI method allows the effect of unobservable parameters to be determined in order to obtain asymptotically correct standard errors for testing t-statistics. In this chapter we use the 2SRI approach to conduct the empirical analysis where from equation (16.2), we introduce the disturbance term δ_i in equation (16.4) as another variable to have equation (16.5) such as:

$$Z_i = \alpha X_i + \varphi P_i + \phi R_i + \upsilon_i \tag{16.5}$$

where R_i represents the residual term estimated from equation (16.3). We perform a robustness evaluation and estimate the average treatment effect on treated producers (ATT), the Probit Endogenous Switching (ESP) and Endogenous Switching Regression (ESR) models are also estimated. Just like Ma and Zhu (2020), we use the ESP model to evaluate the impact of agricultural cooperative membership on pesticides reduction and the ESR model to assess the impact of cooperative membership on CPs use efficacy. To identify those three models (2SRI, ESP, and ESR), and improve parametric estimations, Y_i in equation (16.2) should contain X_i and at least one instrumental variable that affects the decision of participate to an agricultural cooperative, but does not directly affect the treatment of CPs (Takahashi et al. 2019; Terza 2016).

16.3 Empirical Results and Discussion

A preliminary analysis of the main empirical results requires us to perform a variance inflation factor (VIF) test to check out for problems of multicollinearity of the explanatory variables. Thus, the VIF values of the socioeconomic and agricultural variables in equations (16.2) and (16.5) were less than 1.03 and 3.65 respectively. These results indicate that there are no problems of multi-collinearity in our estimated models. The significance of the residual terms in column 4 of Table 16.3 suggests that some unobservable factors influenced the decision of Cameroonian producers to join agricultural cooperatives. Thus, we can conclude that the 2SRI model had strong explanatory power.

16.3.1 Impact Evaluation of Cooperative Membership

Table 16.3 reports the regression results of equations (16.2) and (16.5), which are estimated simultaneously with the 2SRI approach for the impacts of agricultural cooperatives on the usage of CPs.

Table 16.3 Impact of cooperative membership on CPs use: estimates by the 2SRI model

	Cooperative membership	*Pesticide reduction*	*Net return per CFA CFPs*
Membership	–	0.839***	0.1603***
		(0.0540)	(0.0244)
Age	0.418***	−0.103***	−0.00142
	(0.0154)	(0.00481)	(0.00139)
Sex	2.567***	0.296***	0.00943
	(0.175)	(0.0627)	(0.0213)
Education	−0.886***	−0.330***	−0.0181
	(0.117)	(0.0661)	(0.0138)
Year	−0.361***	0.00852	0.00271
	(0.0167)	(0.00637)	(0.00187)
Size	3.540***	2.184***	0.192***
	(0.110)	(0.0365)	(0.0106)
Labor	0.737***	0.107***	−0.0108**
	(0.0361)	(0.0158)	(0.00444)
Residual		55.68***	4.83e-06***
		(1.086)	(3.47e-07)
Constant term	55.68***	−26.94***	11.32***
	(1.086)	(1.212)	(0.0645)
Log-pseudo likelihood	**−4.915e-08**	**−3.206e-09**	**−6730.004208**
Observations	559	559	557

Source: Authors' construction 2021

*** represent the 1% significance level.

After monitoring the endogeneity, the results revealed that, with a significance level of 1%, joining a Cameroonian agricultural cooperative improved the net return per CFA cost of CPs by 160.3 CFA. Indeed, some authors found that there are other source of pesticides reduction. For example, Yuan, Tang, and Shi (2021) conclude that using internet services in rural area in China helps to reduce the overuse of chemical pesticides. In the same line, the research of Zhao, Pan, and Xia (2021) also observes that access to internet services helps reduce pesticide use among vegetable farmers in China. In the light of these studies, our chapter proposes new evidence that rural organization such as agricultural cooperatives can be a competent institutional that reduces both pesticides use in Cameroon. The results of estimations by 2SRI approach of equation (16.2) also show that factors such as farm management experience of the producer and family labor force has a positive impact on farmers' decision to join an agricultural cooperative, while the farm size and the proportion of family labor force had a negative effect. These results agree with existing literature on agricultural cooperatives. Indeed, Li and Ito (2021), Gava et al. (2021), Imami, Valentinov, and Skreli (2021), Deng et al. (2021), and Ndlovu and Masuku (2021) explored the effective role of agricultural cooperatives in enhancing food security and social welfare of farm households in rural areas.

The results of IV1 was positively and statistically significant, demonstrating that the possibility of a Cameroonian producer joining an agricultural cooperative was positively associated with the proportion of other producers who participate in cooperatives in the same country.

16.3.2 Average Treatment Effect on the Treated (ATT)

Table 16.4 presents the results of regression of the ESP and ESR models. In order to be as concise as possible, we will just discuss the ATT results.

The results presented in Table 16.4 show that agricultural cooperative members had a significantly higher probability of reducing the usage of CPs than the non-member farmers and higher application efficacy of CPs. The estimated ATT results in Table 16.4 were positively and statistically significant, which confirmed that agricultural cooperatives could help Cameroonian producers to reduce the overuse of CPs and improve the utilization efficacy of the agrochemicals. Specifically, after joining an agricultural çooperative, the possibility of a Cameroonian producer reducing the overuse of pesticides is 46.4%, and the net return per CFA of CPs increased by 45 CFA. This result led us to conclude that agricultural cooperatives in Cameroon had a reliable impact on reducing the consumption and increasing the application efficacy of CPs.

16.3.3 Mechanism Analysis

Our previous results demonstrated that agricultural cooperatives in Cameroon effectively help their members to reduce the overuse of CPs. To further learn how these producers achieved this, we focused our research exclusively on grain and market garden producers just because the grain, vegetables and other crops were different in production, marketing, and application of CPs. Indeed, there is no important literature that studied the grain and market garden producers at the same time. In total, 560 market garden and grain producers had used at least one service provided by agricultural cooperatives (including production and marketing services). According to whether the producer used the production or marketing services provided

Table 16.4 Results of the robustness control

	ESR model			*ESP model*	
	ATT	*t-value*		*ATT*	*t-value*
Net return per CFA CPs	0.465*** (0.0733)	6.34	**Pesticides reduction**	0.466*** (0.115)	4.06

Source: Authors' construction 2021

Note: Standard errors are in parentheses.

*** represent the 1% significance level.

by agricultural cooperatives, we assessed the effect of cooperative membership on the usage of CPs by the 2SRI approach. The results are presented in Tables 16.5 and 16.6.

The estimated coefficients in columns 3 and 4 of Table 16.5 indicated that after monitoring for other socioeconomic variables, using the production services provided by agricultural cooperatives increased the probability of reducing the consumptions of pesticides by 120%, and increased the net return per CFA cost of CPs by 36 CFA. Furthermore, in relation to the membership coefficients in Table 16.3, which do not distinguish between producers who use production services and those who do not, it is clear that the cooperative's production services have a strongly positive effect on reducing pesticide consumption.

The marketing services' coefficient in column 3 of Table 16.6 was significant at the 1% level, which indicated that using cooperative marketing services significantly reduced the spraying of pesticides.

At the same time, the marketing services had a significant impact on the consumption of chemical pesticides and the net return per CFA CPs. In conclusion, Cameroonian agricultural cooperatives made their producers members lower the usage of CPs by providing services. Production services had reduced the usage of CPs, while marketing services had significantly reduced pesticide usage.

Table 16.5 Impacts of using cooperative production services on the usage of CPs: 2SRI model estimations

	Production services	*Pesticide reduction*	*Net return per CFA CPs*
Production services	–	1.206***	0.366***
		(0.184)	(0.0221)
Age	−0.138***	−0.160***	0.00220
	(0.0317)	(0.0142)	(0.00150)
Sex	−1.693***	1.395***	0.0205
	(0.458)	(0.187)	(0.0215)
Education	1.582***	−0.746***	−0.0371***
	(0.238)	(0.138)	(0.0130)
Year	−0.219***	0.0341**	0.00180
	(0.0577)	(0.0166)	(0.00192)
Size	1.137***	4.130***	0.197***
	(0.158)	(0.104)	(0.0100)
Labor	−0.126	0.288***	−0.0115**
	(0.0805)	(0.0308)	(0.00463)
Residual	–	30.99***	4.66e-06***
		(0.582)	(3.28e-07)
Constant term	−17.01***	−5.841***	11.20***
	(1.271)	(0.723)	(0.0654)
Log-pseudo likelihood	**−1.195e-10**	**−.01038541**	**−6725.484373**
Observations	559	559	557

Source: Authors' construction 2021

Table 16.6 Impacts of using cooperative marketing services on the usage of CPs: 2SRI model approach

	Marketing services	*Pesticide reduction*	*Net return per CFA CFPs*
Marketing services	–	4.466***	0.412***
		(0.168)	(0.0338)
Age	−0.156***	−0.0771***	0.00443*
	(0.0190)	(0.00795)	(0.00239)
Sex	0.724***	0.235	0.0223
	(0.248)	(0.160)	(0.0294)
Education	0.135	−0.426***	−0.0215
	(0.205)	(0.0706)	(0.0183)
Year	−0.131***	0.0171	0.00166
	(0.0227)	(0.0107)	(0.00254)
Size	6.131***	1.510***	0.147***
	(0.111)	(0.0762)	(0.0155)
Labor	0.0694*	0.121***	−0.0122**
	(0.0414)	(0.0175)	(0.00618)
Residual	–	11.42***	0.0912***
		(0.337)	(0.0174)
Constant term	−17.23***	1.095***	11.18***
	(0.975)	(0.274)	(0.103)
Log-pseudo likelihood	**−.0019427**	**−9.929e-12**	**−6634.930033**
Observations	553	553	557

Source: Authors' construction 2021

16.4 Conclusion and Agricultural Policies Implications

The overuse of pesticides by Cameroonian producers leads to amplified environmental pollution and does not always guarantee the quality of certain essential agri-food products. Indeed, the excessive spraying of CPs has led to a progressive deterioration of food safety, which has led to Cameroonian consumers being exposed to diseases and disallowed national producers from playing an important role in international markets. Our study provided empirical evidence of the influence of agricultural cooperatives in improving pesticide management for a more effective environment and food safety, through production and marketing services to their members. Using the 2SRI method and ATT estimators based on the ESP and ESR models, we evaluated the impacts of membership in an agricultural cooperative on pesticides use. We investigated the mechanisms using producer data from a field survey in west region Cameroon. Our empirical results showed that membership in agricultural cooperatives in Cameroon increases the probability of reducing pesticide consumption. According to the ATT estimation results, the net return to cooperative members per FCFA of CPs increased by 44 CFA. The production and marketing services of agricultural cooperatives significantly reduced the overuse of chemicals pesticides.

From the different results of this research, some implications of agricultural policies that can contribute to guarantee the development of rural associations, but especially to define the framework for the implementation of a productive system more respectful of the consumer's health and the environment, emerge. Indeed, it is fundamental that more technical training should be provided to members of rural associations since they can contribute efficiently to the transformation of the agricultural production system, to guarantee the quality of agricultural products by a required level of pesticide residues and finally to disseminate modern agricultural technologies and technics. It is also important to support agricultural cooperatives with production technologies that would allow them to test the level of pesticide residues. In addition, it would be equally interesting to provide agricultural production and marketing services that would facilitate the role of cooperatives in the management of agricultural value chains, through end-to-end supervision of the quality of agricultural products. Finally, it would be important for policy makers to create incentives that would encourage large producers to join rural cooperatives, which could be the backbone of the cooperative. Finally, dedicated technical support to cooperatives to encourage their members to manage pesticides more effectively would lead to a coordinated development between producers and agricultural cooperatives in Cameroon.

Acknowledgements

The authors are especially grateful to all the courageous agricultural producers of the west region of Cameroon for providing all the information in order to facilitate the achievement of this research.

Conflict of Interest

The authors declare that there is no conflict of interest regarding this research.

Bibliography

Abebaw, D, and Mekbib G Haile. 2013. "The impact of cooperatives on agricultural technology adoption: Empirical evidence from Ethiopia". *Food Policy* 38 (février): 82–91. https://doi.org/10.1016/j.foodpol.2012.10.003

Coulibaly, O, D Mbila, DJ Sonwa, Akin Adesina, and J Bakala. 2002. "Responding to economic crisis in sub-Saharan Africa: New farmer-developed pest management strategies in cocoa-based plantations in Southern Cameroon". *Integrated Pest Management Reviews* 7 (3): 165–172.

Deng, Lei, Lei Chen, Jingjie Zhao, and Ruimei Wang. 2021. "Comparative analysis on environmental and economic performance of agricultural cooperatives and smallholder farmers: The case of grape production in Hebei, China". *Plos One* 16 (1): e0245981.

Djoumessi, Y. 2020. *Essais sur la Productivite Agricole en Afrique Sub-Saharienne*. Dschang: Universite de Dschang.

Gava, Oriana, Zahra Ardakani, Adela Delalić, Nour Azzi, and Fabio Bartolini. 2021. "Agricultural cooperatives contributing to the alleviation of rural poverty. The case of Konjic (Bosnia and Herzegovina)". *Journal of Rural Studies* 82: 328–339.

Imami, Drini, Vladislav Valentinov, and Engjell Skreli. 2021. "Food safety and value chain coordination in the context of a transition economy: The role of agricultural cooperatives". *International Journal of the Commons* 15 (1), 21–34.

Ito, Junichi, Zongshun Bao, and Qun Su. 2012. "Distributional effects of agricultural cooperatives in China: Exclusion of smallholders and potential gains on participation". *Food Policy* 37 (6): 700–709.

Kolade, Oluwaseun, and Trudy Harpham. 2014. "Impact of cooperative membership on farmers' uptake of technological innovations in Southwest Nigeria". *Development Studies Research* 1 (1): 340–353. https://doi.org/10.1080/21665095.2014.978981

Li, Xinyi, and Junichi Ito. 2021. "An empirical study of land rental development in rural Gansu, China: The role of agricultural cooperatives and transaction costs". *Land Use Policy* 109: 105621.

Liu, Yuying, Wanglin Ma, Alan Renwick, and Xinhong Fu. 2019. "The role of agricultural cooperatives in serving as a marketing channel: Evidence from low-income regions of Sichuan province in China". *Annals of Public and Cooperative Economics*, 92(02): 207–231. https://doi.org/10.1111/apce.12301

Ma, Wanglin, and Awudu Abdulai. 2016. "Does cooperative membership improve household welfare? Evidence from apple farmers in China". *Food Policy* 58 (janvier): 94–102. https://doi.org/10.1016/j.foodpol.2015.12.002

Ma, Wanglin, and Awudu Abdulai. 2019. "IPM adoption, cooperative membership and farm economic performance". *China Agricultural Economic Review*, 149–162.

Ma, Wanglin, Alan Renwick, Peng Yuan, and Nazmun Ratna. 2018. "Agricultural cooperative membership and technical efficiency of apple farmers in China: An analysis accounting for selectivity bias". *Food Policy* 81: 122–132.

Ma, Wanglin, and Zhongkun Zhu. 2020. "A note: Reducing cropland abandonment in China—Do agricultural cooperatives play a role?" *Journal of Agricultural Economics* 71 (3): 929–935.

Manda, Julius, Makaiko G Khonje, Arega D Alene, Adane H Tufa, Tahirou Abdoulaye, Munyaradzi Mutenje, Peter Setimela, and Victor Manyong. 2020. "Does cooperative membership increase and accelerate agricultural technology adoption? Empirical evidence from Zambia". *Technological Forecasting and Social Change* 158 (septembre): 120160. https://doi.org/10.1016/j.techfore.2020.120160

Minader 2017. "Agriculture". In *Annuaire statistique du Cameroun*, 24. Cameroun: Institut National de la Statistique.

Mojo, Dagne, Christian Fischer, and Terefe Degefa. 2017. "The determinants and economic impacts of membership in coffee farmer cooperatives: Recent evidence from rural Ethiopia". *Journal of Rural Studies* 50: 84–94.

Ndlovu, Confidence, and Mfundo M Masuku. 2021. "The efficacy of agricultural cooperatives towards enhancing food security in rural areas: Mbombela local municipality, Mpumalanga Province". *Technium Social Sciences Journal* 21: 661.

Okolle, Nambangia Justin, Victor Afari-Sefa, Jean-Claude Bidogeza, Precillia Ijang Tata, and Francis Ajebesone Ngome. 2016. "An evaluation of smallholder farmers' knowledge, perceptions, choices and gender perspectives in vegetable pests and diseases control practices in the humid tropics of Cameroon". *International Journal of Pest Management* 62 (3): 165–174.

Sonchieu, Jean, Bitsoga Marie Gracile, and Ngassoum Martin Benoit. 2019. "Characterization of personal clothing worn by pesticide sprayers in Foumbot agricultural area (Cameroon)". *Agricultural Sciences* 10 (08): 1056–1072. https://doi.org/10.4236/as.2019.108080

Takahashi, Kazushi, Rie Muraoka, and Keijirō Ōtsuka. 2019. "Technology adoption, impact, and extension in developing countries' agriculture: A review of the recent literature". *Agricultural Economics, International Association of Agricultural Economists* 51(1): 31–45.

Tarla, DN, IN Manu, ZT Tamedjouong, A Kamga, and DA Fontem. 2015. "Plight of pesticide applicators in Cameroon: Case of tomato (Lycopersicone sculentum Mill). Farmers in Foumbot". *Journal of Agriculture and Environmental Sciences* 4 (2): 87–98.

Tarla, Divine, V Félix Meutchieye Dominic Fontem Assako, and J Kome. 2013. "Exposure of market gardeners during pesticide application in the western highlands of Cameroon". *Scholarly Journal of Agricultural Science* 3 (janvier): 172–177.

Terza, Joseph V. 2016. "Simpler standard errors for two-stage optimization estimators". *The Stata Journal* 16 (2): 368–385.

Varlet, Frédéric, and Dominique Berry. 1997. *Réhabilitation de la protection phytosanitaire des cacaoyers et caféiers du Cameroun. Tome 1: Rapport principal.* Tome 2: Annexes".

Wilson, Clevo, and Clem Tisdell. 2001. "Why farmers continue to use pesticides despite environmental, health and sustainability costs". *Ecological Economics* 39: 449–462.

Ying, Andrew, Ronghui Xu, and James Murphy. 2019. "Two-stage residual inclusion for survival data and competing risks—An instrumental variable approach with application to SEER-Medicare linked data". *Statistics in Medicine* 38 (10): 1775–1801.

Yuan, Fang, Kai Tang, and Qinghua Shi. 2021. "Does internet use reduce chemical fertilizer use? Evidence from rural households in China". *Environmental Science and Pollution Research* 28 (5): 6005–6017. https://doi.org/10.1007/s11356-020-10944-4

Zhang, Shemei, Zhanli Sun, Wanglin Ma, and Vladislav Valentinov. 2020. "The effect of cooperative membership on agricultural technology adoption in Sichuan, China". *China Economic Review* 62 (août): 101334. https://doi.org/10.1016/j.chieco.2019.101334

Zhao, Qiuqian, Yuhe Pan, and Xianli Xia. 2021. "Internet can do help in the reduction of pesticide use by farmers: Evidence from rural China". *Environmental Science and Pollution Research* 28 (2): 2063–2073. https://doi.org/10.1007/s11356-020-10576-8

17 Impact of Urban Growth on Land Surface Temperature Pattern of Urban Landscape Using Space-Borne Images

Gurucharan Karmakar, Apurba Sarkar, Pradip Chouhan, and Margubur Rahaman

17.1 Introduction

Land surface temperature is an important element of climate dynamics and serves as an indicator of environmental health. It increases with land-use patterns, degradation of green coverage, and environmental pollution (Xiao et al., 2008). Vegetation plays a crucial role in regulating terrestrial thermal behavior by influencing the effective radiating temperature of the Earth's surface (Mao et al., 2020; Wang et al., 2015). Additionally, the increase in atmospheric greenhouse gases contributes to higher land surface temperatures (Lu, 2022; Li et al., 2013). However, estimating and validating this parameter is challenging due to the extreme heterogeneity of natural land surfaces (Guha & Govil, 2022). Several factors fundamentally affect the derivation of land surface temperature, including temperature variations with viewing angles, sub-pixel temperature and cover inhomogeneities, surface spectral emissivity, atmospheric temperature and humidity variations, clouds, and large aerosol particles such as dust (Pramanik & Punia, 2019). Urban built-up areas retain higher land surface temperatures compared to surrounding areas, forming urban heat islands (UHI) that impact local climate change and disrupt monsoon patterns due to land cover and land use alterations (Kafy et al., 2021; Wang et al., 2015). The conversion of wetlands, ponds, and vegetation-covered areas into built-up spaces or agricultural land within urban areas contributes to global warming and affects climate change. Urban green spaces, including parks, gardens, and natural environments, play a crucial role in land-use planning. Research suggests that the decline in urban green spaces contributes to increased land surface temperatures in urban settings (Estoque & Murayama, 2017). The World Health Organization recognizes land surface temperature as an indicator of climate change dynamics and urban health issues, emphasizing the need for sustainable urban environments through urban planning that includes the conservation of urban green spaces, natural water bodies, and other environmental aspects (Santra, 2019). Urban green spaces provide recreational opportunities, ecological benefits, aesthetic value, and positive health impacts. They absorb rainwater pollutants, CO_2 gases, provide fresh air, oxygen, water, and

DOI: 10.4324/9781003275916-20

contribute to soil health (Sultana and Satyanarayana, 2020). Urban green spaces help maintain a balance between concrete urban areas and green vegetation cover, mitigating urban heat island effects and reducing urban stress. They improve mental and physical health, create attractive spaces for social interactions, and serve as playgrounds and parks for children (Nurwanda & Honjo, 2018; Soydan, 2020). Urban green infrastructure is vital for preserving the natural environment, promoting ecology, and ensuring the sustainability of metropolitan cities. Most studies examined the changing patterns of land use patterns, urban sprawl, flood assessment, and land surface temperature in Siliguri city (Bose & Chowdhury, 2020; Hoque & Lepcha, 2020; Roy et al., 2021). There were limited studies that focused on the association between land-use patterns and land-surface temperature in Siliguri city (Hoque & Lepcha, 2020). Siliguri city in West Bengal, India, is located in the foothills of the Himalayas and is renowned for its pleasant weather and tourist attractions. However, rapid population growth and economic expansion have led to haphazard urban development without proper planning and sustainability considerations. Previous studies have observed significant urban expansion in Siliguri city (Bose & Chowdhury, 2020), but research on the nexus between changing land use patterns and land surface temperature remains limited. Therefore, this study aims to examine the relationship between changing land use patterns and land surface temperature in Siliguri city using multi-spatial analytical techniques. The study will contribute to understanding the role of changing land use patterns on land surface temperature, along with the associated factors responsible for these changes.

17.2 Study Area

Siliguri, located in the Darjeeling and Jalpaiguri districts of West Bengal, India, is known as the "Gateway of Northeast India" and is famous for its tea, timber, and tourism. Situated on the banks of the Mahananda River at the foothills of the Himalayas, Siliguri is the third largest urban agglomeration in West Bengal, after Kolkata and Asansol. It is spread over an area of 260 km^2 within the Siliguri Corridor. The Siliguri Municipal Corporation administers an area of 41.90 km^2, comprising both Darjeeling and Jalpaiguri districts. Siliguri has a humid subtropical climate, with warm summers, cool winters, and a significant monsoon season. The average annual temperature is 23.7°C, and the city receives an average rainfall of 3340 mm per year. As of the 2011 census, Siliguri has a population of 701,489 in the urban agglomeration, with 513,264 residing within the Municipal Corporation area. The city has a literacy rate of 77.64%, and Bengali is the most commonly spoken language, followed by Hindi, Nepali, Bhojpuri, and Urdu. The majority religion is Hinduism, with Islam as the largest minority religion, along with smaller populations following Christianity and Buddhism. Figure 17.1 shows the location of study area.

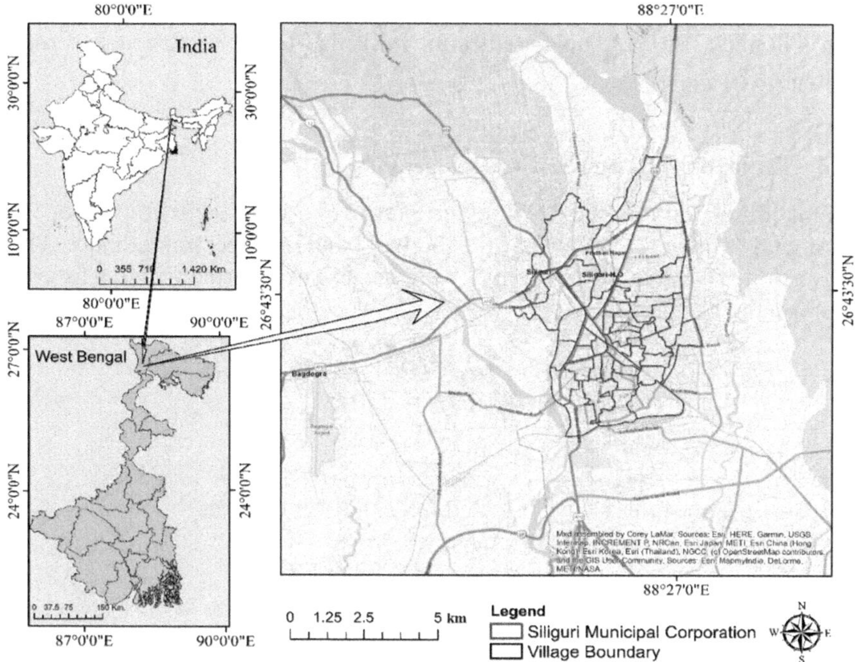

Figure 17.1 Location of the Siliguri Municipal Corporation, 2021

Source: Own

17.3 Methodology

The land use map has been prepared using Landsat imagery to observe the land use dynamic in the area. All nonthermal bands of multispectral Landsat 5, 7, and 8 have prepared false-color composite images (Table 17.1). The composite image has been classified using a maximum likelihood classification algorithm. A training sample of the classification has been collected from Google Earth.

Table 17.1 Data sources used to derive land-use map and drivers variable

Type of data used	*Year*	*path: row*	*Scale/resolution*	*Source*
Landsat TM	1990	139/42,139/41	28.5 m	earthexplorer.usgs.gov
Landsat TM	2000	139/42,139/42	28.5 m	earthexplorer.usgs.gov
Landsat ETM	2010	139/42,139/42	28.5 m	earthexplorer.usgs.gov
Landsat 8 OLI	2020	139/42,139/42	30 m	earthexplorer.usgs.gov
Village boundary map	2011			West Bengal administrative atlas
Siliguri Municipal corporation	2011			Municipality Office

Fifty sample points have been the drive for each class for land use (Sarkar and Chouhan, 2019). The classification task has been performed in Erdas Imagine 13.

17.4 Estimation of Land Surface Temperature

Estimation of Land surface temperature involves a three-steps conversion of the digital number (DN) value of the thermal band into spectral radiance. The conversion DN value of the thermal band has been converted with the following formula;

$$L\lambda = (LMAX\lambda - LMIN\lambda / QCALMAX - QCALMIN) * (QCAL - QCALMIN) + LMIN\lambda$$

where:

Lλ = the atmospherically corrected cell value as radiance,
QCAL = the digital value of imagery,
LMINλ = spectral radiance scales to QCALMIN,
LMAXλ = spectral radiance scales to QCALMAX,
QCALMIN = minimum quantized calibrated pixel value (typically 1),
QCALMAX = maximum quantized calibrated pixel value (typically 255)

For conversion of thermal band the band 10 and 11 digital values of Landsat OLI, have been done with the following formula:

$$L\lambda = MLQcal + AL$$

where:

Lλ = the atmospherically corrected cell value as radiance,
ML = Radiance Multi-band X (X = Band no.),
AL = radiance adds band X (X = Band no.),
QCal = quantized and calibrated standard product pixel values

Step 2 involves the conversion of spectral reliance to at-satellite brightness temperature, namely known as black body temperature.

$$T = -\frac{k2}{\ln(k1+1)} / L\lambda$$

where,

T = at-satellite brightness temperature in Kelvin (K),
Lλ = at-satellite radiance,
K1 and K2 = thermal constant

The values of thermal constant for Landsat TM and Landsat OLI-8 are then converted from Kelvin (K) to °C using the following equation,

$$TC = T - 273.15$$

where,
Tc = degrees Celsius,
T = degrees Kelvin

Step 3 involves the Calculation of Land Surface Emissivity (e)

Mapping emissivity from satellite data is important for surface characterization.

Land surface emissivity has been calculated using the following equation:

$$e = 0.004PV + 0.986$$

where,
PV is the proportion of vegetation

$$PV = \left(NDVI - NDVImin / NDVImax - NDVIMIN\right)2$$

Calculation of Land Surface Temperature (LST)

To acquire LST, the following equation is used:

$$TS = BT/1 + W * \left(BT/P\right) * \ln\left(e\right)$$

where,
BT = at-satellite temperature
W = wavelength of emitted radiance
P = h*c/s (1.438 * 10^{-2} m K)
h = Planck's constant (6.626 * 10^{-34} Js)'
s = Boltzmann constant (1.38 * 10^{-23} J/K)
c = velocity of light (2.998 * 10 8 m/s)

17.5 Calculation of Indexes

17.5.1 Calculation of Normalized Vegetation Index (NDVI)

Normalized Vegetation Index (NDVI) has been derived using flowing formula calculated from the

$$NDVI = \left(NIR - RED\right)/\left(NIR + RER\right)$$

where,
NIR = reflectance value of near infrared band,
RED = reflectance value of red band

17.5.2 Normalized Difference Built-up Index (NDBI)

The Normalized Difference Built-up Index (NDBI) was calculated from the Sentinel 2A images using the following equation:

$$NDBI = (SWIR - NIR) / (SWIR + NIR)$$

where,

SWIR = Reflectance value of shortwave–infrared band,
NIR = Reflectance value of the near-infrared band

17.5.3 Urban Built-Up Index

$$UBI = NDBI - NDVI$$

NDBI = Normalized Difference Built-up Index,
NDVI = Normalized Difference vegetation Index

17.5.4 Enhanced Vegetation Index (EVI)

$$EVI = G * ((NIR - RED)/(NIR + C1 * RED - C2 * B + L))$$

NIR = reflectance value of near infrared band,
RED = reflectance value of red band

In Landsat 8, EVI = 2.5 × ((Band 5 – Band 4) / (Band 5 + 6 × Band 4 – 7.5 × Band 2 + 1)).

Modified Normalized Difference Water Index (MNDWI): MNDWI = (Green − SWIR)/(Green + SWIR)

Green = Reflectance value of shortwave Green band,
NIR = Reflectance value of the near-infrared band

17.6 Results and Discussion

17.6.1 Seasonal Pattern

In the case of pre-monsoon, the highest LST is 31.12 and the highest LST is located in the Nort-East portion of the SMC area and the lowest LST value is 22.21 in the middle and surrounding the city in the year 1990 (Figure 17.2). In 2000, the highest LST is 34.87, and it increased by almost 5°C compared with the previous year. The highest LST zones are in SMC and it grows in its surrounding areas. The lowest LST is nearly the same as the last year, that is 22.17 and it is seen in tea gardens, forests, and some grasslands. In 2010, the highest LST is 33.87 and it is almost the same as in 2000. The lowest LST

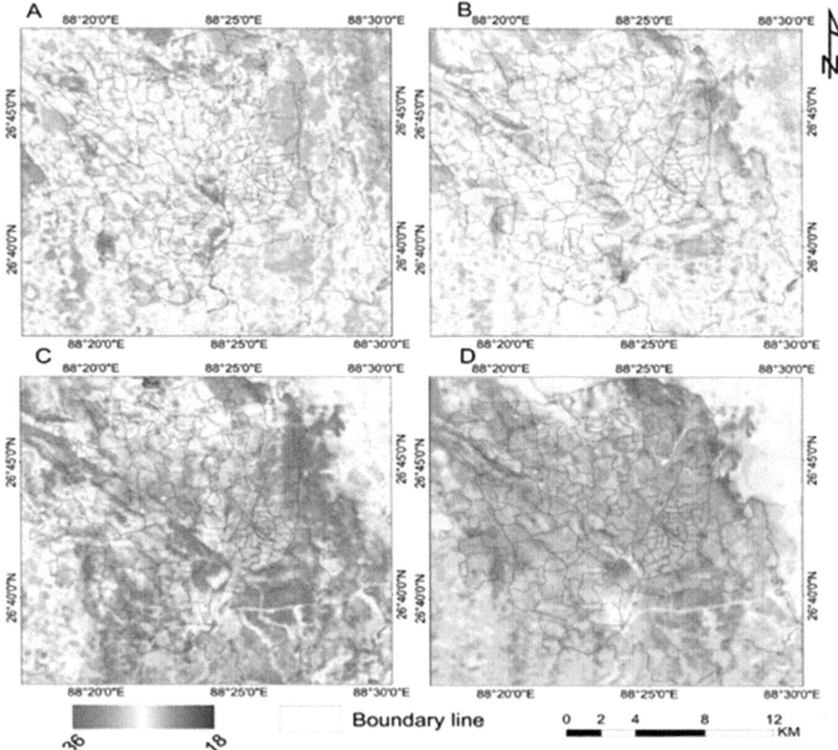

Figure 17.2 Land surface temperature distribution, A(1991), B(2001), C(2011), D(2020)

is 21.36. In 2020, the LST increased very rapidly due to the high rate of urbanization and it is 36.23, which is a 2°C increase from 2010. The highest LST zones are SMC and its surrounding areas and it expands towards the outer regions of SMC. The lowest LST is 24.97; this temperature also increased 5°C. Some patches of lower LST zones are in some North portions, middle of South and some South-East parts. In the case of monsoon, in 1990, the highest LST is 29.69, which is located in the SMC area. The lowest LST is 21.79, which is found in tea gardens and forest areas. In 2000, the LST increased very rapidly, the value is 31.32 and it is located in the core area of SMC. The lowest LST is 23.94, which also increased by almost 2 degrees. The lower LST is in the upper and some lower portions of SMC. In the year 2010, the highest LST is 33.13, which is located in the middle part of SMC and the lowest LST is 22.43 in the outer portion of the SMC area. The highest LST is 34.54 in 2020; the zones are the built-up areas of SMC. The lowest LST is 25.60. In the case of post-monsoon, the highest LST is 26.23 and the built-up areas are expand randomly in 1990. The higher LST zones are now randomly distributed in some upper portions of SMC. In 2000, the highest LST is 28.12 and it is found in SMC and almost all over the

Siliguri region except for some upper patches of SMC, that is tea gardens and some forest cover, and here the LST is lowest at 18.23. In 2010, the highest LST increased by 2 degrees, which is 30.12 and it is found in SMC and surrounding land-use regions (Figure 17.3–17.5).

The lowest LST is 19.39, which is in a very small portion of North Siliguri. In 2020, the highest LST was 32.50 and that increased by almost 3°C from 2010. The highest LST is 20.53 and it is found in tea gardens of some northern patches of SMC and moderately in some South-East portions. In the case of the winter season, the highest LST is 24.77 and it is found in built-up areas. The lowest LST is 16.32 in the outer SMC area, in 1990. In 2000, the highest LST is located in the periphery North-East region and some randomly distributed patches. The lowest LST is 15.99. In 2010, the highest

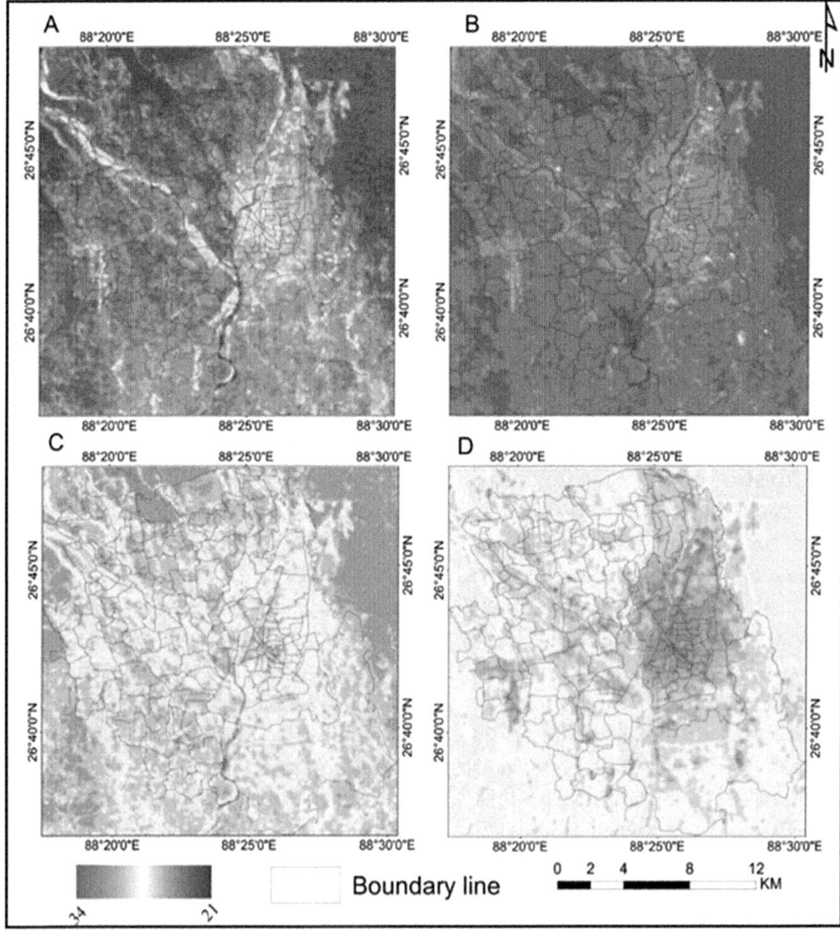

Figure 17.3 LST map of different years in the pre Monsoon season, A(1991), B(2001), C(2011), D(2020)

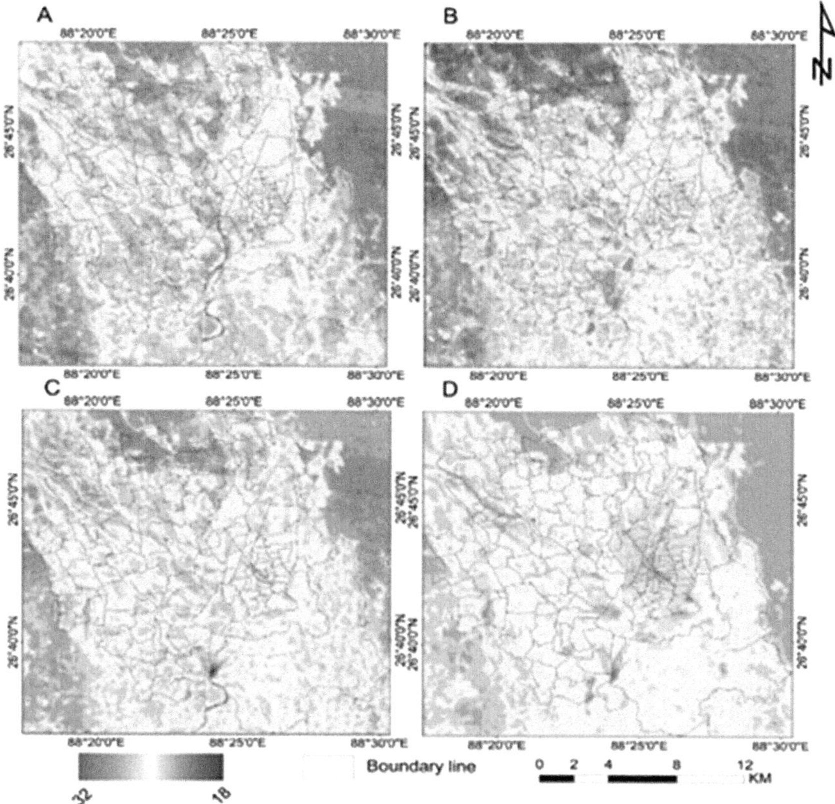

Figure 17.4 LST map of different years in the post-Monsoon season, A(1991), B(2001), C(2011), D(2020)

LST is found in fallow land and the value is 26.46. The lowest LST is 15.75, which is found in tea gardens. In 2020, the highest LST is 28.53, which is an almost 3 degrees increase. The higher LST zones are located in SMC and the periphery of fallow lands. The lowest LST is 19.05, which is found in the tea garden and some forest cover. In the post-monsoon season, the NDVI value of the surrounding rural areas is significantly less. Therefore, a high patch of LST is observed in the north-eastern part of the SMC area and adjacent rural areas. Still, in the monsoon and post-monsoon season, this shifting of high patches of temperature is observed over in the urban area, because in the surrounding rural area, basically over the fellow land and grassland area NDVI value increased, which means greenery is increased. Therefore, this greenery helps to reduce the surface temperature of those areas. The growth of settlements and urban areas has increased the land surface temperature of the whole Siliguri city.

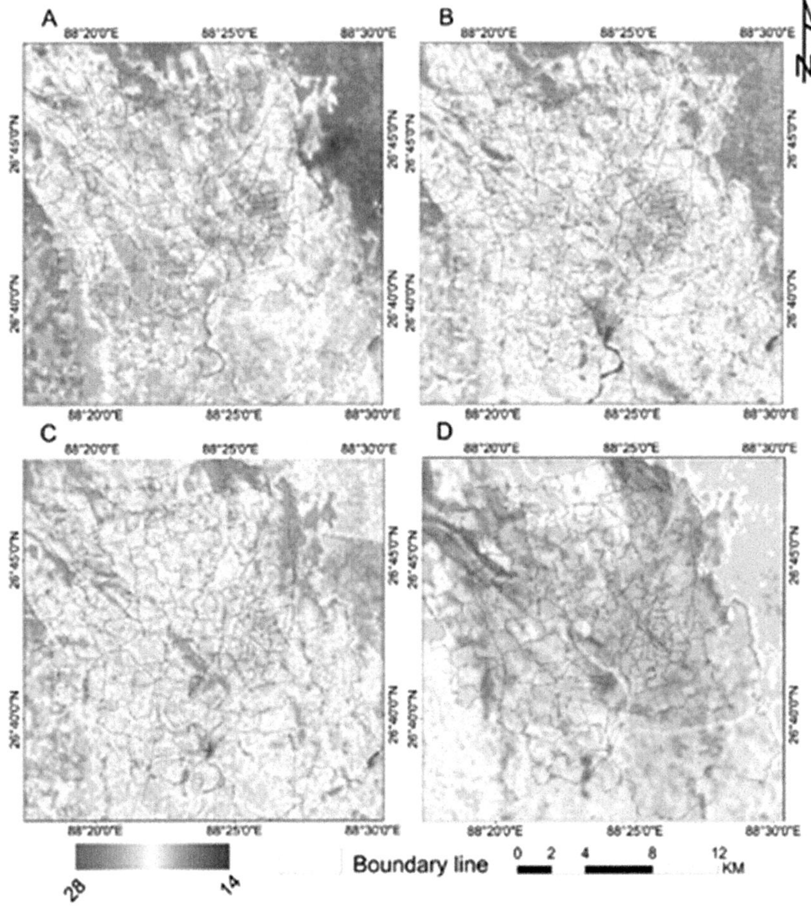

Figure 17.5 LST map of different years in the winter season, A(1991), B(2001), C(2011), D(2020)

17.7 Relationship between LST and Environmental Index

The scatter plot and regression line indicate that NDVI and EVI have a solid significant negative relationship with LST. The coefficient of correlation is −0.600. −0.574 respectively. That signifies with the increase of NDVI and EVI a sharp decline in LST was observed. This inverse relation is crucial for heat management. On the other hand, LST, NDBI, and UBI have a strong positive relationship with LST (Figure 17.6). The correlation coefficient is 0.747, with a high positive slope of 20.5, which indicates an increase in impervious surface will increase positive thermal anomaly from the sounding area, leading to the urban heat island. However, the water body plays a vital role in reducing surface temperature and air temperature through evaporation. In the present

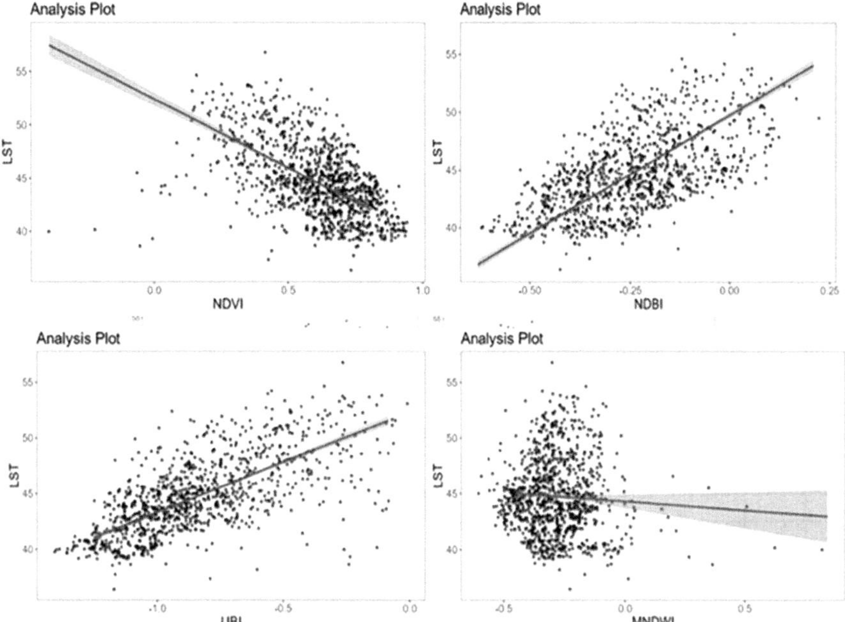

Figure 17.6 Relationship between LST and Environmental Index

study, there is no significant relationship between LST and MNDWI. A fragile negative relation has been observed between LST and MNDWI (r = −0.04715). The Pearson correlation between LST and EVI shows a similar type of negative slope like NDVI (−10.3). On the other hand, LST and UBI have an enormously positive relationship. The coefficient of correlation is 0.706. The scatter plot and regression line indicate that EVI and LST have a significant negative relationship. The coefficient of correlation is −0.574. On the other hand, LST and NDBI have an enormously positive relationship. The coefficient of correlation is 0.747.

Temperature indices were produced to show the spatial distribution of the LST in different temporal periods. It is found that, across the study area, LST values increased from the outskirts to the inner urban area. The lowest value for the temperature is located in the years 1990, which was recorded in pre-monsoon, post-monsoon monsoon, and winter. High-temperature zone clustered in the inner urban areas and in the northeast and southwest directions. It has been observed from Figure 17.6 figure that the temperature trend in the southwest and south and southwest directions follows the direction of urban sprawl. It gradually decreases towards the east because of the dense vegetation cover. Some patches of the high-temperature zone were observed very near the forest cover because the area has a high sand content. The temperature has also trended to decline northward with the increasing elevation. Both Stand NDBI values appear high in the heavy sand content areas for high reflectivity. Significant variation has been found between forest cover water bodies and agricultural land, and barren land. The high temperature was observed on the sand and built-up areas compared to forest and water bodies in both temporal spans. Variation of temperature is also increases over a built-up area and less fluctuation of temperature is observed over forest cover.

17.8 Conclusion

Urbanization impacts negatively on weather conditions. It increases the LST. If we do not control the growth of built-up areas, excessive deforestation, and land alterations, the temperature will increase rapidly. We should ensure immediate tree plantation, agro forestry, and more green spaces in this region to reduce land surface temperature, air pollution, and other local climatic abnormalities. The cooling effect of urban green space plays a crucial role in delineating the density and intensity of cooling in urban heat islands. This chapter assesses the multi-temporal relationship between land surface temperature and vegetation greenness of Siliguri municipal and neighboring areas. LST maintains a robust negative relationship with Urban green spaces for the whole of the study area. The relationship is relatively insignificant for both the high LST zones and low LST zones. An increase in heterogeneous landscape inside the city boundary strongly supports the changing pattern of this relationship.

Bibliography

Bose, A., & Chowdhury, I. R. (2020). Monitoring and modeling of spatio-temporal urban expansion and land-use/land-cover change using markov chain model: A case study in Siliguri Metropolitan area, West Bengal, India. *Modeling Earth Systems and Environment*, *6*, 2235–2249.

Estoque, R.C., & Murayama, Y., 2017. Monitoring surface urban heat island formation in a tropical mountain city using Landsat data (1987–2015). *ISPRS Journal of Photogrammetry and Remote Sensing*, *133*, 18–29. https://doi.org/10.1016/j.isprsjprs.2017.09.008

Guha, S., & Govil, H. (2022). Seasonal impact on the relationship between land surface temperature and normalized difference vegetation index in an urban landscape. *Geocarto International*, *37*(8), 2252–2272.

Hoque, I., & Lepcha, S. K. (2020). A geospatial analysis of land use dynamics and its impact on land surface temperature in Siliguri Jalpaiguri development region, West Bengal. *Applied Geomatics*, *12*(2), 163–178.

Kafy, A. Al, Faisal, A. Al, Shuvo, R. M., et al. (2021). Remote sensing approach to simulate the land use/land cover and seasonal land surface temperature change using machine learning algorithms in a fastest- growing megacity of Bangladesh. *Remote Sensing Applications: Society and Environment*, 21, 100463. https://doi.org/10.1016/j.rsase.2020.100463

Li, C., Li, J., & Wu, J. (2013). Quantifying the speed, growth modes, and landscape pattern changes of urbanization: A hierarchical patch dynamics approach. *Landscape Ecology*, *28*(10), 1875–1888. https://doi.org/10.1007/s10980-013-9933-6

Lu, Q. B. (2022). Major contribution of halogenated greenhouse gases to global surface temperature change. *Atmosphere*, *13*(9), 1419.

Mao, C., Xie, M., & Fu, M., 2020. Thermal response to patch characteristics and configurations of industrial and mining land in a Chinese mining city. *Ecological Indicators*, *112*, 106075. https://doi.org/10.1016/j.ecolind.2020.106075

Nurwanda, A., & Honjo, T., 2018. Analysis of land use change and expansion of surface urban heat island in Bogor city by remote sensing. *ISPRS International Journal of Geo-Information*, *7*. https://doi.org/10.3390/ijgi7050165

Pramanik, S., & Punia, M. (2019). Assessment of green space cooling effects in dense urban landscape: A case study of Delhi, India.*Modeling Earth Systems and Environment*, 5(3), 867–884. https://doi.org/10.1007/s40808-019-00573-3

Roy, S., Bose, A., Singha, N., Basak, D., & Chowdhury, I. R. (2021). Urban waterlogging risk as an undervalued environmental challenge: An Integrated MCDA-GIS based modeling approach. *Environmental Challenges*, *4*, 100194.

Santra, A. (2019). Land surface temperature estimation and urban heat island detection: A remote sensing perspective. In *Environmental information systems: Concepts, methodologies, tools, and applications* (pp. 1538–1560). IGI Global.

Sarkar, A., & Chouhan, P. (2019). Dynamic simulation of urban expansion based on Cellular Automata and Markov Chain Model: A case study in Siliguri Metropolitan Area, West Bengal. *Modeling Earth Systems and Environment*, 5, 1723–1732,

Soydan, O., 2020. Effects of landscape composition and patterns on land surface temperature: Urban heat island case study for Nigde, Turkey. *Urban Climate*, 34, 100688. https://doi.org/10.1016/j.uclim.2020.100688

Sultana, S., & Satyanarayana, A. N. V. (2020). Assessment of urbanisation and urban heat island intensities using landsat imageries during 2000–2018 over a sub-tropical Indian city. *Sustainable Cities and Society*, 52, Article 101846. https://doi.org/10.1016/j.scs.2019.101846

Wang, F., Qin, Z., Song, C., Tu, L., Karnieli, A., Zhao, S., 2015. An improved mono-window algorithm for land surface temperature retrieval from landsat 8 thermal infrared sensor data. *Remote Sensing*, 7, 4268–4289. https://doi.org/10.3390/rs70404268

Xiao, R., Weng, Q., Ouyang, Z., Li, W., Schienke, E. W., & Zhang, Z. (2008). Land surface temperature variation and major factors in Beijing, China. *Photogrammetric Engineering & Remote Sensing*, *74*(4), 451–461.

18 Mapping and Investigating of Land Use and Land Cover Changes Using Geo-Spatial Analysis

A Case Study of Sagardighi Block, Murshidaba

Md. Mustaquim and Woheeul Islam

18.1 Introduction

We regularly use the terms land usage and land cover (LULC) in today's development era. Although we use these phrases interchangeably, it's vital to clarify and understand their meanings in order to use them correctly, meaningfully, and successfully (Giri, 2012). The biotic and abiotic assemblage of the earth's surface and immediate subsurface is referred to as land cover (Meyer and Turner, 1992; Giri, 2012). In contrast, land use is defined as the mode or manner in which land is used or occupied by humans (Giri, 2012).

As the global population expands, so does the demand for food and more economical infrastructure, putting strain on available land. Meeting the demands of a rapidly growing population necessitates greater resources to maintain essential services and maintain a high standard of living. Finally, it exerts pressure on the ground. As a result, modifications in land use are essential for economic and social development. Changes in land use, on the other hand, are not without cost (Wu 2008). The Earth's surface has been dramatically altered by postmodern human activity, which involves a relationship between cultures from one perspective and the land's cultural advancement and carrying capacity from another (Lal 2011). Land has always been altered by humans in order to gain the essentials of life. Nonetheless, increased exploitation has resulted in significant changes in ecosystems and environmental processes at the local, regional, and global levels (Amin and Fazal, 2012). Natural resources and socioeconomic activity will inexorably be impacted by these changes (Islam et al., 2018). The rate of the biological cycle can be affected by changes in land use (Prasad and Ramesh, 2019). Land use/cover changes are key difficulties and challenges for ecologically friendly and sustainable economic growth in any region. As a result, every accessible piece of land should be utilized to the greatest extent possible (Rawat et al., 2013).

Understanding the dynamics of LULC change, as well as the underlying reasons, is essential for developing and implementing successful environmental policies and programs. The purpose of this research was to look into and understand the long-term dynamics of land-use and land-cover change (Tadese

DOI: 10.4324/9781003275916-21

et al., 2020). Land use/cover data from multi-temporal remotely sensed data could be timely and reliable (Rahman, 2007; Rahman et al., 2012). When used in conjunction with RS, remote sensing (RS) will supply us with data on land use/cover, and GIS is an effective data integration and analysis tool (Mhawish and Saba, 2016; Bhalli and Ghaffar, 2015; Prakasam, 2010; Mishra et al., 2020). GIS allows us to combine spatial and non-spatial data, which can help us analyze the change and identify the factors that may be contributing to it. As a result, we may use GIS and remote sensing to map, monitor, and track changes across time.

18.2 Study Area

The Sagardighi CD block is located in Murshidabad's Rarh area. The district is divided into two natural physiographic zones by the Bhagirathi River: Rarh on the west and Bagri on the east. The soil in the Rarh area is primarily clay and lateritic clay-based. The Nabagram plain is formed at the lowest limit of the Rajmahal hills' height in this region as it dips gently down from bordering Jharkhand. Numerous cliffs and bluffs may be seen on the region's eastern slope. Raghunathganj I and Raghunathganj II CD blocks in the north, Lalgola CD block in the east, Nabagram CD block in the south, and Nalhati II CD block in Birbhum district in the west, all fall within the Jangipur sub-division. There are 11 gramme panchayats, 199 gramme sansads, 197 mouzas, and 178 inhabited villages in this panchayat samity. This block is served by the Sagardighi police station. This CD block's headquarters are at Sagardighi. CD block had a total population of 310,461, all of whom were rural, according to the 2011 Indian Census. Males made up 51% of the population, while females made up 49%. Between 2001 and 2011, the population of the Sagardighi block increased by 23.06%. According to the 2011 census, 65.26% of the population in the Sagardighi CD block was literate, with men accounting for 68.34% of the male population and females accounting for 62.05% of the female population. The gap in literacy rates between men and women was 6.29% (Figure 18.1).

18.3 Database and Methods

The USGS Earth Explorer website has made satellite images available for free. Multispectral images acquired in the months before each calendar year were used in this investigation. Table 18.1 contains detailed information about the images.

18.4 Image Processing

The first and most important goal of this chapter is to manage each year's cloud-free satellite images in order to measure land-cover type and establish its trend and pattern of change. The data were downloaded for free from the

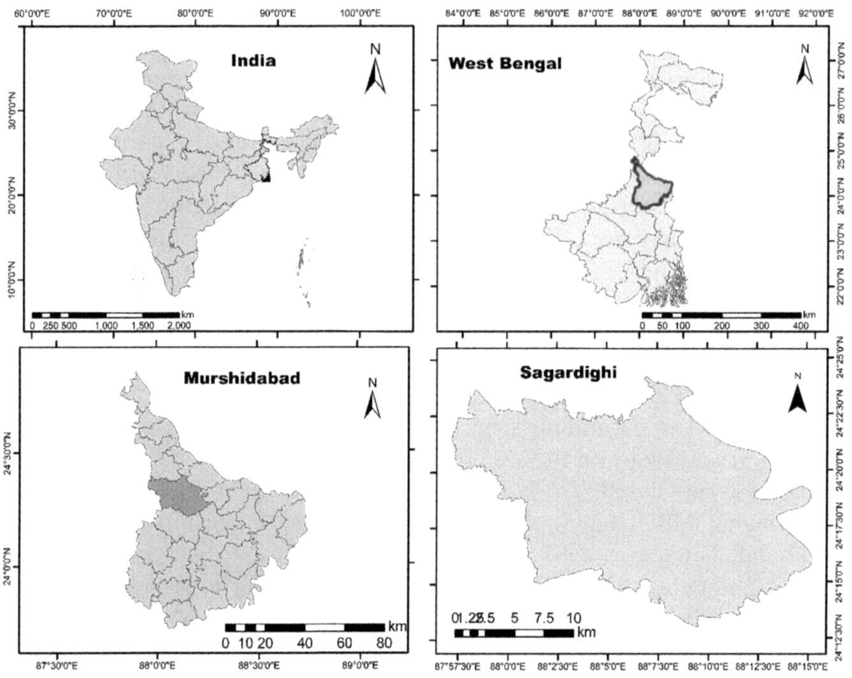

Figure 18.1 Location map of the study area

Table 18.1 Details of satellite images used in the study

Year	*Acquisition date*	*Satellite*	*Sensor*	*Path/Row*	*Resolution (m)*	*Projection*
2001	19-01-2001	Landsat 5	TM	139/43	30	UTM-WGS84
2021	04-02-2021	Landsat 8	OLI/TIRS	139/43	30	UTM-WGS84

Source: USGS Earth Explorer

Table 18.2 LULC class description

LULC classes	*Descriptions*
Agriculture	Land used primarily for growing crops and vegetables
Built-up	Artificial and non-porous structures such as settlements, roads, etc.
Plantation and forest	The area consists of light vegetation where crops are grown and some small trees, and some orchards.
Water bodies	There are rivers, ponds, waterlogging, etc.

USGS Earth Explorer. The layers are then stacked together to create a false color composition. After gathering all of the images, it is blended into a single image. The unified image then requires color modifications, which are done automatically with the ERDAS Imagine 2014 program. The image was then exposed to a subset. The image was then subset according to our area of interest using a shapefile of the research region digitized in ArcGis 10.4 from the Murshidabad District planning map series of national atlases and thematic mapping organizations (NATMO). After performing the tasks above, the image was categorized using the Supervised Classification technique of the Maximum Likelihood Algorithm. Furthermore, in some of the more recent studies, Alam et al. (2021), Saber et al. (2021), Roy and Kasemi (2021), and Chowdhury et al. (2020) used supervised classification with the MCL approach to identify the LULC change (Table 18.2). A set of training pixels was gathered for each land use/cover type in order to generate a class. The required changes were done to construct the land use and land cover map after classifying the satellite pictures. ERDAS Imagine 2014 was used to assess the accuracy; it calculates the kappa coefficient and overall accuracy for each land use map. Accuracy assessment is an important part of studying image classification and land use-land cover classification detection to recognize and evaluate the correct transition (Andualem et al., 2018) To assess overall accuracy and the kappa coefficient, 75 random reference sites were produced for each land use map. As a reference, processed satellite images, local knowledge, and a Google Earth map were all used. For 2001 it was 0.85, and it was 0.88 in 2021, with an overall accuracy of 90 and 96%, respectively

18.5 Analysis and Discussions

The entire Sagardihi block is made up of rural settlements, according to the census of India's rural-urban classification. People here are mostly involved in agriculture and related activities. However, as the population grew and industrialization took hold, the pattern of land use and land cover (LULC) changed, as seen in Figure 18.2, which compares the LULC of 2001 and 2021 in the Sagardighi block. Figure 18.3 shows that built-up area increased from 3.6% to 8.3% during this time period, while plantations and forests increased from.5% to 9.6%. Agricultural land has shrunk considerably from 93.7% to 78.9%.

Furthermore, there is a slight rise in the water body. Agriculture land has diminished, making it neither economically feasible nor ecologically sustainable for agrarian rural communities, according to the magnitude of change. As a result, Table 18.3 reveals the direction of change.

Table 18.3 shows that the majority of agricultural land has been transformed into the built-up (20 km^2) and plantation and forest (29 km^2) categories. This shift in agricultural land cover, on the other hand, reflects a shift in local occupational character. Although we only have census data up to 2011, it shows that cultivators have declined from 29.46% in 2001 to 16.83% in 2011, while agricultural laborers have increased from 51.85% to 64.12%.

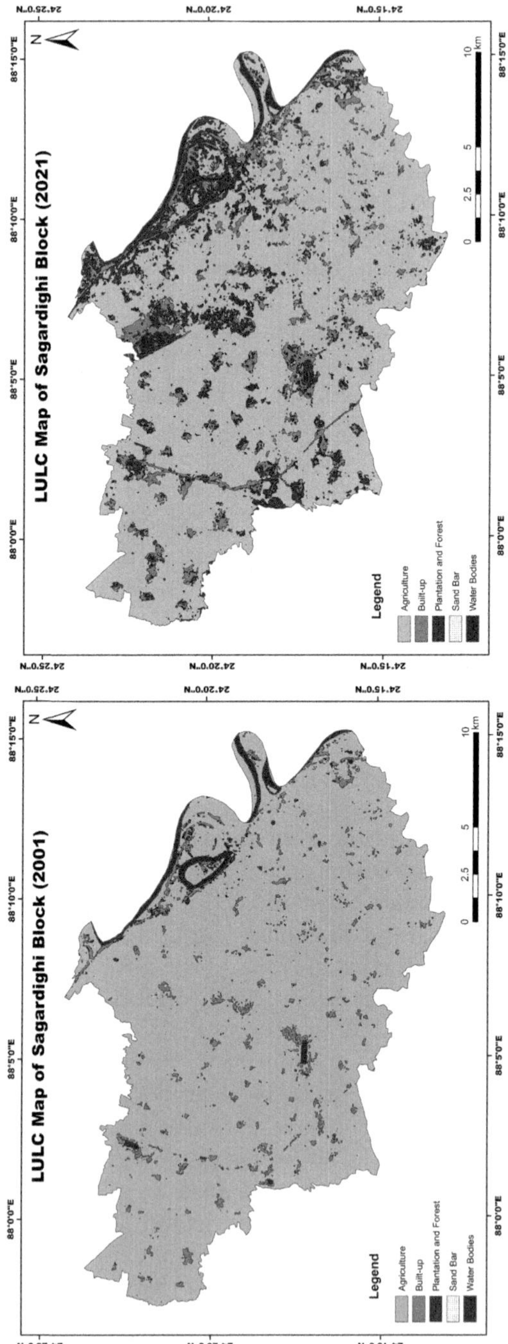

Figure 18.2 LULC map of Sagardighi block of 2001 and 2021

Source: Authors' own

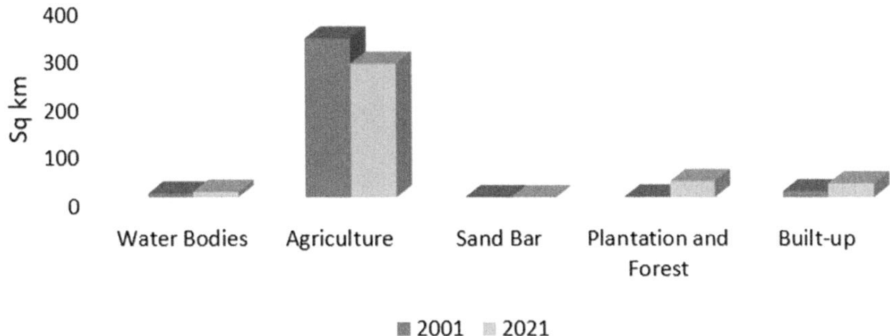

Figure 18.3 Shear of LULC in Sagardighi block (Year 2001 and 2021)

Source: Authors' own

Table 18.3 LULC transition matrix from 2001 to 2021

2001	*2021*					
	Agriculture	*Built-up*	*Plantation and forest*	*Sand bar*	*Water bodies*	*Grand Total*
Agriculture	275.0	20.0	28.8	0.0	5.2	329.1
Built-up	1.2	8.6	2.7		0.1	12.6
Plantation and forest	0.2	0.2	1.1		0.0	1.6
Sand bar	0.0	0.0	0.0		0.0	0.0
Water bodies	0.6	0.1	1.0		6.1	7.8
Grand Total	277	29	34	0	11	351

The population of the Sagardighi block increased from 252,293 to 31,0461 in 2011, necessitating the accommodation of an additional 58,168 people on the same parcel of land. Another significant cause of increasing built-up land is the Sagardighi thermal power station, which has been operational since September 2008 (Figure 18.4), and both of these factors contribute to the reduction of agricultural land.

18.6 Conclusions

Long-term planning and better land management require an understanding of the extent, rate, and degree of change in the study area's spatial pattern. The built-up area has increased by 16 km^2, while agricultural land has decreased by 52 km^2, according to a recent evaluation. The rising population

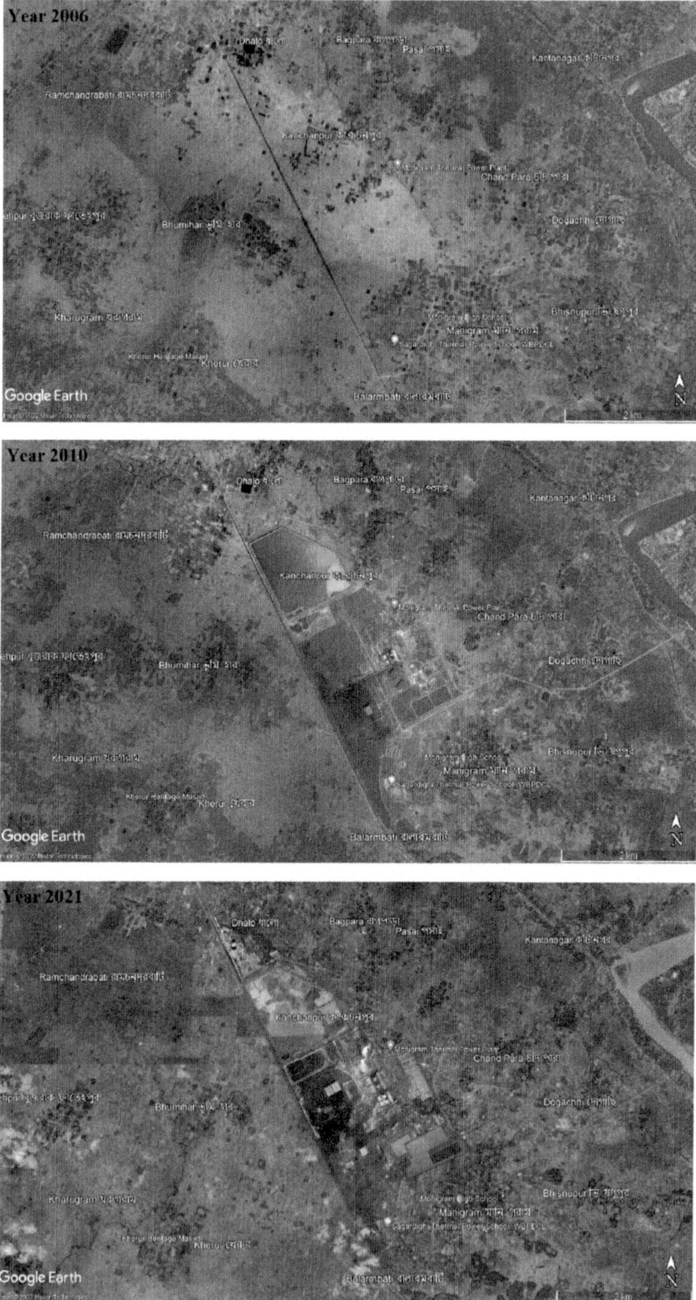

Figure 18.4 Before and after Sagardighi thermal power plant

Source: Authors' own

and the construction of thermal power plants contributed to the development of the built-up region throughout time, resulting in the transformation of a fruitful agricultural field into an impenetrable surface. Plantation and woodland expansion are essentially converting traditional agriculture practices to semi-permanent orchard cropping. Orchard-based farming is less vulnerable to changes in traditional agricultural inputs such as monsoonal rain, fertilizer, insecticides, and labor. This transition away from agricultural land use reflects a change in the occupational character of the community. As a result, any conversion or development inside this region should be carefully weighed in terms of economic, ecological, and social factors before proceeding.

Bibliography

Alam, N., Saha, S., Gupta, S., & Chakraborty, S. (2021). Prediction modelling of riverine landscape dynamics in the context of sustainable management of floodplain: A geospatial approach. *Annals of GIS*, *27*(3), 299–314. https://doi.org/10.1080/19475683.2020.1870558

Amin, A., & Fazal, S. (2012). Land transformation analysis using remote sensing and GIS techniques (A case study). *Journal of Geographic Information System*, *4*, 229–236.

Andualem, T. G., Belay, G., & Guadie, A. (2018). Land use change detection using remote sensing technology. *Journal of Earth Science & Climatic Change*, *9*(10), 1–6.

Bhalli, M.N. Ghaffar, A. (2015). Use of geospatial techniques in monitoring urban expansion and land use change analysis: A case of Lahore, Pakistan. *Journal of Basic & Applied Sciences*, *11*, 265–273.

Chowdhury, M., Hasan, M. E., & Abdullah-Al-Mamun, M. M. (2020). Land use/land cover change assessment of Halda watershed using remote sensing and GIS. *The Egyptian Journal of Remote Sensing and Space Science*, *23*(1), 63–75. https://doi.org/10.1016/j.ejrs.2018.11.003

Giri, C. P. (2012). *Remote sensing of land use and land cover principles and applications*. CRC Press, Taylor & Francis Group.

Islam, K., Jashimuddin, M., Nath, B., & Nath, T. K. (2018). Land use classification and change detection by using multi-temporal remotely sensed imagery: The case of Chunati wildlife sanctuary, Bangladesh. *The Egyptian Journal of Remote Sensing and Space Science*, *21*(1), 37–47. https://doi.org/10.1016/j.ejrs.2016.12.005

Lal, T. (2011). *Population pressure and changes in landuse pattern A study of Akhnoor*. Department of Geography, University of Jammu.

Meyer, W. B., & Turner, B. L. (1992). Human-population growth and global land-use cover change. *Annual Review of Ecology and Systematics*, *23*, 39–61.

Mhawish, Y. M., & Saba, M. (2016). Impact of population growth on land use changes in Wadi Ziqlab of Jordan between 1952 and 2008. *International Journal of Applied Sociology*, *6*(1), 7–14.

Mishra, P. K., Rai, A., & Rai, S. C. (2020). Land use and land cover change detection using geospatial techniques in the Sikkim Himalaya, India. *The Egyptian Journal of Remote Sensing and Space Science*, *23*(2), 133–143. https://doi.org/10.1016/j.ejrs.2019.02.001

Prakasam, C. (2010). Land use and land cover change detection through remote sensing approach: A case study of Kodaikanal taluk, Tamil nadu. *International Journal of Geomatics and Geosciences*, *1*(2), 150–158.

Prasad, G., & Ramesh, M. V. (2019). Spatio-temporal analysis of land use/land cover changes in an ecologically fragile area—Alappuzha District, Southern Kerala, India. *Natural Resources Research*, *28*, 31–42.

Rahman, A. (2007). Application of remote sensing and GIS technique for urban environmental management and sustainable development of Delhi, India. In M. Netzband, W. L. Stefnow, & C. L. Redman (Eds.), *Applied remote sensing for urban planning, governance and sustainability* (pp. 165–197). Springer-Verlag.

Rahman, A., Kumar, S., Fazal, S., & Siddiqui, M. A. (2012). Assessment of land use/ land cover change in the North–West district of Delhi using remote sensing and GIS techniques. *Journal of the Indian Society of Remote Sensing*, *40*, 689–697.

Rawat, J. S., Biswas, V., & Kumar, M. (2013, June). Changes in land use/cover using geospatial techniques: A case study of Ramnagar town area, district Nainital, Uttarakhand, India. *The Egyptian Journal of Remote Sensing and Space Science*, *16*(1), 111–117.

Roy, B., & Kasemi, N. (2021). Monitoring urban growth dynamics using remote sensing and GIS techniques of Raiganj urban agglomeration, India. *The Egyptian Journal of Remote Sensing and Space Science*, *24*(2), 221–230. https://doi.org/10.1016/j.ejrs.2021.02.001

Saber, A., El-Sayed, I., Rabah, M., & Selim, M. (2021). Evaluating change detection techniques using remote sensing data: Case study new administrative capital Egypt. *The Egyptian Journal of Remote Sensing and Space Science*, *24*(3), 635–648. https://doi.org/10.1016/j.ejrs.2021.03.001

Tadese, M., Kumar, L., Koech, R., & Kogo, B. K. (2020). Mapping of land-use/land-cover changes and its dynamics in Awash river basin using remote sensing and GIS. *Remote Sensing Applications: Society and Environment, 19*. *Tehsil* [Doctoral Thesis, University of Jammu]. http://hdl.handle.net/10603/78480

Wu, JunJie, 2008. Land use changes: Economic, social, and environmental impacts. *Choices: The Magazine of Food, Farm, and Resource Issues, Agricultural and Applied Economics Association*, *23*(4), 6–10.

Index

Pages in *italics* refer to figures and pages in **bold** refer to tables.